# LABORATORY MANUAL
## Cat Version

# HOLE'S
# HUMAN
# ANATOMY &
# PHYSIOLOGY
## FIFTEENTH EDITION

### DAVID SHIER    JACKIE BUTLER    RICKI LEWIS

## TERRY R. MARTIN

KISHWAUKEE COLLEGE

Mc
Graw
Hill
Education

**LABORATORY MANUAL TO ACCOMPANY HOLE'S HUMAN ANATOMY & PHYSIOLOGY, FIFTEENTH EDITION (CAT)**

Portfolio Manager: *Amy Reed*
Product Developer: *Michelle Gaseor*
Marketing Manager: *James Connelly*
Content Project Managers: *Jeni McAtee, Christina Nelson*
Buyer: *Sandy Ludovissy*
Design: *Tara McDermott*
Content Licensing Specialist: *Lori Hancock*
Cover Image: *©Tim Tadder/Getty Images*
Compositor: *SPi Global*

mheducation.com/highered

# CONTENTS

*These laboratory exercises are available for instructors via the Instructor Resources in Connect. They are available for students via the eBook in Connect.

# GUIDED TOUR THROUGH A LAB EXERCISE

This laboratory manual was prepared to be used with the textbook *Hole's Human Anatomy & Physiology*, Fifteenth Edition, by David Shier, Jackie Butler, and Ricki Lewis. As with the textbook, the laboratory manual is designed for students with minimal backgrounds in the physical and biological sciences who are pursuing careers in professional health fields.

The laboratory manual contains sixty-three laboratory exercises and reports, which are closely integrated with the chapters of the textbook. (Four of these exercises appear only online.) The exercises are planned to illustrate and review the anatomical and physiological facts and principles presented in the textbook and to help students investigate some of these ideas in greater detail.

Many of the laboratory exercises are short or are divided into several separate procedures. This allows an instructor to select those exercises or parts of exercises that will best meet the needs of a particular program. Also, exercises requiring a minimal amount of laboratory equipment have been included.

The laboratory exercises have a variety of special features that are designed to stimulate interest in the subject matter, to involve students in the learning process, and to guide them through the planned activities. These special features include the following.

## Materials Needed

This section lists the laboratory materials that are required to complete the exercise and to perform the demonstrations and learning extensions.

## ⚠️ Safety

A list of Laboratory Safety Guidelines appears in Appendix 1 of your laboratory manual. Each laboratory session that requires special safety guidelines has a safety section following "Materials Needed." The instructor might require some modifications of these guidelines.

## Purpose of the Exercise

The purpose states the intent of the exercise—that is, what will be accomplished.

## Learning Outcomes

The learning outcomes ① list what a student should be able to do after completing the exercise. Each learning outcome has a matching assessment indicated by a corresponding icon  in the laboratory exercise or the laboratory report.

## Introduction

The introduction briefly describes the subject of the exercise or the ideas that will be investigated.

## Procedure

The procedure provides a set of detailed instructions for accomplishing the planned laboratory activities. Each procedure heading includes a practice icon 🔄 indicating ways to accomplish the learning outcomes. Usually, these instructions are presented in outline form, so that a student can proceed efficiently through the exercise in stepwise fashion. Often the student is referred to particular sections of a textbook for necessary background information or for review of subject matter presented previously.

The procedures include a wide variety of laboratory activities and, from time to time, direct the student to complete various tasks in the laboratory reports.

---

LABORATORY EXERCISE

# 12

## Bone Structure and Classification

### MATERIALS NEEDED

Textbook
Prepared microscope slide of ground compact bone
Human bone specimens, including long, short, flat, irregular, and sesamoid types
Human long bone, sectioned longitudinally
Fresh animal bones, sectioned longitudinally and transversely
Compound light microscope
Dissecting microscope
Fresh chicken bones (radius and ulna from wings)
Vinegar or dilute hydrochloric acid

### ⚠️ SAFETY

- Wear disposable gloves for handling fresh bones and for the demonstration of a bone soaked in vinegar or dilute hydrochloric acid.
- Wash your hands before leaving the laboratory.

### PURPOSE OF THE EXERCISE

To review the way bones are classified and to examine the structure of a long bone.

### 🔄 LEARNING OUTCOMES APⓇR

After completing this exercise, you should be able to

① Locate the major structures of a long bone.
② Distinguish between compact and spongy bone.
③ Differentiate the special characteristics of compact bone tissue.
④ Arrange five groups of bones based on their shapes and identify an example for each group.
⑤ Describe the functions of various structures of a bone.

A bone represents an organ of the skeletal system. As such, it is composed of a variety of tissues, including bone tissue, cartilage, dense connective tissue, blood, and nervous tissue. Bones are not only alive but also multifunctional. They support and protect softer tissues, provide points of attachment for muscles, house blood-producing cells, and store inorganic salts.

Bones are classified according to their shapes as long, short, flat, or irregular. Although the bones of the skeleton vary greatly in size and shape, they have much in common structurally and functionally.

### 🔄 PRACTICE

PROCEDURE—Bone Structure and Classification

1. Review the section entitled "Connective Tissues" in chapter 5 of the textbook and the section entitled "Bone Shape and Structure" in chapter 7 of the textbook.
2. As a review activity, label figures 12.1 and 12.2.
3. Examine the microscopic structure of bone tissue by observing a prepared microscope slide of ground compact bone. Use figure 12.3, a micrograph of bone tissue, to locate the following features:

   **osteon** (Haversian system)—cylinder-shaped unit
   **central canal** (Haversian canal)—contains blood vessels and nerves
   **lamella**—concentric ring of matrix around central canal
   **lacuna**—small chamber for an osteocyte
   **bone extracellular matrix**—collagen and calcium phosphate
   **canaliculus**—minute tube containing cellular process

**LEARN: ACTIVITY**

Examine a fresh chicken bone and a chicken bone that has been soaked for several days in vinegar or overnight in dilute hydrochloric acid. Wear disposable gloves for handling these bones. This acid treatment removes the inorganic salts from the bone extracellular matrix. Rinse the bones in water and note the texture and flexibility of each (fig. 12.4*a*). The bone becomes soft and flexible without the support of the inorganic salts with calcium.

Examine the specimen of chicken bone that has been exposed to high temperature (baked at 121°C [250°F] for 2 hours). This treatment removes the protein and other organic substances from the bone extracellular matrix (fig. 12.4*b*). The bone becomes brittle and fragile without the benefit of the collagen fibers. A living bone with a combination of the qualities of inorganic and organic substances possesses tensile strength.

## Learn: Activity

Learn activities appear in separate boxes. They describe specimens, specialized laboratory equipment, or other materials to enrich the student's laboratory experience.

 **LEARN: LAB IN MOTION**

A long rubber band can be used to simulate muscle locations, origins, insertions, and actions on the human torso model, the skeleton, or a laboratory partner. Hold one end of the rubber band firmly on the origin location of a muscle; then slightly stretch the rubber band and hold the other end on the insertion site. Allow the insertion end to slowly move toward the origin end to simulate the contraction and action of the muscle. A A A

## Learn: Lab in Motion

Some learning activities involve a greater degree of student group or class participation. These learning activities are often more active with physiological concepts.

 **ASSESS**

### CRITICAL THINKING

Assessments involving critical thinking skills are incorporated within many of the laboratory exercises and laboratory reports to enhance valuable critical thinking skills that students will need throughout their lives.

 **ASSESS**

### CRITICAL THINKING

Explain the advantage for melanin granules being located in the deep layer of the epidermis.

_____

_____

_____

_____

_____

## Laboratory Reports

A laboratory report to be completed by the student immediately follows each exercise. These reports include various types of review activities, spaces for sketches of microscopic objects, tables for recording observations and experimental results, and questions dealing with the analysis of such data.

As a result of these activities, students will develop a better understanding of the structural and functional characteristics of their bodies and will increase their skills in gathering information by observation and experimentation. By completing all of the assessments in the laboratory report, students are able to determine if they have accomplished all of the learning outcomes.

## Illustrations

Diagrams from the textbook and diagrams similar to those in the textbook often are used as aids for reviewing subject matter. Other illustrations provide visual instructions for performing steps in procedures or are used to identify parts of instruments or specimens.

Frequent variations exist in anatomical structures among humans. The illustrations in the textbook and the laboratory manual represent normal (normal means the most common variation) anatomy. Variations from normal anatomy do not represent abnormal anatomy unless some function is impaired.

**(c)** Simple columnar epithelium (nonciliated) (from intestine) (400×)

©Victor P.Eroschenko

Micrographs are included to help students identify microscopic structures or to evaluate student understanding of tissues.

In some exercises, the figures include line drawings suitable for students to color with colored pencils. This activity may motivate students to observe the illustrations more carefully and help them to locate the special features represented in the figures. Students can check their work by referring to the corresponding full-color illustrations in the textbook.

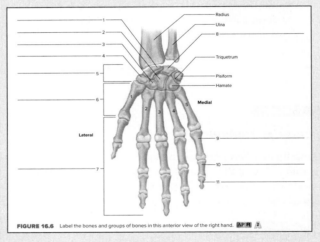

**FIGURE 16.6** Label the bones and groups of bones in this anterior view of the right hand. APR A

**FIGURE 14.5** Label the bones and features of the sagittal section of the skull. APR A A A

This sagittal section of the skull illustrates the more vivid colors used in the all new art program for the entire lab manual including the dissection labs.

# PREFACE

## Changes to this Edition

Many of the changes in this new edition are a result of evaluations and suggestions from anatomy and physiology students. Many suggestions from users and reviewers of the fourteenth edition have been incorporated:

### Global Changes

- New Learn, Practice, and Assess icons matching textbook icons
- Critical thinking applications modified to Assess: Critical Thinking
- Demonstrations modified to a Learn: Activity
- Learning extensions modified to a Learn: Activity
- Some learning extensions modified to a Learn: Lab in Motion
- Safety Guidelines and Periodic Table of Elements moved to Appendices

### Specific Changes to Individual Laboratory Exercises

Lab 6: Updated membrane passage terminology
Lab 9: New figure 9.1*b*
Lab 11: Improved figures 11.2 and 11.3*d*
Lab 14: Revised figures 14.7 and 14.10
Lab 16: Modified figure 16.2*a*
Lab 18: Revised Part E Assessments
Lab 19: Modified Part C Assessments
Lab 20: New figures 20.1 and 20.3; updated Part A and B Assessments
Lab 21: New figures 21.1 and 21.4*b*
Lab 22: New figure 22.1; updated Procedure B
Lab 24: Replaced figures 24.2 and 24.3*b*
Lab 28: Updated terminology in Procedure A and figures 28.1, 28.2, 28.3, 28.5, and 28.6
Lab 32: Modified figure 32.2; new figure 32.3; revised Part A Assessments
Lab 40: Revised figure 40.5
Lab 43: Revised figures 43.6, 43.14, and 43.15
Lab 45: New figures 45.6, 45.9, and 45.10
Lab 46: New figure 46.5
Lab 47: New figures 47.1 and 47.2
Lab 48: New figure 48.4; revised Procedure C
Lab 49: Added figure 49.2
Lab 50: Revised Procedure B and figure 50.4
Lab 51: Updated terminology; new figure 51.2
Lab 52: Updated terminology; revised figure 52.1

## Brand New! Connect Digital Content

Save time with and engage students with auto-graded practice and assessments! Each exercise of the lab manual now has questions available on the Connect platform that can be assigned for a variety of modalities. Fully editable, customizable, auto-graded questions can be mixed and matched to recreate the lab manual work or unique assignments that fit your course. You can generate powerful data related to student performance based on content tags in Connect and help your students learn in the way that suits them best.

The new Connect content that enhances the *Laboratory Manual for Hole's Human Anatomy and Physiology* is available as a bank in the core text's, *Hole's Human Anatomy and Physiology, 15th edition,* Connect shell. Contact your McGraw-Hill Education representative to find out more.

## Acknowledgments

I value all the support and encouragement by the staff at McGraw-Hill Higher Education, including Amy Reed, Jim Connely, Michelle Gaseor, and Fran Simon. Special recognition is granted to Colin Wheatley for his insight, confidence, wisdom, warmth, and friendship.

I am grateful for the professional talent of textbook authors David Shier, Jackie Butler, and Ricki Lewis for all of their coordination efforts and guidance with the laboratory manual. Gratitude is extended to John Hole for his years of dedicated effort in early editions of this classic work and the opportunity he extended to revise his established laboratory manual.

I am particularly thankful to Dr. Norman Jenkins, Dr. David Louis, and Dr. Thomas Choice, retired presidents of Kishwaukee College, and Dr. Laurie Borowicz, president of Kishwaukee College, for their support, and confidence in my endeavors. I am appreciative for the expertise of Womack Photography for numerous contributions. The professional reviews of the nursing procedures were provided by Kathy Schnier. I am also grateful to Laura Anderson, Rebecca Doty, Michele Dukes, Troy Hanke, Jenifer Holtzclaw, Stephen House, Shannon Johnson, Brian Jones, Marissa Kannheiser, Morgan Keen, Marcie Martin, Angele Myska, Sparkle Neal, Bonnie Overton, Susan Rieger, Eric Serna, Robert Stockley, Shatina Thompson, Nancy Valdivia, Marla Van Vickle, Jana Voorhis, Joyce Woo, and DeKalb Clinic for their special contributions. There have been valuable contributions from my students, who have supplied thoughtful suggestions and assisted in the clarification of details.

To my son Ross, I owe gratitude for his keen eye and creative suggestions. Foremost, I am appreciative to Sherrie Martin, my spouse and best friend, for advice, understanding, and devotion throughout the writing and revising.

**Terry R. Martin**
Kishwaukee College
21193 Malta Road
Malta, IL 60150

## Reviewers

I would like to express my sincere gratitude to all users of the laboratory manual who provided suggestions for its improvement. I am especially grateful for the contributions of the reviewers who kindly reviewed the manual. Their thoughtful comments and valuable suggestions are greatly appreciated. They include the following:

Geralyn M. Caplan, *Owensboro Community & Technical College*
Smruti Desai, *Lone Star College–CyFair*
Purti Gadkari, *Wharton County Junior College*
Joseph D. Gar, *West Kentucky Community and Technical College*
Peter Kobella, *Owensboro Community and Technical College*
Timothy Koneval, *Mount Aloysius College*
Deborah J. McCool, *Mount Aloysius College*
Anthony Russo, *Mountain Empire Community College*
Padmaja B. Vedartham, *Lone Star College Cy–Fair*

## About the Author

©J & J Photography

This fifteenth edition is the ninth revision by Terry R. Martin of Kishwaukee College. Terry's teaching experience of over forty years, his interest in students and love for college instruction, and his innovative attitude and use of technology-based learning enhance the solid tradition of John Hole's laboratory manual. Among Terry's awards are the Kishwaukee College Outstanding Educator, Phi Theta Kappa Outstanding Instructor Award, Kishwaukee College ICCTA Outstanding Educator Award, Who's Who Among America's Teachers, Kishwaukee College Faculty Board of Trustees Award of Excellence, Continued Excellence Award for Phi Theta Kappa Advisors, and John C. Roberts Community Service Award. Terry's professional memberships include the National Association of Biology Teachers (NABT), Human Anatomy and Physiology Society (HAPS), former Chicago Area Anatomy and Physiology Society (founding member), and The Nature Conservancy. In addition to writing many publications, he co-produced with Hassan Rastegar a videotape entitled *Introduction to the Human Cadaver and Prosection,* published by Wm. C. Brown Publishers. Terry revised the *Laboratory Manual to Accompany Hole's Essentials of Human Anatomy & Physiology,* Thirteenth Edition, and authored *Laboratory Manual for Human Anatomy & Physiology,* main, cat, and fetal pig, third editions, as well as co-authored the fourth editions with Cynthia Prentice-Craver. A series of seven LabCam videos of anatomy and physiology laboratory processes were produced at Kishwaukee College, published by McGraw-Hill Higher Education. Cadaver dissection experiences have been provided for his students for over thirty years. Terry teaches portions of EMT and paramedic classes. Terry also was a faculty exchange member in Ireland. He locally supports historical preservation, natural areas, scouting, and scholarship.

# TO THE STUDENT

The exercises in this laboratory manual will provide you with opportunities to observe various anatomical parts and to investigate certain physiological phenomena. Such experiences should help you to relate specimens, models, microscope slides, and your body to what you have learned in the lecture and read about in the textbook.

The following list of suggestions and study skills may help make your laboratory activities more effective and profitable.

1. Prepare yourself before attending the laboratory session by reading the assigned exercise and reviewing the related sections of the textbook. It is important to have some understanding of what will be done in the laboratory before you come to class.

2. Bring your laboratory manual and textbook to each laboratory session. These books are closely integrated and will help you complete most of the exercises.

3. Be on time. During the first few minutes of the laboratory meeting, the instructor often will provide verbal instructions. Make special note of any changes in materials to be used or procedures to be followed. Also listen carefully for information about special techniques to be used and precautions to be taken.

4. Keep your work area clean and your materials neatly arranged so that you can locate needed items. This will enable you to proceed efficiently and will reduce the chances of making mistakes.

5. Pay particular attention to the purpose of the exercise, which states what you are to accomplish in general terms, and to the learning outcomes, which list what you should be able to do as a result of the laboratory experience. Then, before you leave the class, review the outcomes and make sure that you can meet all of the assessments.

6. Precisely follow the directions in the procedure and proceed only when you understand them clearly. Do not improvise procedures unless you have the approval of the laboratory instructor. Ask questions if you do not understand exactly what you are supposed to do and why you are doing it.

7. Handle all laboratory materials with care. Some of these materials are fragile and expensive to replace. Whenever you have questions about the proper treatment of equipment, ask the instructor.

8. Treat all living specimens humanely and try to minimize any discomfort they might experience.

9. Although at times you might work with a laboratory partner or a small group, try to remain independent when you are making observations, drawing conclusions, and completing the activities in the laboratory reports.

10. Record your observations immediately after making them. In most cases, such data can be entered in spaces provided in the laboratory reports.

11. Read the instructions for each section of the laboratory report before you begin to complete it. Think about the questions before you answer them. Your responses should be based on logical reasoning and phrased in clear and concise language.

12. At the end of each laboratory period, clean your work area and the instruments you have used. Return all materials to their proper places and dispose of wastes, including glassware or microscope slides that have become contaminated with human blood or body fluids, as directed by the laboratory instructor. Wash your hands thoroughly before leaving the laboratory.

## Study Skills for Anatomy and Physiology

Students have found that certain study skills worked well for them while they were enrolled in Human Anatomy and Physiology. Although everyone has his or her learning style, there are techniques that work well for most students. Using some of the skills listed here can make your course more enjoyable and rewarding.

1. **Time management:** Prepare monthly, weekly, and daily schedules. Include dates of quizzes, exams, and projects on the calendar. On your daily schedule, budget several short study periods. Daily repetition alleviates cramming. Prioritize your tasks, so that you still have time for work and leisure activities. Find an appropriate study atmosphere with minimum distractions.

2. **Note taking:** Look for the main ideas and briefly express them in your own words. Organize, edit, and review your notes soon after the lecture. Add textbook information to your notes as you reorganize them. Underline or highlight with different colors the important points, major headings, and key terms. Study your notes daily, as they provide sequential building blocks of the course content.

3. **Chunking:** Organize information into logical groups or categories. Study and master one chunk of information at a time. For example, study the bones of the upper limb, lower limb, trunk, and head as separate study tasks.

4. **Mnemonic devices:** An *acrostic* is a combination of association and imagery to aid your memory. It is often in the form of a poem, rhyme, or jingle in which the first letter of each word corresponds to the first letters of the words you need to remember. **S**o **L**ong **T**op **P**art, **H**ere **C**omes **T**he **T**humb is an example of such a mnemonic device for remembering the eight carpals in a correct sequence. *Acronyms* are words formed by the first letters of the items to remember. *IPMAT* is an example of this type of mnemonic device to help you remember the phases of the cell cycle in the correct sequence. Try to create some of your own.

5. **Note cards/flash cards:** Make your own. Add labels and colors to enhance the material. Keep them with you; study them often and for short periods. Concentrate on a small number of cards at one time. Shuffle your cards and have someone quiz you on their content. As you become familiar with the material, you can set aside cards that don't require additional mastery.

6. **Recording and recitation:** An auditory learner can benefit by recording lectures and review sessions with a cassette recorder. Many students listen to the taped sessions as they drive or just before they go to bed. Reading your notes aloud can help also. Explain the material to anyone (even if there are no listeners). Talk about anatomy and physiology in everyday conversations.

7. **Study groups:** Small study groups that meet periodically to review course material and compare notes have helped and encouraged many students. However, keep the group on the task at hand. Work as a team and alternate leaders. This group often becomes a support group.

Practice sound study skills during your anatomy and physiology endeavor.

# Correlation of Textbook Chapters and Laboratory Exercises

| Textbook Chapters | Related Laboratory Exercises |
|---|---|
| **Chapter 1** Introduction to Human Anatomy and Physiology | **Exercise 1** Scientific Method and Measurements |
| | **Exercise 2** Body Organization and Terminology |
| **Chapter 2** Chemical Basis of Life | **Exercise 3** Chemistry of Life |
| **Chapter 3** Cells | **Exercise 4** Care and Use of the Microscope |
| | **Exercise 5** Cell Structure and Function |
| | **Exercise 6** Movements Through Membranes |
| | **Exercise 7** Cell Cycle |
| **Chapter 4** Cellular Metabolism | |
| **Chapter 5** Tissues | **Exercise 8** Epithelial Tissues |
| | **Exercise 9** Connective Tissues |
| | **Exercise 10** Muscle and Nervous Tissues |
| **Chapter 6** Integumentary System | **Exercise 11** Integumentary System |
| **Chapter 7** Skeletal System | **Exercise 12** Bone Structure and Classification |
| | **Exercise 13** Organization of the Skeleton |
| | **Exercise 14** Skull |
| | **Exercise 15** Vertebral Column and Thoracic Cage |
| | **Exercise 16** Pectoral Girdle and Upper Limb |
| | **Exercise 17** Pelvic Girdle and Lower Limb |
| **Chapter 8** Joints of the Skeletal System | **Exercise 18** Joint Structure and Movements |
| **Chapter 9** Muscular System | **Exercise 19** Skeletal Muscle Structure and Function |
| | **Exercise 20** Muscles of the Head and Neck |
| | **Exercise 21** Muscles of the Chest, Shoulder, and Upper Limb |
| | **Exercise 22** Muscles of the Deep Back, Abdominal Wall, and Pelvic Floor |
| | **Exercise 23** Muscles of the Hip and Lower Limb |
| | **Exercise 24** Surface Anatomy |
| | **Exercise 54** Cat Dissection: Musculature |
| **Chapter 10** Nervous System I: Basic Structure and Function | **Exercise 25** Nervous Tissue and Nerves |
| **Chapter 11** Nervous System II: Divisions of the Nervous System | **Exercise 26** Brain and Cranial Nerves |
| | **Exercise 27** Dissection of the Sheep Brain |
| | **Exercise 28** Spinal Cord and Meninges |
| | **Exercise 29** Reflex Arc and Reflexes |
| **Chapter 12** Nervous System III: Senses | **Exercise 30** Receptors and General Senses |
| | **Exercise 31** Smell and Taste |
| | **Exercise 32** Ear and Hearing |
| | **Exercise 33** Ear and Equilibrium |
| | **Exercise 34** Eye Structure |
| | **Exercise 35** Visual Tests and Demonstrations |
| **Chapter 13** Endocrine System | **Exercise 36** Endocrine Histology and Diabetic Physiology |
| **Chapter 14** Blood | **Exercise 37** Blood Cells and Blood Typing |
| **Chapter 15** Cardiovascular System | **Exercise 38** Heart Structure |
| | **Exercise 39** Cardiac Cycle |
| | **Exercise 40** Blood Vessel Structure, Arteries, and Veins |
| | **Exercise 41** Pulse Rate and Blood Pressure |
| | **Exercise 55** Cat Dissection: Cardiovascular System |
| **Chapter 16** Lymphatic System and Immunity | **Exercise 42** Lymphatic System |
| **Chapter 17** Digestive System | **Exercise 43** Digestive Organs |
| | **Exercise 44** Action of a Digestive Enzyme |
| | **Exercise 56** Cat Dissection: Digestive System |
| **Chapter 18** Nutrition and Metabolism | |

## Correlation of Textbook Chapters and Online Laboratory Exercises

*These laboratory exercises are available online for instructors via the Instructor Resources in Connect. They are available for students via the eBook in Connect.

# The Use of Animals in Biology Education*

The National Association of Biology Teachers (NABT) believes that the study of organisms, including nonhuman animals, is essential to the understanding of life on Earth. NABT recommends the prudent and responsible use of animals in the life science classroom. NABT believes that biology teachers should foster a respect for life. Biology teachers also should teach about the interrelationship and interdependency of all things.

Classroom experiences that involve nonhuman animals range from observation to dissection. NABT supports these experiences as long as they are conducted within the long-established guidelines of proper care and use of animals, as developed by the scientific and educational community.

As with any instructional activity, the use of nonhuman animals in the biology classroom must have sound educational objectives. Any use of animals, whether for observation or for dissection, must convey substantive knowledge of biology. NABT believes that biology teachers are in the best position to make this determination for their students.

NABT acknowledges that no alternative can substitute for the actual experience of dissection or other use of animals and urges teachers to be aware of the limitations of alternatives. When the teacher determines that the most effective means to meet the objectives of the class do not require dissection, NABT accepts the use of alternatives to dissection, including models and the various forms of multimedia. The association encourages teachers to be sensitive to substantive student objections to dissection and to consider providing appropriate lessons for those students where necessary.

To implement this policy, NABT endorses and adopts the "Principles and Guidelines for the Use of Animals in Precollege Education" of the Institute of Laboratory Animals Resources (National Research Council). Copies of the "Principles and Guidelines" may be obtained from the ILAR (2101 Constitution Avenue, NW, Washington, DC, 20418; 202-334-2590).

---

*Adopted by the Board of Directors in October 1995. This policy supersedes and replaces all previous NABT statements regarding animals in biology education.

# 1

# Scientific Method and Measurements

## MATERIALS NEEDED

Meterstick
Calculator
Human skeleton

## PURPOSE OF THE EXERCISE

To become familiar with the scientific method of investigation, learn how to formulate sound conclusions, and provide opportunities to use the metric system of measurements.

 **LEARNING OUTCOMES**

After completing this exercise, you should be able to

1. Convert English measurements to the metric system, and vice versa.

2. Calculate expected upper limb length and actual percentage of height from recorded upper limb lengths and heights.

3. Apply the scientific method to test the validity of a hypothesis concerning the direct, linear relationship between human upper limb length and height.

4. Design an experiment, formulate a hypothesis, and test it using the scientific method.

Scientific investigation involves a series of logical steps to arrive at explanations for various biological phenomena. It reflects a long history of asking questions and searching for knowledge. This technique, called the *scientific method,* is used in all disciplines of science. It allows scientists to draw logical and reliable conclusions about phenomena.

The scientific method begins with making *observations* related to the topic under investigation. This step commonly involves the accumulation of previously acquired information and/or your observations of the phenomenon. These observations are used to formulate a tentative explanation known as the *hypothesis.* An important attribute of a hypothesis is that it must be testable. The testing of the proposed hypothesis involves designing and performing a carefully controlled *experiment* to obtain data that can be used to support, reject, or modify the hypothesis.

During the experiment to test the proposed hypothesis, it is important to be able to examine only a single changeable factor, known as a *variable.* An *independent variable* is one that can be changed, but is determined before the experiment occurs; a *dependent variable* is determined from the results of the experiment.

An *analysis of data* is conducted using sufficient information collected during the experiment. Data analysis may include organization and presentation of data as tables, graphs, and drawings. From the interpretation of the data analysis, *conclusions* are drawn. (If the data do not support the hypothesis, you must reexamine the experimental design and the data, and if needed, develop a new hypothesis.) The final presentation of the information is made from the conclusions. Results and conclusions are presented to the scientific community for evaluation through peer reviews, presentations at professional meetings, and published articles. If many investigators working independently can validate the hypothesis by arriving at the same conclusions, the explanation can become a *theory.* A theory serves as the explanation from a summary of known experiments and supporting evidence unless it is disproved by new information. The five components of the scientific method are summarized as

**Observations**
↓
**Hypothesis**
↓
**Experiment**
↓
**Analysis of data**
↓
**Conclusions**

Metric measurements are characteristic tools of scientific investigations. The English system of measurements is often used in the United States, so the investigator must make conversions from the English system to the metric system. Table 1.1 provides the conversion factors necessary to change from English units to metric units.

**TABLE 1.1  Metric Measurement System and Conversions**

| Measurement | Unit & Abbreviation | Metric Equivalent | Conversion Factor Metric to English (approximate) | Conversion Factor English to Metric (approximate) |
|---|---|---|---|---|
| **LENGTH** | 1 kilometer (km) | 1,000 ($10^3$) m | 1 km = 0.62 mile | 1 mile = 1.61 km |
| | 1 meter (m) | 100 ($10^2$) cm<br>1,000 ($10^3$) mm | 1 m = 1.1 yards<br>    = 3.3 feet<br>    = 39.4 inches | 1 yard = 0.9 m<br>1 foot = 0.3 m |
| | 1 decimeter (dm) | 0.1 ($10^{-1}$) m | 1 dm = 3.94 inches | 1 inch = 0.25 dm |
| | 1 centimeter (cm) | 0.01 ($10^{-2}$) m | 1 cm = 0.4 inches | 1 foot = 30.5 cm<br>1 inch = 2.54 cm |
| | 1 millimeter (mm) | 0.001 ($10^{-3}$) m<br>0.1 ($10^{-1}$) cm | 1 mm = 0.04 inches | |
| | 1 micrometer (μm) | 0.000001 ($10^{-6}$) m<br>0.001 ($10^{-3}$) mm | | |
| **MASS** | 1 metric ton (t) | 1,000 ($10^3$) kg | 1 t = 1.1 ton | 1 ton = 0.91 t |
| | 1 kilogram (kg) | 1,000 ($10^3$) g | 1 kg = 2.2 pounds | 1 pound = 0.45 kg |
| | 1 gram (g) | 1,000 ($10^3$) mg | 1 g = 0.04 ounce | 1 pound = 454 g<br>1 ounce = 28.35 g |
| | 1 milligram (mg) | 0.001 ($10^{-3}$) g | | |
| **VOLUME (LIQUIDS AND GASES)** | 1 liter (L) | 1,000 ($10^3$) mL | 1 L = 1.06 quarts | 1 gallon = 3.78 L<br>1 quart = 0.95 L |
| | 1 milliliter (mL) | 0.001 ($10^{-3}$) L<br>1 cubic centimeter (cc or $cm^3$) | 1 mL = 0.03 fluid ounce<br>1 mL = 1/5 teaspoon<br>1 mL = 15–16 drops | 1 quart = 946 mL<br>1 fluid ounce = 29.6 mL<br>1 teaspoon = 5 mL |
| **TIME** | 1 second (s) | 1/60 minute | Same | Same |
| | 1 millisecond (ms) | 0.001 ($10^{-3}$) s | Same | Same |
| **TEMPERATURE** | Degrees Celsius (°C) | | °F = 9/5°C + 32 | °C = 5/9 (°F − 32) |

**PROCEDURE A—Using the Steps of the Scientific Method**

1. Refer to Appendix A of the textbook for additional information on the scientific method.
2. Many people have observed a correlation between the length of the upper and lower limbs and the height (stature) of an individual. For example, a person who has long upper limbs (the arm, forearm, and hand combined) tends to be tall. Make some visual observations of other people in your class to observe a possible correlation.
3. From such observations, the following hypothesis can be formulated: The length of a person's upper limb is equal to 0.4 (40%) of the height of the person. To test this hypothesis, perform the following experiment.
4. Use a meterstick (fig. 1.1) to measure an upper limb length of ten subjects. Place the meterstick in the axilla (armpit) and record the length in centimeters to the end of the longest finger (fig. 1.2). Obtain the height of each person in centimeters by measuring them without shoes against a wall (fig. 1.3). The height of each person can also be calculated by multiplying each individual's height in inches by 2.54 to obtain his/her height in centimeters. Record all your measurements in Part A of Laboratory Report 1.

5. The data collected from all of the measurements can now be analyzed. The expected (predicted) correlation between upper limb length and height is determined using the following equation:

$$\text{Height} \times 0.4 = \text{expected upper limb length}$$

The observed (actual) correlation to be used to test the hypothesis is determined by

$$\text{Length of upper limb/height} = \text{actual \% of height}$$

6. A graph is an excellent way to display a visual representation of the data. Plot the subjects' data in Part A of the laboratory report. Plot the upper limb length of each subject on the *x*-axis (independent variable) and the

Metric ruler

**FIGURE 1.1** Metric ruler with metric lengths indicated. A meterstick length would be 100 centimeters (10 decimeters). (The image size is approximately to scale.)

**FIGURE 1.2** Measurement of upper limb length. (©J & J Photography.)

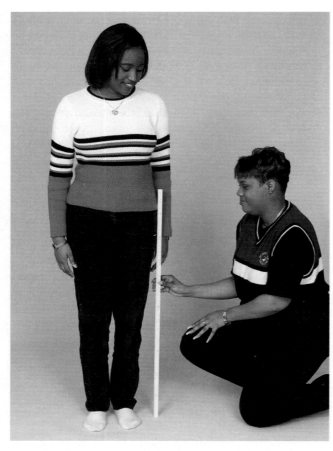

**FIGURE 1.3** Measurement of height. (©J & J Photography)

height of each person on the *y*-axis (dependent variable). A line is already located on the graph that represents a hypothetical relationship of 0.4 (40%) upper limb length compared to height. This is a graphic representation of the original hypothesis.

7. Compare the distribution of all of the points (actual height and upper limb length) that you placed on the graph with the distribution of the expected correlation represented by the hypothesis.

8. Complete Part A of the laboratory report.

 **PRACTICE**

PROCEDURE B—Design an Experiment

 **ASSESS**

**CRITICAL THINKING**

You have probably concluded that there is some correlation of the length of body parts to height. Often when a skeleton is found, it is not complete. It is occasionally feasible to use the length of a single bone to estimate the height of an individual. Observe human skeletons and locate the humerus bone in an upper limb or the femur bone in a lower limb. Use your observations to identify a mathematical relationship between the length of the humerus or femur and height. Formulate a hypothesis that can be tested. Make measurements, analyze data, and develop a conclusion from your experiment. Complete Part B of the laboratory report.

Name _____

Date _____

Section _____

The Ⓐ corresponds to the indicated Learning Outcome(s) found at the beginning of the laboratory exercise.

# Scientific Method
# and Measurements

## PART A ASSESSMENTS

1. Record measurements for the upper limb length and height of ten subjects. Use a calculator to determine the expected upper limb length and the actual percentage (as a decimal or a percentage) of the height for the ten subjects. Record your results in the following table. Ⓐ**2**

| Subject | Measured Upper Limb Length (cm) | Height* (cm) | Height × 0.4 = Expected Upper Limb Length (cm) | Actual % of Height = Measured Upper Limb Length (cm)/Height (cm) |
|---------|--------------------------------|--------------|-----------------------------------------------|----------------------------------------------------------------|
| 1. | | | | |
| 2. | | | | |
| 3. | | | | |
| 4. | | | | |
| 5. | | | | |
| 6. | | | | |
| 7. | | | | |
| 8. | | | | |
| 9. | | | | |
| 10. | | | | |

*The height of each person can be calculated by multiplying each individual's height in inches by 2.54 to obtain his/her height in centimeters. Ⓐ**1**

**2.** Plot the distribution of data (upper limb length and height) collected for the ten subjects on the following graph. The line located on the graph represents the *expected* 0.4 (40%) ratio of upper limb length to measured height (the original hypothesis). (The *x*-axis represents upper limb length and the *y*-axis represents height.) Draw a line of *best fit* through the distribution of points of the plotted data of the ten subjects. Compare the two distributions (expected line and the distribution line drawn for the ten subjects). A

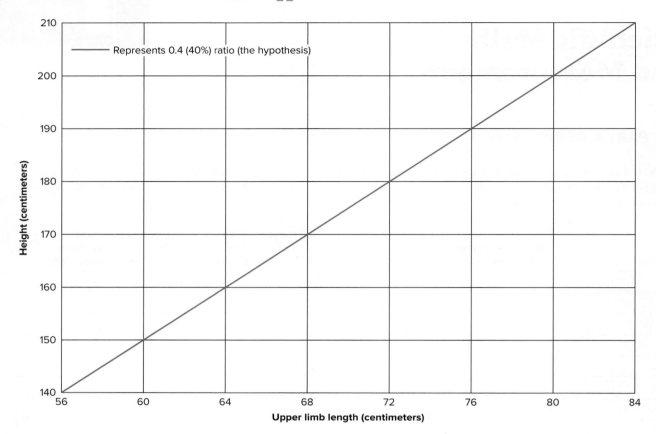

**3.** Does the distribution of the ten subjects' measured upper limb lengths support or reject the original hypothesis? _____ Explain your answer. A

**1.** Describe your observations of a possible correlation between the humerus or femur length and height. **A**

**2.** Write a hypothesis based on your observations. **A**

**3.** Describe the design of the experiment that you devised to test your hypothesis. **A**

**4.** Place your analysis of the data in this space in the form of a table and a graph. **A**
  **a.** Table:

  **b.** Graph:

**5.** Based on an analysis of your data, what can you conclude? Do these conclusions confirm or refute your original hypothesis? **A**

**6.** Discuss your results and conclusions with classmates. What common conclusion can the class formulate about the correlation between the humerus or femur length and height? **A**

# 2

# Body Organization and Terminology

## PURPOSE OF THE EXERCISE

To review the organizational pattern of the human body, to review its organ systems and the organs included in each system, and to become acquainted with the terms used to describe the relative position of body parts, body sections, and body regions.

 **LEARNING OUTCOMES** AP|R

After completing this exercise, you should be able to

①  Locate and name the major body cavities and identify the membranes associated with each cavity.

②  Differentiate the general functions of the organ systems of the human body.

③  Associate the organs included within each system and locate the organs in a dissectible human torso model.

④  Select the terms used to describe the relative positions of body parts.

⑤  Match the terms used to identify body sections and identify the plane along which a particular specimen is cut.

⑥  Label the body regions and associate the terms used to identify body regions.

⑦  Recognize the anatomical terms pertaining to body organization.

**T**he major features of the human body include certain cavities, a set of membranes associated with these cavities, and a group of organ systems composed of related organs. In order to communicate effectively with each other about the body, scientists have devised names to describe these body features. They also have developed terms to represent the relative positions of body parts, imaginary planes passing through these parts, and body regions.

 **PRACTICE**

### PROCEDURE A—Body Cavities and Membranes

1.  Review the sections entitled "Body Cavities" and "Thoracic and Abdominopelvic Membranes" in chapter 1 of the textbook.
2.  As a review activity, label figures 2.1 and 2.2.
3.  Locate the following features on the reference plates near the end of chapter 1 of the textbook and on the dissectible human torso model (fig. 2.3):

**body cavities**
    cranial cavity
    vertebral canal (spinal cavity)
    thoracic cavity
        mediastinum (region between the lungs;
            includes pericardial cavity)
        pleural cavities
    abdominopelvic cavity
        abdominal cavity
        pelvic cavity
**diaphragm**
**smaller cavities within the head**
    oral cavity
    nasal cavity with connected sinuses
    orbital cavity
    middle ear cavity
**membranes and cavities**
    pleural cavity
        parietal pleura
        visceral pleura

pericardial cavity
    parietal pericardium (covered by fibrous
      pericardium)
    visceral pericardium (epicardium)

peritoneal cavity
    parietal peritoneum
    visceral peritoneum

**4.** Complete Part A of Laboratory Report 2.

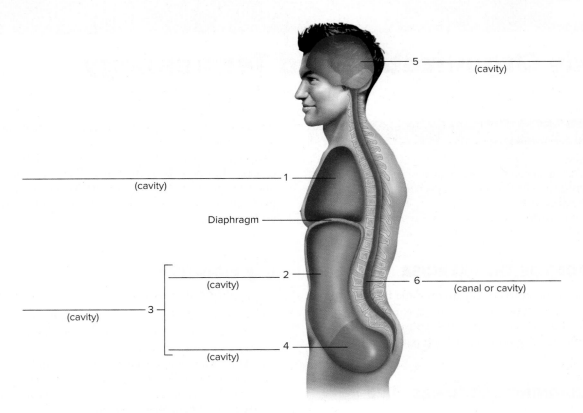

5 _____ (cavity)

1 _____ (cavity)

Diaphragm

2 _____ (cavity)

3 _____ (cavity)

4 _____ (cavity)

6 _____ (canal or cavity)

**FIGURE 2.1**   Label these major body cavities. A AP|R

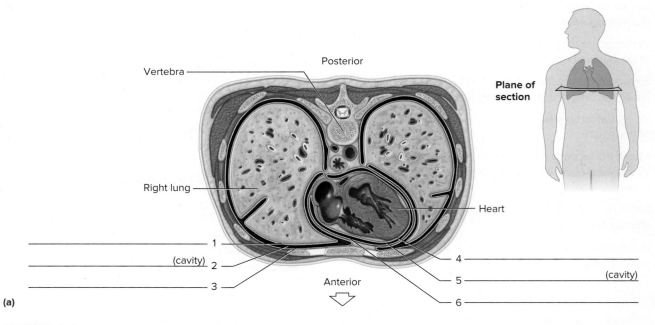

Posterior

Vertebra

Plane of section

Right lung

Heart

1 _____ (cavity)
2 _____
3 _____

4 _____
5 _____ (cavity)
6 _____

Anterior

(a)

**FIGURE 2.2**   Label the thoracic membranes and cavities in (*a*) and the abdominopelvic membranes and cavity in (*b*), as shown in these superior views of transverse sections. A AP|R

Posterior

Kidney

Spleen

Liver

Stomach

Small intestine

7

(cavity) 8

9

Anterior

Plane of section

**(b)**

**FIGURE 2.2** *Continued.* **AP|R**

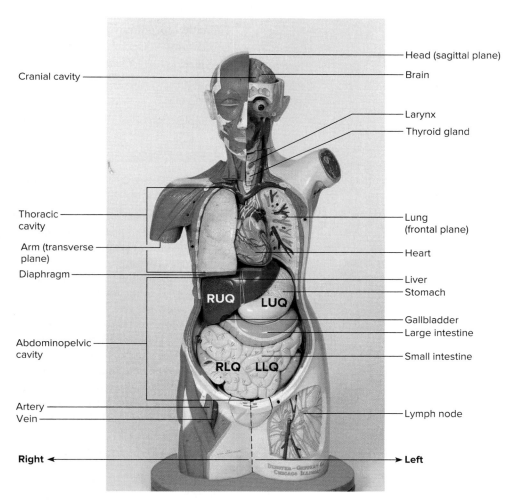

Head (sagittal plane)

Cranial cavity

Brain

Larynx

Thyroid gland

Thoracic cavity

Lung (frontal plane)

Arm (transverse plane)

Heart

Diaphragm

Liver

Stomach

RUQ    LUQ

Gallbladder

Large intestine

Abdominopelvic cavity

Small intestine

RLQ    LLQ

Artery

Vein

Lymph node

**Right**

**Left**

**FIGURE 2.3** Dissectible human torso model with body cavities, abdominopelvic quadrants, body planes, and major organs indicated. The abdominopelvic quadrants include right upper quadrant (RUQ), left upper quadrant (LUQ), right lower quadrant (RLQ), and left lower quadrant (LLQ). **AP|R** (©J & J Photography)

**PROCEDURE B—Organ Systems**

1. Review the section entitled "Organ Systems" in chapter 1 of the textbook.
2. Use the reference plates near the end of chapter 1 of the textbook and the dissectible human torso model (fig. 2.3) to locate the following systems and their major organs:

**integumentary system**
  skin
  accessory organs such as hair and nails

**skeletal system**
  bones
  ligaments

**muscular system**
  skeletal muscles
  tendons

**nervous system**
  brain
  spinal cord
  nerves

**endocrine system**
  pituitary gland
  thyroid gland
  parathyroid glands
  adrenal glands
  pancreas
  ovaries
  testes
  pineal gland
  thymus

**cardiovascular system**
  heart
  arteries
  veins

**lymphatic system**
  lymphatic vessels
  lymph nodes
  thymus
  spleen

**digestive system**
  mouth
  tongue
  teeth
  salivary glands
  pharynx
  esophagus
  stomach
  liver
  gallbladder
  pancreas
  small intestine
  large intestine

**respiratory system**
  nasal cavity
  pharynx
  larynx
  trachea
  bronchi
  lungs

**urinary system**
  kidneys
  ureters
  urinary bladder
  urethra

**male reproductive system**
  scrotum
  testes
  penis
  urethra

**female reproductive system**
  ovaries
  uterine tubes (oviducts; fallopian tubes)
  uterus
  vagina

3. Complete Part B of the laboratory report.

 **PRACTICE**

**PROCEDURE C—Relative Positions, Planes, Sections, and Regions**

1. Observe the person standing in anatomical position (fig. 2.4). Anatomical terminology assumes the body is in anatomical position even though a person is often observed differently.
2. Review the section entitled "Anatomical Terminology" in chapter 1 of the textbook.
3. As a review activity, label figures 2.5, 2.6, and 2.7.
4. Examine the sectioned specimens on the demonstration table and identify the plane along which each is cut. Cylindrical structures, such as a long bone or a blood vessel, may be cut in cross section, or longitudinal section. The same three sections can be demonstrated by three cuts of a banana (fig. 2.8).
5. Complete Parts C, D, E, F, and G of the laboratory report.

 **LEARN: ACTIVITY**

Use different colored pencils to distinguish body regions in figure 2.7.

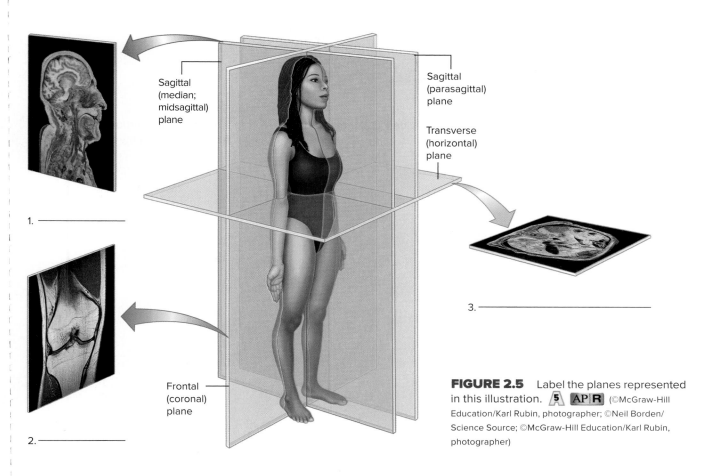

**FIGURE 2.4** Anatomical position with directional terms indicated. The body is standing erect, face forward, with upper limbs at the sides and palms forward. When the palms are forward (supinated), the radius and ulna in the forearm are nearly parallel. This results in an anterior view of the body. Terms of relative position are used to describe the location of one body part with respect to another. **AP|R** (©J & J Photography)

Superior

Right ←——→ Left

Proximal

Medial ←——→ Lateral

Distal

Proximal

Inferior

Distal

Sagittal (median; midsagittal) plane

Sagittal (parasagittal) plane

Transverse (horizontal) plane

Frontal (coronal) plane

1. ———————

2. ———————

3. ———————

**FIGURE 2.5** Label the planes represented in this illustration. **5 AP|R** (©McGraw-Hill Education/Karl Rubin, photographer; ©Neil Borden/ Science Source; ©McGraw-Hill Education/Karl Rubin, photographer)

(a)

(b)

**FIGURE 2.6** The abdominopelvic cavity is subdivided into either nine regions or four quadrants for purposes of locating organs, injuries, pain, or performed medical procedures. Label (*a*) the 9 regions and (*b*) the 4 quadrants of the abdominopelvic cavity.

(nose) _____ 1

(mouth) _____ 2

(neck) _____ 3

(shoulder) _____ 4

(armpit) _____ 5

(breast) _____ 6

(arm) _____ 7

(anterior elbow) _____ 8

(abdomen) _____ 9

(forearm) _____ 10

(wrist) _____ 11

(palm) _____ 12

(finger or toe) _____ 13

(reproductive) _____ 14

(leg) _____ 15

(ankle) _____ 16

17 _____ (head)

18 _____ (forehead)

19 _____ (eye)

20 _____ (cheek)

21 _____ (chin)

22 _____ (middle thorax)

23 _____ (chest)

24 _____ (navel)

25 _____ (groin)

26 _____ (hip)

27 _____ (anterior knee)

28 _____ (foot)

(a)

**FIGURE 2.7**　Label these diagrams with terms used to describe body regions: (*a*) anterior regions;
(*b*) posterior regions. 　6　AP|R

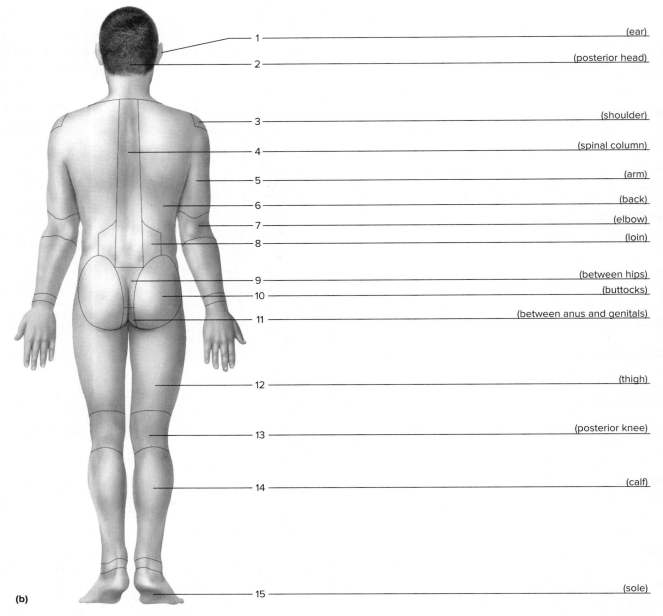

1 —————————————— (ear)

2 —————————————— (posterior head)

3 —————————————— (shoulder)

4 —————————————— (spinal column)

5 —————————————— (arm)

6 —————————————— (back)

7 —————————————— (elbow)

8 —————————————— (loin)

9 —————————————— (between hips)

10 —————————————— (buttocks)

11 —————————————— (between anus and genitals)

12 —————————————— (thigh)

13 —————————————— (posterior knee)

14 —————————————— (calf)

15 —————————————— (sole)

(b)

**FIGURE 2.7**   *Continued.*  **AP|R**

(a)                              (b)                              (c)

**FIGURE 2.8**   Three possible cuts of a banana: (*a*) cross section; (*b*) oblique section; (*c*) longitudinal section. Sections through an organ, as a body tube, frequently produce views similar to the cut banana.  ((*a,b,c*) ©J & J Photography)

LABORATORY REPORT

The ⚠ corresponds to the indicated Learning Outcome(s) found at the beginning of the laboratory exercise.

# Body Organization and Terminology

## PART A ASSESSMENTS

Match the body cavities in column A with the organs contained in the cavities in column B. Place the letter of your choice in the space provided. ⚠

| Column A | Column B |
|---|---|
| **a.** Abdominal cavity | _____ **1.** Liver |
| **b.** Cranial cavity | _____ **2.** Lungs |
| **c.** Pelvic cavity | _____ **3.** Spleen |
| **d.** Thoracic cavity | _____ **4.** Stomach |
| **e.** Vertebral canal (spinal cavity) | _____ **5.** Brain |
| | _____ **6.** Internal reproductive organs |
| | _____ **7.** Gallbladder |
| | _____ **8.** Urinary bladder |
| | _____ **9.** Mediastinum |
| | _____ **10.** Spinal cord |
| | _____ **11.** Heart |
| | _____ **12.** Small intestine |

## PART B ASSESSMENTS

Match the organ systems in column A with the general functions in column B. Place the letter of your choice in the space provided. ⚠

**Column A**

**a.** Cardiovascular system
**b.** Digestive system
**c.** Endocrine system
**d.** Integumentary system
**e.** Lymphatic system
**f.** Muscular system
**g.** Nervous system
**h.** Reproductive system
**i.** Respiratory system
**j.** Skeletal system
**k.** Urinary system

**Column B**

_____ **1.** Main system that secretes hormones

_____ **2.** Provides an outer covering of the body for protection

_____ **3.** Produces gametes (eggs and sperm)

_____ **4.** Stimulates muscles to contract and interprets information from sensory organs

_____ **5.** Provides a framework and support for soft tissues and produces blood cells in red marrow

_____ **6.** Exchanges gases between air and blood

_____ **7.** Transports excess fluid from tissues to blood

_____ **8.** Involves contractions and creates most body heat

_____ **9.** Removes liquid and wastes from blood and transports them to the outside of the body

_____ **10.** Converts food molecules into absorbable forms

_____ **11.** Transports nutrients, wastes, and gases throughout the body

**PART C ASSESSMENTS**

Indicate whether each of the following sentences makes correct or incorrect usage of the word in boldface type (assume that the body is in the anatomical position as observed in fig. 2.4). If the sentence is incorrect, in the space provided supply a term that will make it correct. **A**

1. The mouth is **superior** to the nose. _____

2. The stomach is **inferior** to the diaphragm. _____

3. The trachea is **anterior** to the spinal cord. _____

4. The larynx is **posterior** to the esophagus. _____

5. The heart is **medial** to the lungs. _____

6. The kidneys are **inferior** to the lungs. _____

7. The hand is **proximal** to the elbow. _____

8. The knee is **proximal** to the ankle. _____

9. The thumb is the **lateral** digit of the hand. _____

10. A **peripheral** nerve passes from the spinal cord into the limbs. _____

11. The spleen and gallbladder are **ipsilateral**. _____

12. The popliteal surface region is **anterior** to the patellar region. _____

**PART D ASSESSMENTS**

Match the body regions in column A with the locations in column B. Place the letter of your choice in the space provided. **6**

| Column A | | Column B |
|----------|--|----------|
| **a.** Antecubital | _____ | **1.** Wrist |
| **b.** Axillary | _____ | **2.** Reproductive organs |
| **c.** Brachial | _____ | **3.** Armpit |
| **d.** Buccal | _____ | **4.** Posterior region of elbow |
| **e.** Carpal | _____ | **5.** Buttocks |
| **f.** Cephalic | _____ | **6.** Finger or toe |
| **g.** Cervical | _____ | **7.** Neck |
| **h.** Crural | _____ | **8.** Arm |
| **i.** Cubital | _____ | **9.** Cheek |
| **j.** Digital | _____ | **10.** Leg |
| **k.** Genital | _____ | **11.** Head |
| **l.** Gluteal | _____ | **12.** Anterior region of elbow |

Match the body regions in column A with the locations in column B. Place the letter of your choice in the space provided. **6** **A**

**Column A**

a. Inguinal
b. Lumbar
c. Mammary
d. Occipital
e. Palmar
f. Pectoral
g. Pedal
h. Perineal
i. Plantar
j. Popliteal
k. Sternal
l. Tarsal

**Column B**

_____ **1.** Ankle

_____ **2.** Breasts

_____ **3.** Between anus and reproductive organs

_____ **4.** Sole of foot

_____ **5.** Middle of thorax

_____ **6.** Chest

_____ **7.** Posterior region of knee

_____ **8.** Foot

_____ **9.** Inferior posterior region of head

_____ **10.** Abdominal wall near thigh (groin)

_____ **11.** Lower back

_____ **12.** Palm

**CRITICAL THINKING**

State the quadrant of the abdominopelvic cavity where the pain or sound would be located for each of the six conditions listed. In some cases, there may be more than one correct answer, and pain is sometimes referred to another region. This phenomenon, called *referred pain*, occurs when pain is interpreted as originating from some area other than the parts being stimulated. When referred pain is involved in the patient's interpretation of the pain location, the proper diagnosis of the ailment is more challenging. For the purpose of this exercise, assume the pain is interpreted as originating from the organ involved. **3**

1. Stomach ulcer _____

2. Appendicitis _____

3. Bowel sounds _____

4. Gallbladder attack _____

5. Kidney stone in left ureter _____

6. Ruptured spleen _____

## PART G ASSESSMENTS

Locate the 16 anatomical terms pertaining to "Body Organization and Terminology." [7]

| E | N | D | O | C | R | I | N | E | B | C | R |
|---|---|---|---|---|---|---|---|---|---|---|---|
| Q | L | A | N | I | M | O | D | B | A | X | E |
| C | R | A | N | I | A | L | U | P | E | L | P |
| S | L | A | T | T | I | G | A | S | Y | A | R |
| C | R | U | L | M | V | E | R | Q | O | R | O |
| I | L | G | L | U | T | E | A | L | K | O | D |
| C | A | C | F | H | V | W | L | L | T | G | U |
| A | T | O | C | S | B | H | U | I | W | T | C |
| R | N | X | N | B | U | C | C | A | L | K | T |
| O | O | A | D | H | L | A | S | R | A | T | I |
| H | R | L | F | L | A | R | U | S | K | O | V |
| T | F | W | G | Z | R | A | M | L | A | P | E |

Find and circle words pertaining to Laboratory Exercise 2. The words may be horizontal, vertical, diagonal, or backward.

8 body surface regions (buccal, coxal, gluteal, oral, otic, palmar, sural, tarsal)

2 organ system names; 3 body cavities; 3 planes of the body

# Chemistry of Life

## MATERIALS NEEDED

*For pH Tests:*

Chopped fresh red cabbage
Beaker (250 mL)
Distilled water
Tap water
Vinegar
Baking soda
Laboratory scoop for measuring
Pipets for measuring
Full-range pH test papers
7 assorted common liquids clearly labeled in closed
    bottles on a tray or in a tub
Droppers labeled for each liquid

*For Organic Tests:*

Test tubes
Test-tube rack
Test-tube clamps
China marker
Hot plate
Beaker for hot water bath (500 mL)
Pipets for measuring
Benedict's solution
Biuret reagent (or 10% NaOH and 1% $CuSO_4$)
Iodine-potassium-iodide (IKI) solution
Sudan IV dye
Egg albumin
10% glucose solution
Clear carbonated soft drink
10% starch solution
Potatoes for potato water
Distilled water
Vegetable oil
Brown paper
Numbered unknown organic samples

## ⚠ SAFETY

- Review all Laboratory Safety Guidelines in Appendix 1 of your laboratory manual.
- Clean laboratory surfaces before and after laboratory procedures using soap and water.
- Use extreme caution when working with chemicals.
- Wear safety goggles.
- Wear disposable gloves while working with the chemicals.
- Take precautions to prevent chemicals from contacting your skin.
- Do not mix any of the chemicals together unless instructed to do so.
- Clean up any spills immediately and notify the instructor at once.
- Wash your hands before leaving the laboratory.

## PURPOSE OF THE EXERCISE

To review the organization of atoms and molecules, types of chemical interactions, and basic categories of organic compounds and to differentiate between types of organic compounds.

##  LEARNING OUTCOMES

After completing this exercise, you should be able to

1. Associate and illustrate the basic organization of atoms and molecules.
2. Measure pH values of various substances through testing methods.
3. Determine categories of organic compounds with basic colorimetric tests.
4. Discover the organic composition of an unknown solution.

The complexities of the human body arise from the organization and interactions of chemicals. Organisms are made of matter, and the most basic unit of matter is the chemical element. The smallest unit of an element is an atom, and two or more of those can unite to form a molecule. All processes that occur within the body involve chemical reactions—interactions between atoms and molecules. We breathe to supply oxygen to our cells for energy. We eat and drink to bring chemicals into our bodies that our cells need. Water fills and bathes all of our cells and allows an amazing array of reactions to occur, all of which are designed to keep us alive. Chemistry forms the basis of life and thus forms the foundation of anatomy and physiology.

### PROCEDURE A—Matter, Molecules, Bonding, and pH

1. Review the section entitled "Structure of Matter" in chapter 2 of the textbook. Use the Periodic Table of Elements, located in Appendix 2 of the laboratory manual, as a reference.
2. Review figures 3.1, 3.2, and 3.3, which show molecular diagrams, formation of an ionic bond, and formation of a covalent bond, respectively.
3. Complete Part A of Laboratory Report 3.

 **PRACTICE**

### PROCEDURE B—The pH Scale

1. Review the section entitled "Acid and Base Concentrations" in chapter 2 of the textbook.
2. Review figure 3.4, which shows the pH scale. Note the range of the scale and the pH value considered neutral.
3. **Cabbage water tests.** Many tests can determine a pH value, but among the more interesting are colorimetric tests in which an indicator chemical changes color when it reacts. Many plant pigments, especially anthocyanins (which give plants color, from blue to red), can be used as colorimetric pH indicators. One that works well is the red pigment in red cabbage.
   a. Prepare cabbage water to be used as a general pH indicator. Fill a 250 mL beaker to the 100 mL level with chopped red cabbage. Add water to make 150 mL. Place the beaker on a hot plate and simmer the mixture until the pigments come out of the cabbage and the water turns deep purple. Allow the water to cool. (You may proceed to step 4 while you wait for this to finish.)
   b. Label three clean test tubes: one for water, one for vinegar, and one for baking soda.
   c. Place 2 mL of cabbage water into each test tube.
   d. To the first test tube, add 2 mL of distilled water and swirl the mixture. Record the color in Part B of the laboratory report.
   e. Repeat this procedure for test tube 2, adding 2 mL of vinegar and swirling the mixture. Record the results in Part B of the laboratory report.
   f. Repeat this procedure for test tube 3, adding one laboratory scoop of baking soda. Swirl the mixture. Record the results in Part B of the laboratory report.
4. **Testing with pH paper.** Many commercial pH indicators are available. A simple one to use is pH paper, which comes in small strips. Don gloves for this procedure so your skin secretions do not contaminate the paper, and to protect you from the chemicals you are testing.
   a. Test the pH of distilled water by dropping one or two drops of distilled water onto a strip of pH paper, and note the color. Compare the color to the color guide on the container. Note the results of the test as soon

Hydrogen (H)    Helium (He)    Lithium (Li)

**FIGURE 3.1**    The single electron of a hydrogen atom is located in its first shell. The two electrons of a helium atom fill its first shell. Two of the three electrons of a lithium atom are in the first shell, and one is in the second shell. **AP|R**

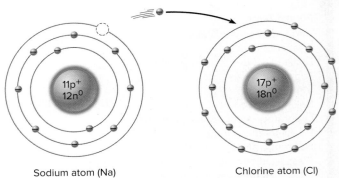

Sodium atom (Na)    Chlorine atom (Cl)

**(a)** Separate atoms
If a sodium atom loses an electron to a chlorine atom, the sodium atom becomes a sodium ion ($Na^+$), and the chlorine atom becomes a chloride ion ($Cl^-$).

Sodium ion ($Na^+$)    Chloride ion ($Cl^-$)

Sodium chloride

**(b)** Bonded ions
These oppositely charged particles attract electrically and join by an ionic bond.

**FIGURE 3.2**    Formation of an ionic bond. **AP|R**

as color appears, as drying of the paper will change the results. Record the pH value in Part B of the laboratory report.
   b. Repeat this procedure for tap water. Record the results.
   c. Individually test the various common substances found on your lab table. Record your results.
   d. Complete Part B of the laboratory report.

**FIGURE 3.3** A hydrogen molecule forms when two hydrogen atoms share a pair of electrons and join by a covalent bond. AP|R

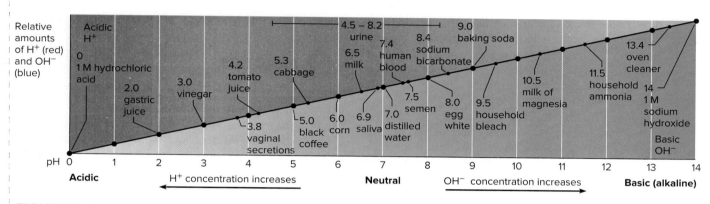

**FIGURE 3.4** As the concentration of hydrogen ions (H⁺) increases, a solution becomes more acidic, and the pH value decreases. As the concentration of ions that combine with hydrogen ions (such as hydroxide ions) increases, a solution becomes more basic (alkaline), and the pH value increases. The pH values of some common substances are shown.

## PRACTICE

### PROCEDURE C—Organic Molecules

In this section, you will perform some simple tests to check for the presence of the following categories of organic molecules (biomolecules): protein, sugar, starch, and lipid. Specific color changes will occur if the target compound is present. Please note the original color of the indicator being added so you can tell if the color really changed. For example, Biuret reagent and Benedict's solution are both initially blue, so an end color of blue would indicate no change. Use caution when working with these chemicals, and follow all directions. Carefully label the test tubes and avoid contaminating any of your samples. Only the Benedict's test requires heating and a time delay for accurate results. Do not heat any other tubes. To prepare for the Benedict's test, fill a 500 mL beaker half full with water and place it on the hot plate. Turn the hot plate on and bring the water to a boil. To save time, start the water bath before doing the Biuret test.

1. **Biuret test for protein.** In the presence of protein, Biuret reagent reacts with peptide bonds and changes to violet or purple. A pinkish color indicates that shorter polypeptides are present. The color intensity is proportional to the number of peptide bonds; thus, the intensity reflects the length of polypeptides (amount of protein).

a. Label six test tubes as follows: 1W, 1E, 1G, 1D, 1S, and 1P.

b. To each test tube, add 2 mL of one of the samples as follows:
   1W—2 mL distilled water
   1E—2 mL egg albumin
   1G—2 mL 10% glucose solution
   1D—2 mL carbonated soft drink
   1S—2 mL 10% starch solution
   1P—2 mL potato juice

c. To each, add 2 mL of Biuret reagent, swirl the tube to mix it, and note the final color. [*Note*: If Biuret reagent is not available, add 2 mL of 10% NaOH (sodium hydroxide) and about 10 drops of 1% CuSO4 (copper sulfate).] **Be careful—NaOH is very caustic.**

d. Record your results in Part C of the laboratory report.

e. Mark a "+" on any tubes with positive results and retain them for comparison while testing your unknown sample. Put remaining test tubes aside so they will not get confused with future trials.

2. **Benedict's test for sugar (monosaccharides).** In the presence of sugar, Benedict's solution changes from its initial blue to a green, yellow, orange, or reddish color, depending on the amount of sugar present. Orange and red indicate greater amounts of sugar.

a. Label six test tubes as follows: 2W, 2E, 2G, 2D, 2S, and 2P.

b. To each test tube, add 2 mL of one of the samples as follows:

2W—2 mL distilled water

2E—2 mL egg albumin

2G—2 mL 10% glucose solution

2D—2 mL carbonated soft drink

2S—2 mL 10% starch solution

2P—2 mL potato juice

c. To each, add 2 mL of Benedict's solution, swirl the tube to mix it, and place all tubes into the boiling water bath for 3 to 5 minutes.

d. Note the final color of each tube after heating and record the results in Part C of the laboratory report.

e. Mark a "+" on any tubes with positive results and retain them for comparison while testing your unknown sample. Put remaining test tubes aside so they will not get confused with future trials.

3. **Iodine test for starch.** In the presence of starch, iodine turns a dark purple or blue-black color. Starch is a long chain formed by many glucose units linked together side by side. This regular organization traps the iodine molecules and produces the dark color.

a. Label six test tubes as follows: 3W, 3E, 3G, 3D, 3S, and 3P.

b. To each test tube, add 2 mL of one of the samples as follows:

3W—2 mL distilled water

3E—2 mL egg albumin

3G—2 mL 10% glucose solution

3D—2 mL carbonated soft drink

3S—2 mL 10% starch solution

3P—2 mL potato juice

c. To each, add 0.5 mL of IKI (iodine solution) and swirl the tube to mix it.

d. Record the final color of each tube in Part C of the laboratory report.

e. Mark a "+" on any tubes with positive results and retain them for comparison while testing your unknown sample. Put remaining test tubes aside so they will not get confused with future trials.

4. **Tests for lipids.**

a. Label two separate areas of a piece of brown paper as "water" or "oil."

b. Place a drop of water on the area marked "water" and a drop of vegetable oil on the area marked "oil."

c. Let the spots dry several minutes; then record your observations in Part C of the laboratory report. Upon drying, oil leaves a translucent stain (grease spot) on brown paper; water does not leave such a spot.

d. In a test tube, add 2 mL of water and 2 mL of vegetable oil and observe. Shake the tube vigorously; then let it sit for 5 minutes and observe again. Record your observations in Part C of the laboratory report.

e. Sudan IV is a dye that is lipid-soluble but not water-soluble. If lipids are present, the Sudan IV will stain them pink or red. Add a small amount of Sudan IV to the test tube that contains the oil and water. Swirl it; then let it sit for a few minutes. Record your observations in Part C of the laboratory report.

## PRACTICE

### PROCEDURE D—Identifying Unknown Compounds

Now you will apply the information you gained with the tests in the previous section. You will retrieve an unknown sample that contains none, one, or any combination of the following types of organic compounds: protein, sugar, starch, or lipid. You will test your sample using each test from the previous section and record your results in Part D of the laboratory report.

1. Label four test tubes (1, 2, 3, 4).

2. Add 2 mL of your unknown sample to each test tube.

3. **Test for protein.** Add 2 mL of Biuret reagent to tube 1. Swirl the tube and observe the color. Record your observations in Part D of the laboratory report.

4. **Test for sugar.** Add 2 mL of Benedict's solution to tube 2. Swirl the tube and place it in a boiling water bath for 3 to 5 minutes. Record your observations in Part D of the laboratory report.

5. **Test for starch.** Add several drops of iodine solution to tube 3. Swirl the tube and observe the color. Record your observations in Part D of the laboratory report.

6. **Test for lipid.** Add 2 mL of water to tube 4. Swirl the tube and note if there is any separation.

7. Add a small amount of Sudan IV to test tube 4, swirl the tube, and record your observations in Part D of the laboratory report.

8. Based on the results of these tests, determine if your unknown sample contains any organic compounds and, if so, what they are. Record and explain your identification in Part D of the laboratory report.

Name _____

Date _____

Section _____

The ⚠ corresponds to the indicated Learning Outcome(s) found at the beginning of the laboratory exercise.

LABORATORY
REPORT

3

# Chemistry of Life

## ♻ PART A ASSESSMENTS

Match the terms in column A with the descriptions in column B. Place the letter of your choice in the space provided. ⚠

| Column A | Column B |
|---|---|
| **a.** Atomic number | |
| **b.** Atomic weight (mass) | _____ **1.** The number of protons plus the number of neutrons |
| **c.** Base | _____ **2.** A combination of two or more atoms of different elements |
| **d.** Catalyst (enzyme) | _____ **3.** A small, negatively charged particle that orbits the nucleus |
| **e.** Compound | _____ **4.** An atom that has gained or lost electrons, and thus carries an electrical charge |
| **f.** Covalent bond | |
| **g.** Electrolyte | _____ **5.** Atoms are held together by sharing electrons |
| **h.** Electron | _____ **6.** The most fundamental substances of matter |
| **i.** Elements | _____ **7.** A substance that combines with hydrogen ions |
| **j.** Ion | _____ **8.** A molecule that influences the rate of chemical reactions but is not consumed in the reaction |
| **k.** Isotopes | |
| **l.** Nucleus | _____ **9.** Two atoms with the same atomic number but different atomic weights |
| | _____ **10.** A substance that releases ions in water |
| | _____ **11.** The number of protons in an atom |
| | _____ **12.** The area of an atom where protons and neutrons are located |

## MOLECULES AND BONDING

After reviewing figures 3.1, 3.2, and 3.3, complete the following:

1. The atomic number of hydrogen is _____, and it has _____ electron(s) in its outer shell. ⚠

2. The atomic number of chlorine is _____, and it has _____ electron(s) in its outer shell. ⚠

3. Using the style shown in figure 3.2, draw an atom of hydrogen and one of chlorine, clearly indicating the numbers and positions of the protons, neutrons, and electrons. ⚠

4. Is the hydrogen atom stable? ⚠ _____

   Is the chlorine atom stable? ⚠ _____

**5.** What type of bond is likely to form the compound HCl (hydrochloric acid)? ⚠ _____

**6.** Repeat your drawings below, but also show and explain how this bond of HCl would form.

Clearly indicate the positions of protons, neutrons, and electrons. ⚠

## PART B ASSESSMENTS

**1.** Results from cabbage water tests: ②

| Substance | Cabbage Water | Distilled Water | Vinegar | Baking Soda |
|---|---|---|---|---|
| Color | | | | |
| Acid, base, or neutral? | | | | |

**2.** Results from pH paper tests: ②

| Substance Tested | Distilled Water | Tap Water | Sample 1:_____ | Sample 2:_____ | Sample 3:_____ | Sample 4:_____ | Sample 5:_____ | Sample 6:_____ | Sample 7:_____ |
|---|---|---|---|---|---|---|---|---|---|
| pH value | | | | | | | | | |

**3.** Are the pH values the same for distilled water and tap water? ② _____

**4.** If not, what might explain this difference? _____

_____

**5.** Draw the pH scale here, and indicate the following values: 0, 7 (neutral), and 14. Now label the scale with the names and indicate the location of the pH values for each substance you tested.

**PART C ASSESSMENTS**

1. **Biuret test results for protein.** Enter your results from the Biuret test for protein. A color change to purple indicates that protein is present; pink indicates that short polypeptides are present. **3**

| Tube | Contents | Color | Protein Present (+) or Absent (−) |
|------|----------|-------|-----------------------------------|
| 1W | Distilled water | | |
| 1E | Egg albumin | | |
| 1G | Glucose solution | | |
| 1D | Soft drink | | |
| 1S | Starch solution | | |
| 1P | Potato juice | | |

2. **Benedict's test results for most sugars (monosaccharides).** Enter your results from the Benedict's test for sugars. A color change to green, yellow, orange, or red indicates that sugar is present. Note the color after the mixture has been heated for 3 to 5 minutes. **3**

| Tube | Contents | Color | Sugar Present (+) or Absent (−) |
|------|----------|-------|----------------------------------|
| 2W | Distilled water | | |
| 2E | Egg albumin | | |
| 2G | Glucose solution | | |
| 2D | Soft drink | | |
| 2S | Starch solution | | |
| 2P | Potato juice | | |

3. **Iodine test results for starch.** Enter your results from the iodine test for starch. A color change to dark blue or black indicates that starch is present. **3**

| Tube | Contents | Color | Starch Present (+) or Absent (−) |
|------|----------|-------|-----------------------------------|
| 3W | Distilled water | | |
| 3E | Egg albumin | | |
| 3G | Glucose solution | | |
| 3D | Soft drink | | |
| 3S | Starch solution | | |
| 3P | Potato juice | | |

4. **Lipid test results.** What did you observe when you allowed the drops of oil and water to dry on the brown paper?

_____

What did you observe when you mixed the oil and water together? _____

_____

What did you observe when you added the Sudan IV dye? **3** _____

**ASSESS**

**CRITICAL THINKING**

If a person were on a low-carbohydrate, high-protein diet, which of the six substances tested would the person want to increase?

_____

Explain your answer.

_____

_____

_____

_____

**PART D ASSESSMENTS**

1. What is the number of your sample? _____

2. Record the results of your tests on the unknown here: **A**

| Test Performed | Results |
| --- | --- |
| Biuret test for protein | |
| Benedict's test for sugars | |
| Iodine test for starch | |
| Water test for lipid | |
| Sudan IV test for lipid | |

3. Based on these results, what organic compound(s) does your unknown sample contain? **A**

_____

4. Explain your answer. **A**

_____

_____

_____

# Care and Use of the Microscope

## MATERIALS NEEDED

Compound light microscope
Lens paper
Microscope slides
Coverslips
Transparent plastic millimeter ruler
Prepared slide of letter *e*
Slide of three colored threads
Medicine dropper
Dissecting needle (needle probe)
Specimen examples for wet mounts
Methylene blue (dilute) or iodine-potassium-
   iodide stain
Stereomicroscope (dissecting microscope)

## PURPOSE OF THE EXERCISE

To become familiar with the major parts of a compound light microscope and their functions and to make use of the microscope to observe small objects.

 **LEARNING OUTCOMES**

After completing this exercise, you should be able to

1. Locate and identify the major parts of a compound light microscope and differentiate the functions of these parts.

2. Calculate the total magnification produced by various combinations of eyepiece and objective lenses.

3. Demonstrate proper use of the microscope to observe and measure small objects.

4. Prepare a simple microscope slide and sketch the objects you observed.

The human eye cannot perceive objects less than 0.1 mm in diameter, so a microscope is an essential tool for the study of small structures such as cells. The microscope used for this purpose is the *compound light microscope*.

It is called compound because it uses two sets of lenses: an eyepiece, or ocular, lens system and an objective lens system. The eyepiece lens system magnifies, or compounds, the image reaching it after the image is magnified by the objective lens system. Such an instrument can magnify images of small objects up to about 1,000 times.

 **PRACTICE**

### PROCEDURE A—Microscope Basics

1. Familiarize yourself with the following list of rules for care of the microscope:

   **a.** Keep the microscope under its *dustcover* and in a cabinet when it is not being used.

   **b.** Handle the microscope with great care. It is an expensive and delicate instrument. To move it or carry it, hold it by its *arm* with one hand and support its *base* with the other hand (fig. 4.1).

   **c.** Always store the microscope with the scanning or lowest-power objective in place. Always start with this objective when using the microscope.

   **d.** To clean the lenses, rub them gently with *lens paper* or a high-quality cotton swab. If the lenses need additional cleaning, follow the directions in the "Lens-Cleaning Technique" section that follows.

   **e.** If the microscope has a substage lamp, be sure the electric cord does not hang off the laboratory table where someone might trip over it. The bulb life can be extended if the lamp is cool before the microscope is moved.

   **f.** Never drag the microscope across the laboratory table.

   **g.** Never remove parts of the microscope or try to disassemble the eyepiece or objective lenses.

   **h.** If your microscope is not functioning properly, report the problem to your laboratory instructor immediately.

2. Observe a compound light microscope and study figure 4.1 to learn the names of its major parts. Your microscope might be equipped with a binocular body and two eyepieces (fig. 4.2). The lens system of a compound microscope includes three parts—the condenser, objective lens, and eyepiece (ocular).

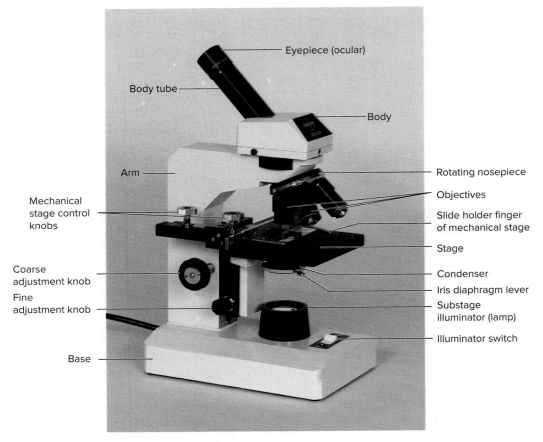

**FIGURE 4.1** Major parts of a compound light microscope with a monocular (one eyepiece) body and a mechanical stage. Some microscopes are equipped with a binocular (two eyepieces) body. (©J & J Photography)

 **LENS-CLEANING TECHNIQUE**

1. Moisten one end of a high-quality cotton swab with one drop of lens cleaner. Keep the other end dry.

2. Clean the optical surface with the wet end. Dry it with the other end, using a circular motion.

3. Use a hand aspirator to remove lingering dust particles.

4. Start with the scanning objective and work upward in magnification, using a new cotton swab for each objective.

5. When cleaning the eyepiece, do not open the lens unless it is absolutely necessary.

6. Use alcohol for difficult cleaning.

Light enters this system from a *substage illuminator (lamp)* or *mirror* and is concentrated and focused by a *condenser* onto a microscope slide (fig. 4.3). The condenser, which contains a set of lenses, usually is kept in its highest position possible.

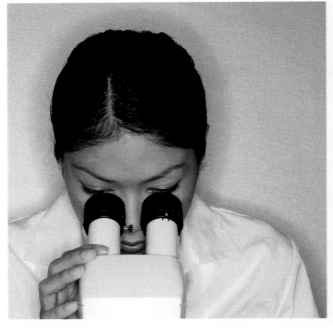

**FIGURE 4.2** Scientist using a compound light microscope equipped with a binocular body and two eyepieces. (©Stockbyte/Getty Images RF)

**FIGURE 4.3** Microscope, showing the path of light through it. (©Nikon Instruments)

← Light path

The *iris diaphragm*, located between the light source and the condenser, can be used to increase or decrease the intensity of the light entering the condenser and

to control the contrast of the image. Locate the lever that operates the iris diaphragm beneath the *stage* and move it back and forth. Note how this movement causes the size of the opening in the diaphragm to change. (Some microscopes have a revolving plate called a *disc diaphragm* beneath the stage instead of an iris diaphragm. Disc diaphragms have different-sized holes to admit varying amounts of light.) Which way do you move the diaphragm to increase the light intensity? _____ Which way to decrease it? _____

After light passes through a specimen mounted on a microscope slide, it enters an *objective lens system*. This lens projects the light upward into the *body tube*, where it produces a magnified image of the object being viewed.

The *eyepiece (ocular) lens* system then magnifies this image to produce another image, seen by the eye. Typically, the eyepiece lens magnifies the image ten times (10×). Look for the number in the metal of the eyepiece that indicates its power (fig. 4.4). What is the eyepiece power of your microscope? _____

The objective lenses are mounted in a *rotating nosepiece* so that different magnifications can be achieved by rotating any one of several objective lenses into position above the specimen. Commonly, this set of

(a)

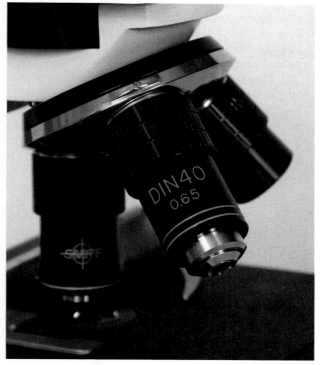

(b)

**FIGURE 4.4** The powers of this 10× eyepiece (*a*) and this 40× objective (*b*) are marked in the metal. DIN is an international optical standard on quality optics. The 0.65 on the 40× objective is the numerical aperture, a measure of the light-gathering capabilities. ((*a,b*) ©J & J Photography) lenses includes a scanning objective (4×), a low-power objective (10×), and a high-power objective, also called a high-dry-power objective (about 40×). Sometimes an oil immersion objective (about 100×) is present. Look for the number printed on each objective lens that indicates its power. What are the objective lens powers of your microscope? _____

To calculate the *total magnification* achieved when using a particular objective, multiply the power of the eyepiece by the power of the objective used. Thus, the 10× eyepiece and the 40× objective produce a total magnification of 10 × 40, or 400×. See a summary of microscope lenses in table 4.1.

**TABLE 4.1  Microscope Lenses**

| Objective Lens Name | Common Objective Lens Magnification | Common Eyepiece Lens Magnification | Total Magnification |
|---|---|---|---|
| Scan | 4× | 10× | 40× |
| Low-power (LP) | 10× | 10× | 100× |
| High-power (HP) | 40× | 10× | 400× |
| Oil immersion | 100× | 10× | 1,000× |

*Note*: If you wish to observe an object under LP, HP, or oil immersion, locate and then center and focus the object first under scan magnification.

3. Complete Part A of Laboratory Report 4.
4. Turn on the substage illuminator and look through the eyepiece. You will see a lighted, circular area called the *field of view*.

   You can measure the diameter of this field of view by focusing the lenses on the millimeter scale of a transparent plastic ruler. To do this, follow these steps:

   a. Place the ruler on the microscope stage in the spring clamp of a slide holder finger on a *mechanical stage* or under the *stage (slide) clips*. (*Note*: If your microscope is equipped with a mechanical stage, it may be necessary to use a short section cut from a transparent plastic ruler. The section should be several millimeters long and can be mounted on a microscope slide for viewing.)

   b. Center the millimeter scale in the beam of light coming up through the condenser, and rotate the scanning objective into position.

   c. While you watch from the side to prevent the lens from touching anything, raise the stage until the objective is as close to the ruler as possible, using the *coarse adjustment knob* (fig. 4.5). (*Note*: The adjustment knobs on some microscopes move the body and objectives downward and upward for focusing.)

d. Look into the eyepiece and use the *fine adjustment knob* to raise the stage until the lines of the millimeter scale come into sharp focus.

e. Adjust the light intensity by moving the *iris diaphragm lever* so that the field of view is brightly illuminated but comfortable to your eye. At the same time, take care not to overilluminate the field because transparent objects tend to disappear in bright light.

f. Position the millimeter ruler so that its scale crosses the greatest diameter of the field of view. In addition, move the ruler so that one of the millimeter marks is against the edge of the field of view.

g. In millimeters, measure the distance across the field of view.

5. Complete Part B of the laboratory report.
6. Most microscopes are designed to be *parfocal*. This means that when a specimen is in focus with a lower-power objective, it will be in focus (or nearly so) when a higher-power objective is rotated into position. Always

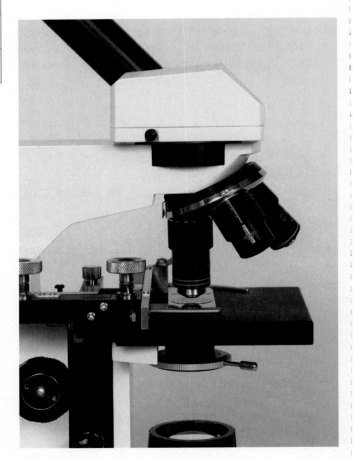

**FIGURE 4.5**  When you focus using a particular objective, you can prevent it from touching the specimen by watching from the side. (©J & J Photography)

center the specimen in the field of view before changing to higher objectives.

Rotate the low-power objective into position, and then look at the millimeter scale of the transparent plastic ruler. If you need to move the low-power objective to sharpen the focus, use the fine adjustment knob.

Adjust the iris diaphragm so that the field of view is properly illuminated. Once again, adjust the millimeter ruler so that the scale crosses the field of view through its greatest diameter, and position the ruler so that a millimeter mark is against one edge of the field. Measure the distance across the field of view in millimeters.

7. Rotate the high-power objective into position while you watch from the side, and then observe the millimeter scale on the plastic ruler. *All focusing using high-power magnification should be done only with the fine adjustment knob.* If you use the coarse adjustment knob with the high-power objective, you can accidentally force the objective into the coverslip and break the slide. This is because the *working distance* (the distance from the objective lens to the slide on the stage) is much shorter when using higher magnifications.

Adjust the iris diaphragm for proper illumination. When using higher magnifications, more illumination usually will help you to view the objects more clearly. Try to measure the distance across the field of view in millimeters.

8. Locate the numeral 4 (or 9) on the plastic ruler and focus on it using the scanning objective. Note how the number appears in the field of view. Move the plastic ruler to the right and note which way the image moves. Slide the ruler away from you and again note how the image moves.

9. Observe a letter *e* slide in addition to the numerals on the plastic ruler. Note the orientation of the letter *e* using the scan, LP, and HP objectives. As you increase magnifications, note that the amount of the letter *e* shown is decreased. Move the slide to the left, and then away from you, and note the direction in which the observed image moves. If a *pointer* is visible in your field of view, you can manipulate the pointer to a location (structure) within the field of view by moving the slide or rotating the eyepiece.

10. Examine the slide of the three colored threads using the low-power objective and then the high-power objective. Focus on the location where the three threads cross. By using the fine adjustment knob, determine the order from top to bottom by noting which color is in focus at different depths. The other colored threads will still be visible, but they will be blurred. Be sure to notice whether the stage or the body tube moves up and down with the adjustment knobs of the microscope being used for this depth determination. The vertical depth of the specimen clearly in focus is called the *depth of field*

(*focus*). Whenever specimens are examined, continue to use the fine adjustment focusing knob to determine relative depths of structures clearly in focus within cells, giving a three-dimensional perspective. The depth of field is less at higher magnifications.

 **ASSESS**

### CRITICAL THINKING

What was the sequence of the three colored threads from top to bottom? Explain how you came to that conclusion.

_____

_____

_____

11. Complete Parts C and D of the laboratory report.

**PRACTICE**

### PROCEDURE B—Slide Preparation

1. Prepare several temporary *wet mounts* using any small, transparent objects of interest, and examine the specimens using the low-power objective and then a high-power objective to observe their details. To prepare a wet mount, follow these steps (fig. 4.6):

   **a.** Obtain a precleaned microscope slide.

   **b.** Place a tiny, thin piece of the specimen you want to observe in the center of the slide, and use a medicine dropper to put a drop of water over it. Consult with your instructor if a drop of stain might enhance the image of any cellular structures of your specimen. If the specimen is solid, you might want to tease some of it apart with dissecting needles. In any case, the specimen must be thin enough that light can pass through it. Why is it necessary for the specimen to be so thin?

   _____

   _____

   **c.** Cover the specimen with a coverslip. Try to avoid trapping bubbles of air beneath the coverslip by slowly lowering it at an angle into the drop of water.

   **d.** Remove any excess water from the edge of the coverslip with absorbent paper. If your microscope has an inclination joint, do not tilt the microscope while observing wet mounts because the fluid will flow.

   **e.** Place the slide under the stage (slide) clips or in the slide holder on a mechanical stage, and position the slide so that the specimen is centered in the light beam passing up through the condenser.

   **f.** Focus on the specimen using the scanning objective first. Next, focus using the low-power objective, and then examine it with the high-power objective.

① Obtain a precleaned microscope slide.

② Place the specimen in the center of the slide and add drop of water to specimen.

Specimen

Paper

Needle probe

③ Slowly lower coverslip to avoid bubbles.

④ Remove excess water with absorbent paper.

**FIGURE 4.6**  Steps in the preparation of a wet mount.

2. If an oil immersion objective is available, use it to examine the specimen. To use the oil immersion objective, follow these steps:

   **a.** Center the object you want to study under the high-power field of view.

   **b.** Rotate the high-power objective away from the microscope slide, place a small drop of immersion oil on the coverslip, and swing the oil immersion objective into position. To achieve sharp focus, use the fine adjustment knob only.

   **c.** You will need to open the iris diaphragm more fully for proper illumination. More light is needed because the oil immersion objective covers a very small lighted area of the microscope slide.

   **d.** The oil immersion objective must be very close to the coverslip to achieve sharp focus, so care must be taken to avoid breaking the coverslip or damaging the objective lens. For this reason, never lower the objective when you are looking into the eyepiece. Instead, always raise the objective to achieve focus, or prevent the objective from touching the coverslip by watching the microscope slide and coverslip from the side if the objective needs to be lowered. Usually when using the oil immersion objective, only the fine adjustment knob needs to be used for focusing.

3. When you have finished working with the microscope, remove the microscope slide from the stage and wipe any oil from the objective lens with lens paper or a high-quality cotton swab. Swing the scanning objective or the low-power objective into position. Wrap the electric cord around the base of the microscope and replace the dustcover.

4. Complete Part E of the laboratory report.

## LEARN: ACTIVITY

A stereomicroscope (dissecting microscope) (fig. 4.7) is useful for observing the details of relatively large, opaque specimens. Although this type of microscope achieves less magnification than a compound microscope, it has the advantage of producing a three-dimensional image rather than the flat, two-dimensional image of the compound light microscope. In addition, the image produced by the stereomicroscope is positioned in the same manner as the specimen, rather than being reversed and inverted, as it is by the compound light microscope.

Observe the stereomicroscope. The eyepieces can be pushed apart or together to fit the distance between your eyes. Focus the microscope on the end of your finger. Which way does the image move when you move your finger to the right? _____

When you move it away? _____

If the instrument has more than one objective, change the magnification to a higher power. Use the instrument to examine various small, opaque objects available in the laboratory.

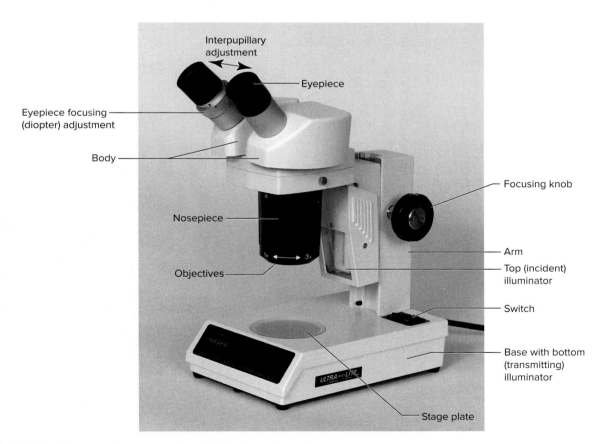

**FIGURE 4.7** A stereomicroscope, also called a dissecting microscope. (©J & J Photography)

# Notes

The $\triangle$ corresponds to the indicated Learning Outcome(s) found at the beginning of the laboratory exercise.

# Care and Use of the Microscope

## PART A ASSESSMENTS

Revisit Procedure A, number 2; then complete the following:

1. What total magnification will be achieved if the 10× eyepiece and the 10× objective are used? **2** _____

2. What total magnification will be achieved if the 10× eyepiece and the 100× objective are used? **2** _____

## PART B ASSESSMENTS

Revisit Procedure A, number 2; then complete the following:

1. Sketch the millimeter scale as it appears under the scanning objective magnification. (The circle represents the field of view through the microscope.)

2. In millimeters, what is the diameter of the scanning field of view? **3** _____

3. Microscopic objects often are measured in *micrometers*. A micrometer equals 1/1,000 of a millimeter and is symbolized by μm. In micrometers, what is the diameter of the scanning power field of view? **3** _____

4. If a circular object or specimen extends halfway across the scanning field, what is its diameter in millimeters? **3** _____

5. In micrometers, what is its diameter? **3** _____

## PART C ASSESSMENTS

Complete the following:

1. Sketch the millimeter scale as it appears using the low-power objective.

2. What do you estimate the diameter of this field of view to be in millimeters? **3**

_____

3. How does the diameter of the scanning power field of view compare with that of the low-power field?

_____

4. Why is it more difficult to measure the diameter of the high-power field of view than that of the low-power field? 🔺3

_____

_____

5. What change occurred in the intensity of the light in the field of view when you exchanged the low-power objective

for the high-power objective? _____

_____

6. Sketch the numeral 4 (or 9) as it appears through the scanning objective of the compound microscope.

7. What has the lens system done to the image of the numeral? (Is it right side up, upside down, or what?) _____

8. When you moved the ruler to the right, which way did the image move? _____

9. When you moved the ruler away from you, which way did the image move? _____

## PART D ASSESSMENTS

Match the names of the microscope parts in column A with the descriptions in column B. Place the letter of your choice in the space provided. 🔺1

| Column A | Column B |
|---|---|
| **a.** Adjustment knob (coarse) | _____ **1.** Increases or decreases the light intensity |
| **b.** Arm | |
| **c.** Condenser | _____ **2.** Platform that supports a microscope slide |
| **d.** Eyepiece (ocular) | _____ **3.** Concentrates light onto the specimen |
| **e.** Field of view | |
| **f.** Iris diaphragm | _____ **4.** Causes stage (or objective lens) to move upward or downward |
| **g.** Nosepiece | _____ **5.** After light passes through the specimen, it next enters this |
| **h.** Objective lens system | lens system |
| **i.** Stage | |
| **j.** Stage (slide) clip | _____ **6.** Holds a microscope slide in position |
| | _____ **7.** Contains a lens at the top of the body tube |
| | _____ **8.** Serves as a handle for carrying the microscope |
| | _____ **9.** Part to which the objective lenses are attached |
| | _____ **10.** Circular area seen through the eyepiece |

## PART E ASSESSMENTS

Prepare sketches of the objects you observed using the microscope. For each sketch, include the name of the object, the magnification you used to observe it, and its estimated dimensions in millimeters and micrometers. 🔺3 🔺4

# 5

# Cell Structure and Function

## MATERIALS NEEDED

Textbook
Animal cell model
Clean microscope slides
Coverslips
Flat toothpicks
Medicine dropper
Methylene blue (dilute) or iodine-potassium-
    iodide stain
Prepared microscope slides of human tissues
Compound light microscope
Single-edged razor blade
Plant materials such as leaves, soft stems, fruits, onion
    peel, and vegetables
Cultures of *Amoeba* and *Paramecium*

## ⚠ SAFETY

- Review all Laboratory Safety Guidelines in Appendix 1 of your laboratory manual.
- Clean laboratory surfaces before and after laboratory procedures.
- Wear disposable gloves for the wet-mount procedures of the cells lining the inside of the cheek.
- Work only with your own materials when preparing the slide of cheek cells. Observe the same precautions as with all body fluids.
- Dispose of laboratory gloves, slides, coverslips, and toothpicks as instructed.
- Use the biohazard container to dispose of items used during the cheek cells procedure.
- Take precautions to prevent cellular stains from contacting your clothes and skin.
- Wash your hands before leaving the laboratory.

## PURPOSE OF THE EXERCISE

To review the structure and functions of major cellular components and to observe examples of human cells.

##  LEARNING OUTCOMES AP|R

After completing this exercise, you should be able to

1. Name and locate the components of a cell.
2. Differentiate the functions of cellular components.
3. Prepare a wet mount of cells lining the inside of the cheek; stain the cells; and identify the cell membrane, nucleus, and cytoplasm.
4. Examine cells on prepared slides of human tissues and identify their major components.

Cells are the "building blocks" from which all parts of the human body are formed. Their arrangement and interactions result in the shape, organization, and construction of the body, and cells are responsible for carrying on its life processes. Under the light microscope, with a properly applied stain to make structures visible, the *cell (plasma) membrane,* the *cytoplasm,* and a *nucleus* are easily seen. The cytoplasm is composed of a clear fluid, the *cytosol,* and numerous *cytoplasmic organelles* suspended in the cytosol.

The cell membrane, composed of lipids and proteins, forms the cell boundary and functions in various methods of membrane transport. Chromatin within the nucleus contains fine strands of DNA and protein. Various cytoplasmic organelles, including mitochondria, endoplasmic reticulum, and Golgi apparatus, provide specialized metabolic functions.

##  PRACTICE

### PROCEDURE—Cell Structure and Function

1. Review the section entitled "A Composite Cell" in chapter 3 of the textbook.
2. Observe the animal cell model and identify its structures.
3. As a review activity, label figure 5.1 and study figure 5.2.
4. Complete Part A of Laboratory Report 5.
5. Prepare a wet mount of cells lining the inside of the cheek. To do this, follow these steps:
   a. Gently scrape (force is not necessary and should be avoided) the inner lining of your cheek with the broad end of a flat toothpick.

**FIGURE 5.1** Label the structures of this composite cell. The structures are not drawn to scale. **A** **AP|R**

**b.** Stir the toothpick in a drop of water on a clean microscope slide and dispose of the toothpick as directed by your instructor.

**c.** Cover the drop with a coverslip.

**d.** Observe the cheek cells by using the microscope. Compare your image with figure 5.3. To report what you observe, sketch a single cell in the space provided in Part B of the laboratory report.

**Extracellular side of membrane**

Glycolipid

Glycoprotein

Double layer (bilayer) of phospholipid molecules

Transmembrane protein

Integral proteins

Peripheral proteins

Cholesterol molecules

Hydrophobic fatty acid "tail"

Hydrophilic phosphate "head"

**Cytoplasmic side of membrane**

Phospholipid bilayer

**FIGURE 5.2** Structures of the cell membrane. **AP|R**

Cell membrane

Folds of the cell membrane

Nucleus

Cytoplasm

**FIGURE 5.3** Iodine-stained cell from inner cheek, as viewed through the compound light microscope using the high-power objective (400×). (©Ed Reschke)

6. Prepare a second wet mount of cheek cells, but this time, add a drop of dilute methylene blue or iodine-potassium-iodide stain to the cells. Cover the liquid with a cover-slip and observe the cells with the microscope. Add to your sketch any additional structures you observe in the stained cells.

7. Answer the questions in Part B of the laboratory report.

8. Observe all safety guidelines for disposal of materials used during this procedure.

9. Using the microscope, observe each of the prepared slides of human tissues. To report what you observe, sketch a single cell of each type in the space provided in Part C of the laboratory report.

10. Complete Parts C and D of the laboratory report.

 **ASSESS**

**CRITICAL THINKING**

The cells lining the inside of the cheek are frequently removed for making observations of basic cell structure. The cells are from stratified squamous epithelium. Explain why these cells are used instead of outer body surface tissue.

_____

_____

_____

_____

_____

**LEARN: ACTIVITY**

Investigate the microscopic structure of various plant materials. To do this, prepare tiny, thin slices of plant specimens, using a single-edged razor blade. (*Take care not to injure yourself with the blade.*) Keep the slices in a container of water until you are ready to observe them. To observe a specimen, place it into a drop of water on a clean microscope slide and cover it with a coverslip. Use the microscope and view the specimen using low- and high-power magnifications. Observe near the edges where your section of tissue is most likely to be one cell thick. Add a drop of dilute methylene blue or iodine-potassium-iodide stain, and note if any additional structures become visible. How are the microscopic structures of the plant specimens similar to the human tissues you observed?

_____

How are they different? _____
_____

 **LEARN: ACTIVITY**

Prepare a wet mount of the *Amoeba* and *Paramecium* by putting a drop of culture on a clean glass slide. Gently cover with a clean coverslip. Observe the movements of the *Amoeba* with pseudopodia and the *Paramecium* with cilia. Try to locate cellular components such as the cell membrane, nuclear envelope, nucleus, mitochondria, and contractile vacuoles. Describe the movement of the *Amoeba*.

_____

_____

Describe the movement of the *Paramecium*.

_____

_____

The Ⓐ corresponds to the indicated Learning Outcome(s) found at the beginning of the laboratory exercise.

# Cell Structure and Function

### PART A ASSESSMENTS

Match the cellular components in column A with the descriptions in column B. Place the letter of your choice in the space provided. ① ②

| Column A | Column B |
|---|---|
| **a.** Chromatin | _____ **1.** Loosely coiled fibers containing protein and DNA within nucleus |
| **b.** Cytoplasm | _____ **2.** Location of ATP production for cellular energy |
| **c.** Endoplasmic reticulum | _____ **3.** Small RNA-containing particles for the synthesis of proteins |
| **d.** Golgi apparatus | _____ **4.** Membranous sac that stores or transports substances |
| **e.** Lysosome | _____ **5.** Dense body of RNA and protein within the nucleus |
| **f.** Microtubule | _____ **6.** Slender tubes that provide movement in cilia and flagella |
| **g.** Mitochondrion | _____ **7.** Composed of membrane-bound canals and sacs for tubular transport throughout the cytoplasm |
| **h.** Nuclear envelope | |
| **i.** Nucleolus | _____ **8.** Occupies space between cell membrane and nucleus |
| **j.** Nucleus | _____ **9.** Flattened, membranous sacs that package a secretion |
| **k.** Ribosome | _____ **10.** Membranous sac that contains digestive enzymes |
| **l.** Vesicle | _____ **11.** Separates nuclear contents from cytoplasm |
| | _____ **12.** Spherical organelle that contains chromatin and nucleolus |

### PART B ASSESSMENTS

Complete the following:

1. Sketch a single cheek cell that has been stained. Label the cellular components you recognize. (The circle represents the field of view through the microscope.) Ⓐ

Magnification _____ ×

2. After comparing the wet mount and the stained cheek cells, describe the advantage gained by staining cells.

_____

_____

**3.** Are the stained cheek cells nearly the same size and shape? _____ Propose an explanation for your answer.

_____

_____

## PART C ASSESSMENTS

Complete the following:

**1.** Sketch a single cell of each type you observed in the prepared slides of human tissues. Name the tissue, indicate the magnification used, and label the cellular components you recognize. 4

_____ ×

Tissue _____

_____ ×

Tissue _____

_____ ×

Tissue _____

_____ ×

Tissue _____

**2.** What do the various types of cells in these tissues have in common? _____

_____

**3.** What are the main differences you observed among these cells? _____

_____

Electron micrographs represent extremely thin slices of cells. The micrograph in figure 5.4 contains a section of a nucleus and some cytoplasm. Compare the organelles shown in this micrograph with organelles of the animal cell model and figure 5.1.

Identify the structures indicated by the arrows in figure 5.4. 🔺**1**

1. _____

2. _____

3. _____

4. _____

5. _____

Answer the following questions after observing the transmission electron micrograph in figure 5.4.

6. What cellular structures were visible in the transmission electron micrograph that were not apparent in the cells you observed using the compound light microscope? _____

_____

7. Before they can be observed by using a transmission electron microscope, cells are sliced into very thin sections. What disadvantage does this procedure present in the study of cellular parts? _____

_____

**Terms:**
Chromatin
Mitochondria
Nuclear envelope
Nucleolus
Rough endoplasmic reticulum

**FIGURE 5.4** Transmission electron micrograph of cellular components (23,000×). The view is only a portion of the cell. Identify the numbered cellular structures, using the terms provided. 🔺**1** **AP|R** (©Keith R. Porter/ Science Source)

# Notes

# Movements Through Membranes

## MATERIALS NEEDED

*For Procedure A—Diffusion*:
Textbook
Petri dish
White paper
Forceps
Potassium permanganate crystals
Millimeter ruler (thin and transparent)

*For Procedure B—Osmosis*:
Thistle tube
Molasses (or Karo dark corn syrup)
Selectively permeable (semipermeable) membrane
 (presoaked dialysis tubing of 1 5/16" or greater
 diameter)
Ring stand and clamp
Beaker
Rubber band
Millimeter ruler

*For Procedure C—Hypertonic, Hypotonic,
 and Isotonic Solutions*:
Test tubes
Marking pen
Test-tube rack
10 mL graduated cylinder
Medicine dropper
Uncoagulated animal blood
Distilled water
0.9% NaCl (aqueous solution)
3.0% NaCl (aqueous solution)
Clean microscope slides
Coverslips
Compound light microscope

*For Procedure D—Filtration*:
Glass funnel
Filter paper
Support stand and ring
Beaker
Powdered charcoal or ground black pepper
1% glucose (aqueous solution)
1% starch (aqueous solution)
Test tubes
10 mL graduated cylinder
Water bath (boiling water)
Benedict's solution

Iodine-potassium-iodide solution
Medicine dropper

*For Alternative Osmosis Activity*:
Fresh chicken egg
Beaker
Laboratory balance
Spoon
Vinegar
Corn syrup (Karo)

 **SAFETY**

- Clean laboratory surfaces before and after
  laboratory procedures.
- Wear disposable gloves when handling chemicals
  and animal blood.
- Wear safety glasses when using chemicals.
- Dispose of laboratory gloves and blood-
  contaminated items as instructed.
- Wash your hands before leaving the laboratory.

## PURPOSE OF THE EXERCISE

To demonstrate some of the passive mechanisms by which
substances move through cell membranes.

 **LEARNING OUTCOMES** AP|R

After completing this exercise, you should be able to

(1) Demonstrate the process of diffusion and identify
 examples of diffusion.

(2) Explain diffusion by preparing and interpreting a graph.

(3) Demonstrate the process of osmosis and identify
 examples of osmosis.

(4) Distinguish among hypertonic, hypotonic, and isotonic
 solutions and observe the effects of these solutions on cells.

(5) Demonstrate the process of filtration and identify
 examples of filtration.

(6) Recognize the anatomical terms pertaining to
 movements through membranes.

A cell membrane functions as a gateway through which chemical substances and small particles may enter or leave a cell. These substances move through the membrane by passive mechanisms that do not require cellular energy, such as diffusion, osmosis, and filtration or by active mechanisms that use cellular energy, such as active transport, phagocytosis, or pinocytosis.

 **PRACTICE**

### PROCEDURE A—Diffusion AP|R

1. Review the section entitled "Diffusion" in chapter 3 of the textbook.
2. To demonstrate *diffusion*, refer to figure 6.1 as you follow these steps:
   a. Place a petri dish, half filled with water, on a piece of white paper that has a millimeter ruler positioned on the paper. Wait until the water surface is still. Allow approximately 3 minutes. *Note:* The petri dish should remain level. A second millimeter ruler may be needed under the petri dish as a shim to obtain a level amount of the water inside the petri dish.
   b. Using forceps, place one crystal of potassium permanganate near the center of the petri dish and near the millimeter ruler (fig. 6.1).
   c. Measure the radius of the purple circle at 1-minute intervals for 10 minutes and record the results in Part A of Laboratory Report 6.
3. Complete Part A of the laboratory report.

**FIGURE 6.1** To demonstrate diffusion, place one crystal of potassium permanganate in the center of a petri dish containing water. Place the crystal near the millimeter ruler (positioned under the petri dish). (©J & J Photography)

 **LEARN: ACTIVITY**

Repeat the demonstration of diffusion using a petri dish filled with ice-cold water and a second dish filled with very hot water. At the same moment, add a crystal of potassium permanganate to each dish and observe the circle as before. What difference do you note in the rate of diffusion in the two dishes? How do you explain this difference?

_____

_____

_____

_____

_____

_____

_____

 **PRACTICE**

### PROCEDURE B—Osmosis AP|R

1. Review the section entitled "Osmosis" in chapter 3 of the textbook.
2. To demonstrate *osmosis*, refer to figure 6.2 as you follow these steps:
   a. One person plugs the tube end of a thistle tube with a finger.
   b. Another person then fills the bulb with molasses until it is about to overflow at the top of the bulb. Allow the molasses to enter the first centimeter of the stem, leaving the rest filled with trapped air.
   c. Cover the bulb opening with a single-thickness piece of moist selectively permeable (semipermeable) membrane. Dialysis tubing that has been soaked for 30 minutes can easily be cut open because it becomes pliable.
   d. Tightly secure the membrane in place with several wrappings of a rubber band.
   e. Immerse the bulb end of the tube in a beaker of water. If leaks are noted, repeat the procedures.
   f. Support the upright portion of the tube with a clamp on a support stand. Folded paper between the stem and clamp will protect the thistle tube stem from breakage.
   g. Mark the meniscus level of the molasses in the tube. *Note:* The best results will occur if the mark of the molasses is a short distance up the stem of the thistle tube when the experiment starts.
   h. Measure the level changes after 10 and 30 minutes and record the results in Part B of the laboratory report.
3. Complete Part B of the laboratory report.

(a)                               (b)                               (c)

**FIGURE 6.2** (*a*) Fill the bulb of the thistle tube with molasses; (*b*) tightly secure a piece of selectively permeable (semipermeable) membrane over the bulb opening; and (*c*) immerse the bulb in a beaker of water. *Note*: These procedures require the participation of two people.

 **ALTERNATIVE PROCEDURE**

Eggshell membranes possess selectively permeable properties. To demonstrate osmosis using a natural membrane, soak a fresh chicken egg in vinegar for about 24 hours to remove the shell. Use a spoon to carefully handle the delicate egg. Place the egg in a hypertonic solution (corn syrup) for about 24 hours. Remove the egg, rinse it, and using a laboratory balance weigh the egg to establish a baseline weight. Place the egg in a hypotonic solution (distilled water). Remove the egg and weigh it every 15 minutes for an elapsed time of 75 minutes. Explain any weight changes that were noted during this experiment.

_____

_____

_____

_____

_____

 **PRACTICE**

**PROCEDURE C—Hypertonic, Hypotonic, and Isotonic Solutions** AP R

1. Review the section entitled "Osmosis" in chapter 3 of the textbook.

2. To demonstrate the effect of *hypertonic, hypotonic,* and *isotonic* solutions on animal cells, follow these steps:
   a. Place three test tubes in a rack and mark them *tube 1, tube 2,* and *tube 3*. (*Note*: One set of tubes can be used to supply test samples for the entire class.)
   b. Using 10 mL graduated cylinders, add 3 mL of distilled water to tube 1; add 3 mL of 0.9% NaCl to tube 2; and add 3 mL of 3.0% NaCl to tube 3.
   c. Place 3 drops of fresh, uncoagulated animal blood into each of the tubes, and gently mix the blood with the solutions. Wait 5 minutes.
   d. Using three separate medicine droppers, remove a drop from each tube and place the drops on three separate microscope slides marked *1, 2,* and *3*.
   e. Cover the drops with coverslips and observe the blood cells, using the high power of the microscope.
3. Complete Part C of the laboratory report.

 **ALTERNATIVE PROCEDURE**

Various substitutes for blood can be used for Procedure C. Onion cells, cucumber cells, and cells lining the inside of the cheek represent three possible options.

**PRACTICE**

### PROCEDURE D—Filtration

1. Review the section entitled "Filtration" in chapter 3 of the textbook.
2. To demonstrate *filtration*, follow these steps:
   a. Place a glass funnel in the ring of a support stand over an empty beaker. Fold a piece of filter paper in half and then in half again. Open one thickness of the filter paper to form a cone. Wet the cone, and place it in the funnel. The filter paper is used to demonstrate how movement across membranes is limited by the size of the molecules, but it does not represent a working model of biological membranes.
   b. Prepare a mixture of 5 cc (approximately 1 teaspoon) powdered charcoal (or ground black pepper) and equal amounts of 1% glucose solution and 1% starch solution in a beaker. Pour some of the mixture into the funnel until it nearly reaches the top of the filter-paper cone. Care should be taken to prevent the mixture from spilling over the top of the filter paper. Collect the filtrate in the beaker below the funnel (fig. 6.3).
   c. Test some of the filtrate in the beaker for the presence of glucose. To do this, place 1 mL of filtrate in a clean test tube and add 1 mL of Benedict's solution. Place the test tube in a water bath of boiling water for 2 minutes and then allow the liquid to cool slowly. If the color of the solution changes to green, yellow, or red, glucose is present (fig. 6.4).
   d. Test some of the filtrate in the beaker for the presence of starch. To do this, place a few drops of filtrate in a test tube and add 1 drop of iodine-potassium-iodide solution. If the color of the solution changes to blue-black, starch is present.
   e. Observe any charcoal in the filtrate.
3. Complete Parts D and E of the laboratory report.

**FIGURE 6.3** Apparatus used to illustrate filtration.

**FIGURE 6.4** Heat the filtrate and Benedict's solution in a boiling water bath for 2 minutes.

Name _____

Date _____

Section _____

The  corresponds to the indicated Learning Outcome(s) found at the beginning of the laboratory exercise.

# Movements Through Membranes

## PART A ASSESSMENTS

Complete the following:

**1.** Enter data for changes in the movement of the potassium permanganate. Ⓐ1

| Elapsed Time | Radius of Purple Circle in Millimeters |
|---|---|
| Initial | _____ |
| 1 minute | _____ |
| 2 minutes | _____ |
| 3 minutes | _____ |
| 4 minutes | _____ |
| 5 minutes | _____ |
| 6 minutes | _____ |
| 7 minutes | _____ |
| 8 minutes | _____ |
| 9 minutes | _____ |
| 10 minutes | _____ |

**2.** Prepare a graph that illustrates the diffusion distance of potassium permanganate in 10 minutes. Ⓐ2

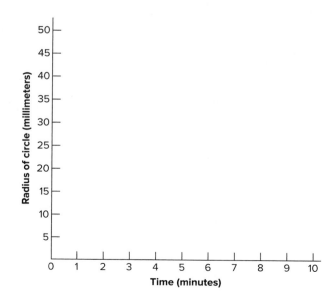

**3.** Interpret your graph. [2] _____

_____

**4.** Define *diffusion.* _____

_____

## ⟳ ASSESS _____

### CRITICAL THINKING

By answering yes or no, indicate which of the following provides an example of diffusion. [1]

1. A perfume bottle is opened, and soon the odor can be sensed in all parts of the room. _____

2. A tea bag is dropped into a cup of hot water, and, without being stirred, all of the liquid becomes the color of the tea leaves. _____

3. A person blows air molecules into a balloon by forcefully exhaling. _____

## ⟳ PART B ASSESSMENTS _____

Complete the following:

1. What was the change in the level of molasses in 10 minutes? [3] _____

2. What was the change in the level of molasses in 30 minutes? [3] _____

3. How do you explain this change? [3] _____

_____

**4.** Define *osmosis.* _____

_____

## ⟳ ASSESS _____

### CRITICAL THINKING

By answering yes or no, indicate which of the following involves osmosis. [3]

1. A fresh potato is peeled, weighed, and soaked in a strong salt solution. The next day, it is discovered that the potato has lost weight. _____

2. Air molecules escape from a punctured tire as a result of high pressure inside. _____

3. Plant seeds soaked in water swell and become several times as large as before soaking. _____

Complete the following:

1. In the spaces, sketch a few blood cells from each of the test tubes, and indicate the magnification. Then, draw an arrow(s) indicating the movement of water (into the cell, out of the cell, or both equally). 🅰

Tube 1                     Tube 2                     Tube 3
(distilled water)          (0.9% NaCl)                (3.0% NaCl)

_____×               _____×               _____×

2. Based on your results, which tube contained a solution hypertonic to the blood cells? _____

   Give the reason for your answer. 🅰 _____

   _____

3. Which tube contained a solution hypotonic to the blood cells? _____

   Give the reason for your answer. 🅰 _____

   _____

4. Which tube contained a solution isotonic to the blood cells? _____

   Give the reason for your answer. 🅰 _____

   _____

 **PART D ASSESSMENTS**

Complete the following:

1. Which of the substances in the mixture you prepared passed through the filter paper into the filtrate? 🄱 _____

   _____

2. What evidence do you have for your answer to question 1? 🄱 _____

   _____

3. What force was responsible for the movement of substances through the filter paper? 🄱 _____

   _____

4. What substances did not pass through the filter paper? 🄱 _____

**5.** What factor prevented these substances from passing through? 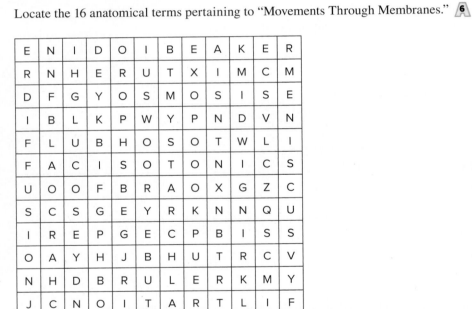 _____

_____

**6.** Define *filtration*. _____

_____

## ⟳ ASSESS

### CRITICAL THINKING

By answering yes or no, indicate which of the following involves filtration:

1. Oxygen molecules move into a cell and carbon dioxide molecules leave a cell because of differences in the concentrations of these substances on either side of the cell membrane. _____

2. Blood pressure forces water molecules and small, dissolved solutes from the blood outward through the thin wall of a blood capillary. _____

3. Noninstant coffee is made using a coffeemaker. _____

## ⟳ PART E ASSESSMENTS

Locate the 16 anatomical terms pertaining to "Movements Through Membranes."

| E | N | I | D | O | I | B | E | A | K | E | R |
|---|---|---|---|---|---|---|---|---|---|---|---|
| R | N | H | E | R | U | T | X | I | M | C | M |
| D | F | G | Y | O | S | M | O | S | I | S | E |
| I | B | L | K | P | W | Y | P | N | D | V | N |
| F | L | U | B | H | O | S | O | T | W | L | I |
| F | A | C | I | S | O | T | O | N | I | C | S |
| U | O | O | F | B | R | A | O | X | G | Z | C |
| S | C | S | G | E | Y | R | K | N | N | Q | U |
| I | R | E | P | G | E | C | P | B | I | S | S |
| O | A | Y | H | J | B | H | U | T | R | C | V |
| N | H | D | B | R | U | L | E | R | K | M | Y |
| J | C | N | O | I | T | A | R | T | L | I | F |

Find and circle words pertaining to Laboratory Exercise 6. The words may be horizontal, vertical, diagonal, or backward.

10 terms (beaker, charcoal, glucose, iodine, meniscus, mixture, ring, ruler, starch, tube)

3 passive mechanisms of membrane passage; 3 tonicity terms

# 7

# Cell Cycle

## MATERIALS NEEDED

Textbook
Models of animal cells during mitosis
Microscope slides of whitefish mitosis (blastula)
Compound light microscope
Microscope slide of human chromosomes from
    leukocytes in mitosis
Oil immersion objective

## PURPOSE OF THE EXERCISE

To review the phases in the cell cycle and to observe cells in various phases of their life cycles.

 ## LEARNING OUTCOMES AP|R

After completing this exercise, you should be able to

1. Describe the cell cycle and locate structures involved with the process.

2. Identify and sketch the phases (stages) in the life cycle of a particular cell.

3. Arrange into a correct sequence a set of models or drawings of cells in various phases of their life cycles.

4. Recognize the anatomical terms pertaining to the cell cycle.

The cell cycle consists of the series of changes a cell undergoes from the time it is formed until it divides. Typically, a newly formed diploid cell with 46 chromosomes grows to a certain size and then divides to form two new cells *(daughter cells)*, each with 46 chromosomes. This cell division process involves two major steps: (1) division of the cell's nuclear parts, *mitosis,* and (2) division of the cell's cytoplasm, *cytokinesis.* Before the cell divides, it must synthesize biochemicals and other contents. This period of preparation is called *interphase.* The extensive period of interphase is divided into three phases. The S phase, when DNA synthesis

occurs, is between two gap phases (G1 and G2), when cell growth occurs and cytoplasmic organelles duplicate. Eventually, some specialized cells such as skeletal muscle cells and most nerve cells cease further cell division but remain alive.

A special type of cell division, called *meiosis,* occurs in the reproductive system to produce haploid gametes with 23 chromosomes. Meiosis is not included in this laboratory exercise.

 ## PRACTICE

### PROCEDURE—Cell Cycle

1. Review the section entitled "The Cell Cycle" in chapter 3 of the textbook.
2. As a review activity, study the various phases of the cell's life cycle represented in figures 7.1–7.3.
3. Label the structures indicated in figure 7.4.
4. Using figure 7.1 as a guide, observe the animal mitosis models, and review the major events in a cell's life cycle represented by each of them. Be sure you can arrange these models in correct sequence if their positions are changed. The acronym IPMAT can help you arrange the correct order of phases in the cell cycle. This includes interphase followed by the four phases of mitosis. Cytokinesis overlaps anaphase and telophase.
5. Complete Part A of Laboratory Report 7.
6. Obtain a slide of the whitefish mitosis (blastula).
   a. Examine the slide using the high-power objective of a microscope. The tissue on this slide was obtained from a developing embryo (blastula) of a fish, and many of the embryonic cells are undergoing mitosis. The chromosomes of these dividing cells are darkly stained (see fig. 7.3).
   b. Search the tissue for cells in various phases of cell division. There are several sections on the slide. If you cannot locate different phases in one section, examine the cells of another section because the phases occur randomly on the slide.
   c. Each time you locate a cell in a different phase, sketch it in an appropriate circle in Part B of the laboratory report.

**Early Interphase**
of daughter cells—
a time of normal cell
growth and function.

Restriction
checkpoint

Cleavage
furrow

Nuclear
envelope

Chromatin
fibers

Centrioles

**Late Interphase**
Cell has passed the
restriction checkpoint
and completed DNA
replication, as well as
replication of centrioles
and mitochondria, and
synthesis of extra
membrane.

Aster

Microtubule

**Prophase**
Chromosomes condense and
become visible. Nuclear
envelope and nucleolus
disperse. Spindle apparatus
forms.

Centromere

Spindle fiber

**Late Prophase**

Sister
chromatids

Chromosomes

Nuclear
envelopes

**Telophase and Cytokinesis**
Nuclear envelopes begin to
reassemble around two daughter
nuclei. Chromosomes decondense.
Spindle disappears. Division of
the cytoplasm into two cells
advances.

**Anaphase**
Sister (replicated) chromatids
separate to opposite poles of
cell. Events begin which lead
to cytokinesis.

**Metaphase**
Chromosomes align along
equator, or metaphase plate,
of cell.

- ◻ Mitosis
- ◼ Cytokinesis
- ◻ G₁ phase ⎫
- ◻ S phase ⎬ Interphase
- ◻ G₂ phase ⎭

**FIGURE 7.1**  The cell cycle: interphase, mitosis, and cytokinesis.  **AP|R**  ((a,b,c,d,e) ©Ed Reschke)

**FIGURE 7.2** Cell in interphase (400×). The nucleoplasm contains a fine network of chromatin. (©Ed Reschke)

Labels: Cell membrane, Cytoplasm, Nuclear envelope, Nucleoplasm

**FIGURE 7.3** Cell in prophase (400×). (©Ed Reschke)

Labels: Nuclear envelope, Chromosomes, Centrosome, Cell membrane

 **ASSESS**

**CRITICAL THINKING**

Which phase (stage) of the cell cycle was the most numerous in the blastula? _____

Explain your answer. _____

_____

_____

7. Complete Parts C and D of the laboratory report.

 **LEARN: ACTIVITY**

Using the oil immersion objective of a microscope, see if you can locate some human chromosomes by examining a prepared slide of human chromosomes from leukocytes. The cells on this slide were cultured in a special medium and were stimulated to undergo mitosis. The mitotic process was arrested in metaphase by exposing the cells to a chemical called colchicine, and the cells were caused to swell osmotically. As a result of this treatment, the chromosomes were spread apart. A complement of human chromosomes should be visible when they are magnified about 1,000×. Each chromosome is double-stranded and consists of two chromatids joined by a common centromere (fig. 7.5).

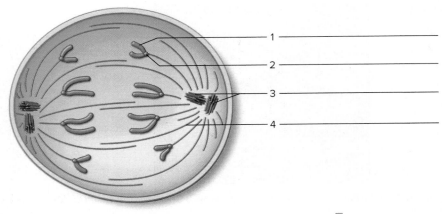

**FIGURE 7.4** Label the structures indicated in the dividing cell during anaphase. [A]

(a)

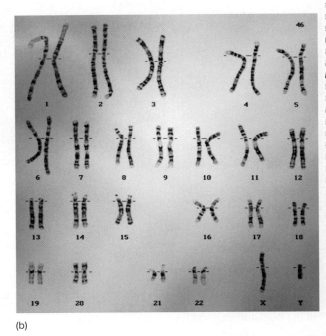

(b)

**FIGURE 7.5** (*a*) A complement of human chromosomes of a female (1,000×). (*b*) A *karyotype* can be constructed by arranging the homologous chromosome pairs together in a chart. A karyotype can aid in the diagnosis of genetic conditions and abnormalities. The completed karyotype indicates a normal male. **AP|R** (©James Cavallini/Science Source; ©Randy Allbritton/Getty Images RF)

The ⓐ corresponds to the indicated Learning Outcome(s) found at the beginning of the laboratory exercise.

# Cell Cycle

## PART A ASSESSMENTS

Complete the table by listing the major events of each phase in the second column. ⓐ

| Phase | Major Events Occurring |
|---|---|
| Interphase (G$_1$, S, and G$_2$) | |
| Mitosis Prophase | |
| Metaphase | |
| Anaphase | |
| Telophase | |
| Cytokinesis | |

Sketch an interphase cell and cells in different phases (stages) of mitosis to illustrate the whitefish cell's life cycle. Label the major cellular structures represented in the sketches and indicate cytokinesis locations. (The circles represent fields of view through the microscope.) 🄰

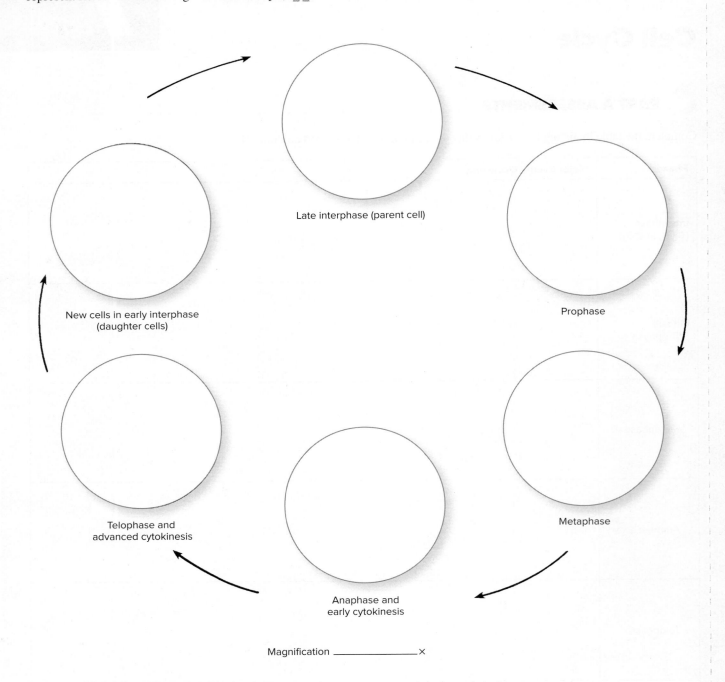

Late interphase (parent cell)

New cells in early interphase
(daughter cells)

Prophase

Telophase and
advanced cytokinesis

Metaphase

Anaphase and
early cytokinesis

Magnification ＿＿＿＿＿×

## PART C ASSESSMENTS

**1.** Identify the mitotic phase represented by each of the micrographs in figure 7.6*a–d.*  **3**

 a. _____  c. _____

 b. _____  d. _____

**2.** Identify the structures indicated by numbers in figure 7.6 by placing the correct numbers in the spaces provided.  **1**

 _____ Cell membrane     _____ Metaphase plate

 _____ Centrosome      _____ Nuclear envelope

 _____ Cleavage furrow     _____ Spindle fiber

(a)

(b)

(c)

(d)

**FIGURE 7.6** Identify the mitotic phase and structures of the cell in each of these micrographs of the whitefish blastula (250×). *((a,b,c,d)* ©Ed Reschke)

Locate the 18 anatomical terms pertaining to the "Cell Cycle."

| | | | | | | | | | | |
|---|---|---|---|---|---|---|---|---|---|---|
| E | S | A | H | P | A | T | E | M | W | P | R |
| C | E | N | T | R | O | M | E | R | E | L | E |
| E | S | A | H | P | R | E | T | N | I | A | B |
| N | U | C | L | E | O | L | U | S | B | T | I |
| T | K | P | A | N | A | P | H | A | S | E | F |
| R | E | L | U | B | U | T | O | R | C | I | M |
| I | D | K | S | Y | N | T | H | E | S | I | S |
| O | E | S | A | H | P | O | L | E | T | X | M |
| L | S | I | S | E | N | I | K | O | T | Y | C |
| E | P | R | O | P | H | A | S | E | Z | V | K |
| R | E | T | S | A | N | I | E | L | C | U | N |
| U | B | F | G | D | S | P | I | N | D | L | E |

Find and circle words pertaining to Laboratory Exercise 7. The words may be horizontal, vertical, diagonal, or backward.

7 terms (aster, cytokinesis, fiber, mitosis, nuclei, plate, synthesis)

6 cellular structures involved during mitosis; 5 phases of the cell cycle

# 8

# Epithelial Tissues

## MATERIALS NEEDED

Textbook
Compound light microscope
Prepared slides of the following epithelial tissues:
    Simple squamous epithelium (lung)
    Simple cuboidal epithelium (kidney)
    Simple columnar epithelium (small intestine)
    Pseudostratified (ciliated) columnar epithelium
      (trachea)
    Stratified squamous epithelium (esophagus)
    Transitional epithelium (urinary bladder)
Colored pencils

## PURPOSE OF THE EXERCISE

To review the characteristics of epithelial tissues and to observe examples.

##  LEARNING OUTCOMES AP|R

After completing this exercise, you should be able to

1. Differentiate the special characteristics of each type of epithelial tissue.

2. Sketch and label the characteristics of epithelial tissues that you were able to observe.

3. Indicate a location and function of each type of epithelial tissue.

4. Identify the major types of epithelial tissues on microscope slides.

**A** tissue is composed of a layer or group of cells similar in size, shape, and function. A study of tissues is called *histology.* Within the human body, there are four major types of tissues: (1) *epithelial*, which cover the body's external and internal surfaces and compose most glands; (2) *connective*, which bind and support parts; (3) *muscle*, which make movement possible; and (4) *nervous*, which conduct impulses from one part of the body to another and help to control and coordinate body activities.

Epithelial tissues are tightly packed single (simple) to multiple (stratified) layers of cells that provide protective barriers. The underside of this tissue contains an acellular basement membrane layer composed of adhesive cellular secretions and collagen through which the epithelial cells anchor to underlying connective tissue. The cells readily divide and lack blood vessels. Epithelial cells always have a free (apical) surface exposed to the outside or to an open space internally and a basal surface that attaches to the basement membrane. The functions of epithelial cells include protection, filtration, secretion, and absorption. Many epithelial cell shapes are used to name and identify the variations. Many of the prepared slides contain more than the tissue to be studied, so be certain that your view matches the correct tissue. Also be aware that stained colors of all tissues might vary.

##  PRACTICE

### PROCEDURE—Epithelial Tissues AP|R

1. Review the section entitled "Epithelial Tissues" in chapter 5 of the textbook.
2. Complete Part A of Laboratory Report 8.
3. Use the microscope to observe the prepared slides of types of epithelial tissues. As you observe each tissue, look for its special distinguishing features as described in the textbook, such as cell size, shape, and arrangement. Compare your prepared slides of epithelial tissues to the micrographs in figure 8.1. As you observe each type of epithelial tissue, prepare a labeled sketch of a representative portion of the tissue in Part B of the laboratory report.
4. Complete Part B of the laboratory report.
5. Test your ability to recognize each type of epithelial tissue. To do this, have a laboratory partner select one of the prepared slides, cover its label, and focus the microscope on the tissue. Then see if you can correctly identify the tissue. 4

(a) Simple squamous epithelium (side view) (from lung) (250×)

(b) Simple cuboidal epithelium (from kidney) (165×)

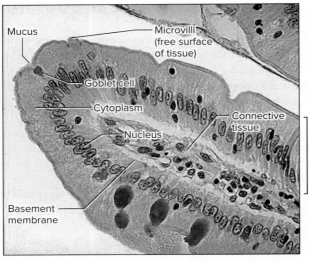

(c) Simple columnar epithelium (nonciliated) (from intestine) (400×)

(d) Pseudostratified columnar epithelium with cilia (from trachea) (1,000×)

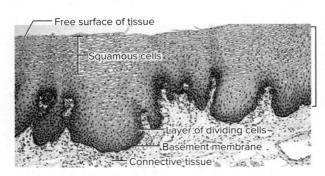

(e) Stratified squamous epithelium (nonkeratinized) (from esophagus) (100×)

(f) Transitional epithelium (stretched) (from urinary bladder) (675×)

**FIGURE 8.1** Micrographs of epithelial tissues (a–f). Note: A bracket to the right of a micrograph indicates the tissue layer.

AP|R ((a,d,e,f) ©Ed Reschke; (b,c) ©Victor P.Eroschenko)

Name _____

Date _____

Section _____

The ⒜ corresponds to the indicated Learning Outcome(s) found at the beginning of the laboratory exercise.

# Epithelial Tissues

**8**

### PART A ASSESSMENTS

Match the tissues in column A with the characteristics in column B. Place the letter of your choice in the space provided. (Some answers may be used more than once.) ⒜ ⒜

**Column A**

a. Simple columnar epithelium
b. Simple cuboidal epithelium
c. Simple squamous epithelium
d. Pseudostratified columnar epithelium
e. Stratified squamous epithelium
f. Transitional epithelium

**Column B**

_____ **1.** Consists of several layers of cells, allowing an expandable lining

_____ **2.** Commonly possesses cilia that move dust and mucus out of the respiratory airways

_____ **3.** Single layer of flattened cells

_____ **4.** Nuclei located at different levels within a single row of aligned cells

_____ **5.** Forms walls of capillaries and air sacs of lungs

_____ **6.** Appear layered (stratified) but are a single layer of cells (simple)

_____ **7.** Deeper cells cuboidal, or columnar; older cells flattened nearest the free surface

_____ **8.** Forms inner lining of urinary bladder

_____ **9.** Lines kidney tubules and ducts of salivary glands

_____ **10.** Forms lining of stomach and intestines

_____ **11.** Elongated cells with elongated nuclei located near basement membrane

_____ **12.** Forms lining of oral cavity, esophagus, anal canal, and vagina

# PART B ASSESSMENTS

In the space that follows, sketch a few cells of each type of epithelium you observed. For each sketch, label the major characteristics, indicate the magnification used, write an example of a location in the body, and provide a function.
Ⓐ Ⓐ Ⓐ

|  |  |
|---|---|
| Simple squamous epithelium (_____×)<br><br>Location: _____<br>Function: _____ | Simple cuboidal epithelium (_____ ×)<br><br>Location: _____<br>Function: _____ |
| Simple columnar epithelium (_____×)<br><br>Location: _____<br>Function: _____ | Pseudostratified columnar epithelium with cilia (_____×)<br><br>Location: _____<br>Function: _____ |
| Stratified squamous epithelium (_____ ×)<br><br>Location: _____<br>Function: _____ | Transitional epithelium (_____×)<br><br>Location: _____<br>Function: _____ |

## ASSESS

### CRITICAL THINKING

As a result of your observations of epithelial tissues, which one(s) provide(s) the best protection? Explain your answer.

_____

_____

_____

## LEARN: ACTIVITY

Use colored pencils to differentiate various cellular structures in Part B. Select a different color for a nucleus, cytoplasm, cell membrane, basement membrane, goblet cell, and cilia whenever visible.

# Connective Tissues

## PURPOSE OF THE EXERCISE

To review the characteristics of connective tissues and to observe examples of the major types.

 ## LEARNING OUTCOMES AP|R

After completing this exercise, you should be able to

1. Differentiate the special characteristics of each of the major types of connective tissue.

2. Sketch and label the characteristics of connective tissues that you were able to observe.

3. Indicate a location and function of each type of connective tissue.

4. Identify the major types of connective tissues on microscope slides.

Connective tissues contain a variety of cell types and occur in all regions of the body. They bind structures together, provide support and protection, fill spaces, store fat, and produce blood cells.

Connective tissue cells are often widely scattered in an abundance of extracellular matrix. The matrix consists of fibers and a ground substance of various densities and consistencies. Many of the prepared slides contain more than the tissue to be studied, so be certain that your view matches the correct tissue. Additional study of bone and blood will be found in Laboratory Exercises 12 and 37.

 ## PRACTICE

### PROCEDURE—Connective Tissues AP|R

1. Review the section entitled "Connective Tissues" in chapter 5 of the textbook.
2. Complete Part A of Laboratory Report 9.
3. Use a microscope to observe the prepared slides of various connective tissues. As you observe each tissue, look for its special distinguishing features as described in the textbook. Compare your prepared slides of connective tissues to the micrographs in figure 9.1. As you observe each type of connective tissue, prepare a labeled sketch of a representative portion of the tissue in Part B of the laboratory report.
4. Complete Part B of the laboratory report.
5. Test your ability to recognize each of these connective tissues by having a laboratory partner select a slide, cover its label, and focus the microscope on this tissue. Then see if you can correctly identify the tissue.

**(a)** Areolar tissue (from beneath the skin) (800×)

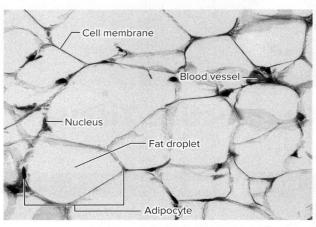

**(b)** Adipose tissue (from subcutaneous layer) (400×)

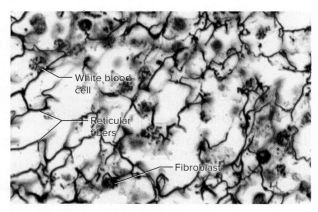

**(c)** Reticular connective tissue (from spleen) (1,000×)

**(d)** Dense regular connective tissue (from tendon) (500×)

**(e)** Elastic connective tissue (from artery wall) (160×)

**(f)** Hyaline cartilage (from costal cartilage of ribs) (160×)

**FIGURE 9.1** Micrographs of connective tissues *(a–j)*. **AP|R** ((a, b, j) ©McGraw-Hill Education/Al Telser, photographer; (c, e, f, g, i) ©Ed Reschke; (d) ©Ed Reschke/Photolibrary/Getty Images; (h) ©Victor P. Eroschenko)

(g) Elastic cartilage (from ear) (400×)

(h) Fibrocartilage (from intervertebral discs) (400×)

(i) Compact bone (from skeleton) (200×)

(j) Blood (400×)

**FIGURE 9.1** *Continued.* AP|R

# Notes

LABORATORY REPORT

The Ⓐ corresponds to the indicated Learning Outcome(s) found at the beginning of the laboratory exercise.

# Connective Tissues

## ⟳ PART A ASSESSMENTS

Match the tissues in column A with the characteristics in column B. Place the letter of your choice in the space provided. (Some answers may be used more than once.) 🄰 🄰

| Column A | Column B |
|---|---|
| **a.** Adipose tissue | _____ **1.** Forms framework of outer ear |
| **b.** Areolar tissue | _____ **2.** Functions as heat insulator beneath skin |
| **c.** Blood | _____ **3.** Contains large amounts of fluid and transports nutrients, wastes, and gases |
| **d.** Bone (compact) | _____ **4.** Cells in solid matrix arranged around central canal |
| **e.** Dense connective tissue (regular) | _____ **5.** Binds skin and fills spaces between organs |
| **f.** Elastic cartilage | _____ **6.** Main tissue of tendons and ligaments |
| **g.** Elastic connective tissue | _____ **7.** Provides stored energy supply in fat droplets in cytoplasm |
| **h.** Fibrocartilage | _____ **8.** Forms the ends of many long bones |
| **i.** Hyaline cartilage | _____ **9.** Pads between vertebrae that are shock absorbers |
| **j.** Reticular connective tissue | _____ **10.** Matrix contains collagen fibers and mineral salts |
| | _____ **11.** Occurs in some ligament attachments between vertebrae and larger artery walls |
| | _____ **12.** Forms supporting tissue in walls of liver and spleen |

In the space that follows, sketch a small section of each of the types of connective tissues you observed. For each sketch, label the major characteristics, indicate the magnification used, write an example of a location in the body, and provide a function. A  A  A

Areolar tissue (_____×)

Location: _____

Function: _____

Adipose tissue (_____ ×)

Location: _____

Function: _____

Reticular connective tissue (_____ ×)

Location: _____

Function: _____

Dense connective tissue (regular type) (_____×)

Location: _____

Function: _____

Elastic connective tissue (_____ ×)

Location: _____

Function: _____

Hyaline cartilage (_____×)

Location: _____

Function: _____

Elastic cartilage (____×)

Location: _____

Function: _____

Fibrocartilage (____×)

Location: _____

Function: _____

Bone (compact) (____×)

Location: _____

Function: _____

Blood (____×)

Location: _____

Function: _____

## ASSESS

### CRITICAL THINKING

Abdominal impact injuries in the region of the LUQ often involve the spleen. Explain the structural tissue characteristics within the spleen that make it so vulnerable to serious injury. **3**

_____

_____

_____

_____

## LEARN: ACTIVITY

Use colored pencils to differentiate various cellular structures in Part B. Select a different color for the cells, fibers, and ground substance whenever visible.

# Notes

74

# 10

# Muscle and Nervous Tissues

## MATERIALS NEEDED

Textbook
Compound light microscope
Prepared slides of the following:
    Skeletal muscle tissue
    Smooth muscle tissue
    Cardiac muscle tissue
    Nervous tissue (spinal cord smear and/or
       cerebellum)
Colored pencils

## PURPOSE OF THE EXERCISE

To review the characteristics of muscle and nervous tissues and to observe examples of these tissues.

## LEARNING OUTCOMES AP|R

After completing this exercise, you should be able to

① Differentiate the special characteristics of each type of muscle tissue and nervous tissue.

② Sketch and label the characteristics of the different muscle and nervous tissues that you were able to observe.

③ Indicate an example location and function of each type of muscle tissue and nervous tissue.

④ Identify three types of muscle tissues and nervous tissues on microscope slides.

**M**uscle tissues are characterized by the presence of elongated cells or muscle fibers that can contract in response to specific stimuli. As they shorten, these fibers pull at their attached ends and cause body parts to move. The three types of muscle tissues are *skeletal*, *smooth*, and *cardiac*.

Nervous tissues occur in the brain, spinal cord, and peripheral nerves. They consist of *neurons* (nerve cells), the impulse-conducting cells of the nervous system, and *neuroglia,* which perform supportive and protective functions for neurons.

## PRACTICE

**PROCEDURE—Muscle and Nervous Tissues** AP|R

1. Review the sections entitled "Muscle Tissues" and "Nervous Tissues" in chapter 5 of the textbook.
2. Complete Part A of Laboratory Report 10.
3. Using the microscope, observe each of the types of muscle tissues on the prepared slides. Look for the special features of each type, as described in the textbook. Compare your prepared slides of muscle tissues to the micrographs in figure 10.1 and the muscle tissue characteristics in table 10.1. As you observe each type of muscle tissue, prepare a labeled sketch of a representative portion of the tissue in Part B of the laboratory report.
4. Observe the prepared slide of nervous tissue and identify neurons (nerve cells), neuron cellular processes, and neuroglia. Compare your prepared slide of nervous tissue to the micrograph in figure 10.1. Prepare a labeled sketch of a representative portion of the tissue in Part B of the laboratory report. AP|R

**TABLE 10.1 Muscle Tissue Characteristics**

| Characteristic | Skeletal Muscle | Smooth Muscle | Cardiac Muscle |
|---|---|---|---|
| Appearance of cells | Unbranched and relatively parallel | Spindle-shaped | Branched and connected in complex networks |
| Striations | Present and obvious | Absent | Present but faint |
| Nucleus | Multinucleated | Uninucleated | Uninucleated (usually) |
| Intercalated discs | Absent | Absent | Present |
| Control | Voluntary (usually) | Involuntary | Involuntary |

5. Complete Part B of the laboratory report.
6. Test your ability to recognize each of these muscle and nervous tissues by having your laboratory partner select a slide, cover its label, and focus the microscope on this tissue. Then see if you correctly identify the tissue. 4

**(a)** Skeletal muscle (400×)

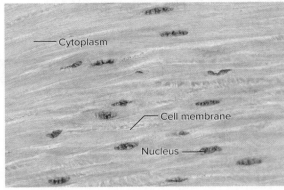

**(b)** Smooth muscle (from small intestine) (400×)

**(c)** Cardiac muscle (from heart) (400×)

**(d)** Nervous tissue (350×)

**FIGURE 10.1**    Micrographs of muscle and nervous tissues *(a–d)*.    AP|R    *((a–d)* ©McGraw-Hill Education/Al Telser, photographer)

The Ⓐ corresponds to the indicated Learning Outcome(s) found at the beginning of the laboratory exercise.

# Muscle and Nervous Tissues

## 🔄 PART A ASSESSMENTS

Match the tissues in column A with the characteristics in column B. Place the letter of your choice in the space provided. (Some answers may be used more than once.) Ⓐ1 Ⓐ3

| Column A | Column B |
|---|---|
| **a.** Cardiac muscle | _____ **1.** Coordinates, regulates, and integrates body functions |
| **b.** Nervous tissue | _____ **2.** Contains intercalated discs |
| **c.** Skeletal muscle | _____ **3.** Muscle that lacks striations |
| **d.** Smooth muscle | _____ **4.** Striated and involuntary |
| | _____ **5.** Striated and voluntary |
| | _____ **6.** Contains neurons and neuroglia |
| | _____ **7.** Muscle attached to bones |
| | _____ **8.** Muscle that composes heart |
| | _____ **9.** Moves food through the digestive tract |
| | _____ **10.** Conducts impulses along cellular processes |
| | _____ **11.** Muscle under conscious control |
| | _____ **12.** Muscle of blood vessels and urinary bladder |

In the space that follows, sketch a few cells or fibers of each of the three types of muscle tissues and of nervous tissue as they appear through the microscope. For each sketch, label the major structures of the cells or fibers, indicate the magnification used, write an example of a location in the body, and provide a function. 🅰 🅰 🅰

Skeletal muscle tissue (_____×)
Location: _____
Function: _____

Smooth muscle tissue (_____×)
Location: _____
Function: _____

Cardiac muscle tissue (_____×)
Location: _____
Function: _____

Nervous tissue (_____×)
Location: _____
Function: _____

 **LEARN: ACTIVITY**

   Use colored pencils to differentiate various cellular structures in Part B.

# 11

# Integumentary System

## MATERIALS NEEDED

Textbook
Skin model
Hand magnifier or dissecting microscope
Forceps
Microscope slide and coverslip
Compound light microscope
Prepared microscope slide of human scalp or axilla
Prepared slide of dark (heavily pigmented) human skin
Prepared slide of thick skin (plantar or palmar)
Tattoo slide

## PURPOSE OF THE EXERCISE

To observe the structures and tissues of the integumentary system and to review the functions of these parts.

##  LEARNING OUTCOMES AP|R

After completing this exercise, you should be able to

1. Locate and name the structures of the integumentary system.
2. Describe the major functions of these structures.
3. Distinguish the locations and tissues among the epidermis, dermis, and subcutaneous layers.
4. Identify and sketch the layers of the skin and associated structures observed on the prepared slide.

The integumentary system includes the skin, hair, nails, sebaceous glands, and sweat glands. These structures provide a protective covering for deeper tissues, aid in regulating body temperature, retard water loss, house sensory receptors, synthesize various chemicals, and excrete small quantities of wastes.

The skin consists of two distinct layers. The outer layer, the *epidermis,* consists of keratinized stratified squamous epithelium. The inner layer, the *dermis,* consists of a thicker layer of mainly dense connective tissue. Beneath the dermis is the *subcutaneous layer* (not considered a true layer of the skin), composed of adipose and areolar connective tissues.

##  PRACTICE

### PROCEDURE—Integumentary System

1. Review the sections entitled "Skin and Its Tissues" and "Accessory Structures of the Skin" in chapter 6 of the textbook.
2. As a review activity, label figures 11.1 and 11.2, and study figure 11.3. Locate as many of these structures as possible on a skin model.
3. Complete Part A of Laboratory Report 11.
4. Use the hand magnifier or dissecting microscope and proceed as follows:
   a. Observe the skin, hair, and nails on your hand.
   b. Compare the type and distribution of hairs on the front and back of your forearm.
5. Use low-power magnification of the compound light microscope and proceed as follows:
   a. Pull out a single hair with forceps and mount it on a microscope slide under a coverslip.
   b. Observe the root and shaft of the hair and note the scalelike parts that make up the shaft.
6. Complete Part B of the laboratory report.
7. As vertical sections of human skin are observed, remember that the lenses of the microscope invert and reverse images. It is important to orient the position of the epidermis, dermis, and subcutaneous (hypodermis) layers using scan magnification before continuing with additional observations. Compare all of your skin observations to figure 11.3. Use low-power magnification of the compound light microscope and proceed as follows:
   a. Observe the prepared slide of human scalp or axilla.
   b. Locate the epidermis, dermis, and subcutaneous layer; a hair follicle; an arrector pili muscle; a sebaceous gland; and a sweat gland.
   c. Focus on the epidermis with high power and locate the stratum corneum, stratum granulosum, stratum spinosum, and stratum basale. Note how the shapes of the cells in these layers differ.
   d. Observe the dense connective tissue (irregular type) that makes up the bulk of the dermis along with some areolar connective tissue.
   e. Observe the adipose tissue that composes most of the subcutaneous layer along with some areolar connective tissue.

**8.** Observe the prepared slide of dark (heavily pigmented) human skin with low-power magnification. The pigment is most abundant in the epidermis. Focus on this region with the high-power objective. The pigment-producing cells, or melanocytes, are located among the deeper layers of epidermal cells. Differences in skin color are primarily due to the amount of pigment (melanin) produced by these cells.

 **ASSESS**

### CRITICAL THINKING

Explain the advantage for melanin granules being located in the deep layer of the epidermis.

_____

_____

_____

_____

_____

**9.** Observe the prepared slide of thick skin from the palm of a hand or the sole of a foot (fig. 11.2). Locate the stratum lucidum. Note how the stratum corneum compares to your observation of the human scalp.

**10.** Complete Part C of the laboratory report.

**11.** Using low-power magnification, locate a hair follicle sectioned longitudinally through its bulblike base. Also locate a sebaceous gland close to the follicle and find a sweat gland (fig. 11.3). Observe the detailed structure of these parts with high-power magnification.

**12.** Complete Parts D and E of the laboratory report.

 **LEARN: ACTIVITY**

Observe a vertical section of human skin through a tattoo, using low-power magnification. Note the location of the dispersed ink granules within the upper portion of the dermis. From a thin vertical section of a tattoo, it is not possible to determine the figure or word of the entire tattoo as seen on the surface of the skin because only a small, thin segment of the image is in view at any one time. Compare this to the location of melanin granules found in dark (heavily pigmented) skin. Suggest reasons why a tattoo is permanent but a suntan is not.

_____

_____

_____

**FIGURE 11.1**  Label this vertical section of the skin and subcutaneous layer.

Labels in figure:
- Dermal papilla
- Touch receptor
- 4
- Pressure receptor
- 5
- Hair root
- 6
- 7
- Nerve cell process
- 8
- Muscle layer below skin

**FIGURE 11.2** Label the features associated with this hair follicle. **1** **AP|R**

1 _____
Pore
Merocrine sweat gland duct
2 _____
3 _____
4 _____
5 _____
6 _____
Region of cell division
Hair papilla

**FIGURE 11.3**
Features of human skin are indicated in these micrographs (a–e). **AP|R** ((a) ©Victor B. Eichler, Ph.D.; (b, d) ©McGraw-Hill Education/Al Telser, photographer; (c) ©Image Source/Getty Images RF; (e) ©doc-stock/Alamy Stock Photo)

Stratum corneum
Stratum granulosum
Stratum spinosum
Stratum basale
Melanin
Dermal papilla
Basement membrane
Dermis

**(a)** Epidermis with melanin (400×)

Stratum corneum
Stratum lucidum
Stratum granulosum
Stratum spinosum
Stratum basale
Dermal papilla
Basement membrane
Dermis

**(b)** Strata of thick skin (500×)

Epidermis

Dermis

Hair follicle

Arrector pili muscle

Sebaceous gland

**(c)** Skin layers (30×)

Dense irregular connective tissue of dermis

Hair follicle

Hair root

Hair bulb

Region of cell division

Adipose tissue

**(d)** Hair follicle (175×)

Epidermis

Dermis

Sweat pore

Hair follicle opening

Dermal papilla

Duct of merocrine sweat gland

Hair root

Hair follicle

Sebaceous glands

Dense irregular connective tissue

Merocrine (eccrine) sweat gland

**(e)** Skin layers with glands (30×)

**FIGURE 11.3** *Continued.* AP|R

Name _____

Date _____

Section _____

The [A] corresponds to the indicated Learning Outcome(s) found at the beginning of the laboratory exercise.

LABORATORY
REPORT

11

# Integumentary System

## PART A ASSESSMENTS

Match the structures in column A with the descriptions and functions in column B. Place the letter of your choice in the space provided. [A1] [A2]

**Column A**

a. Apocrine sweat gland
b. Arrector pili muscle
c. Dermis
d. Epidermis
e. Hair follicle
f. Keratin
g. Melanin
h. Merocrine sweat gland
i. Sebaceous gland
j. Sebum
k. Stratum basale
l. Stratum corneum

**Column B**

_____ 1. Oily secretion that helps to waterproof body surface

_____ 2. Outermost layer of epidermis

_____ 3. Become active at puberty in axillary and groin regions

_____ 4. Epidermal pigment

_____ 5. Inner layer of skin

_____ 6. Responds to elevated body temperature

_____ 7. General name of entire superficial layer of the skin

_____ 8. Gland that secretes an oily mixture

_____ 9. Tough protein of nails and hair

_____ 10. Cell division and deepest layer of epidermis

_____ 11. Tubelike part that contains the root of the hair

_____ 12. Causes hair to stand on end and goose bumps to appear

## PART B ASSESSMENTS

Complete the following:

1. How does the skin of your palm differ from that on the back (posterior) of your hand? [A1] _____

_____

2. Describe the differences you observed in the type and distribution of hair on the front (anterior) and back (posterior) of your forearm. [A1] _____

_____

3. Explain how a hair is formed. [A2] _____

_____

4. What cells produce the pigment in hair? [A2] _____

## PART C ASSESSMENTS

Complete the following:

1. Distinguish the locations and tissues among the epidermis, dermis, and subcutaneous layer. ⓐ _____

_____

_____

2. How do the cells of stratum corneum and stratum basale differ? ③ _____

_____

_____

3. State the specific location of melanin observed in dark (heavily pigmented) skin. ③ _____

_____

4. What special qualities, due to the presence of fibers, does the connective tissue of the dermis have? ③ _____

_____


## PART D ASSESSMENTS

Complete the following:

1. What part of the hair extends from between the bulb and the shaft? ① _____

2. In which layer of skin are sebaceous glands found? ① _____

3. How are sebaceous glands associated with hair follicles and what do they secrete? ① _____

_____

4. The ducts of apocrine sweat glands open into _____ . ①


## PART E ASSESSMENTS

Using the scanning objective, sketch a vertical section of human skin. Label the skin layers and a hair follicle, a sebaceous gland, and a sweat gland. ④

# 12

# Bone Structure and Classification

## MATERIALS NEEDED

Textbook
Prepared microscope slide of ground compact bone
Human bone specimens, including long, short, flat,
  irregular, and sesamoid types
Human long bone, sectioned longitudinally
Fresh animal bones, sectioned longitudinally and
  transversely
Compound light microscope
Dissecting microscope
Fresh chicken bones (radius and ulna from wings)
Vinegar or dilute hydrochloric acid

## ⚠ SAFETY

- Wear disposable gloves for handling fresh bones
  and for the demonstration of a bone soaked in
  vinegar or dilute hydrochloric acid.
- Wash your hands before leaving the laboratory.

## PURPOSE OF THE EXERCISE

To review the way bones are classified and to examine the
structure of a long bone.

## LEARNING OUTCOMES AP|R

After completing this exercise, you should be able to

1. Locate the major structures of a long bone.
2. Distinguish between compact and spongy bone.
3. Differentiate the special characteristics of compact
   bone tissue.

4. Arrange five groups of bones based on their shapes and
   identify an example for each group.
5. Describe the functions of various structures of a bone.

A bone represents an organ of the skeletal system. As
such, it is composed of a variety of tissues, including
bone tissue, cartilage, dense connective tissue, blood, and
nervous tissue. Bones are not only alive but also multifunctional. They support and protect softer tissues, provide points
of attachment for muscles, house blood-producing cells, and
store inorganic salts.

Bones are classified according to their shapes as long,
short, flat, or irregular. Although the bones of the skeleton
vary greatly in size and shape, they have much in common
structurally and functionally.

## PRACTICE

### PROCEDURE—Bone Structure and Classification

1. Review the section entitled "Connective Tissues" in
   chapter 5 of the textbook and the section entitled "Bone
   Shape and Structure" in chapter 7 of the textbook.
2. As a review activity, label figures 12.1 and 12.2.
3. Examine the microscopic structure of bone tissue by
   observing a prepared microscope slide of ground compact bone. Use figure 12.3, a micrograph of bone tissue,
   to locate the following features:

   **osteon** (Haversian system)—cylinder-shaped unit
   **central canal** (Haversian canal)—contains blood
     vessels and nerves
   **lamella**—concentric ring of matrix around central
     canal
   **lacuna**—small chamber for an osteocyte
   **bone extracellular matrix**—collagen and calcium
     phosphate
   **canaliculus**—minute tube containing cellular
     process

Epiphyseal plates

Space occupied by red marrow

Endosteum

1 _____

2 _____

3 _____

4 _____

5 _____

6 _____

7 _____

8 _____

9 _____

**FIGURE 12.1** Label the major structures of this long bone (femur). Ⓐ1 Ⓐ2 AP|R

 **ASSESS**

### CRITICAL THINKING

Explain how bone cells embedded in a solid ground substance obtain nutrients and eliminate wastes. Ⓐ3

_____

_____

_____

_____

**4.** Observe the individual bone specimens and arrange them into groups according to the following shapes and examples: Ⓐ4

**long**—femur; humerus; phalanges

**short**—carpals; tarsals; sesamoid (round) bones

**flat**—ribs; some cranial bones, such as frontal bone

**irregular**—vertebrae; some facial bones, such as sphenoid

**5.** Complete Part A of Laboratory Report 12.

**6.** Examine the sectioned long bones and locate the following:

**epiphysis**
  proximal—nearest limb attachment to torso
  distal—farthest from limb attachment to torso
**epiphyseal plate**—growth zone of hyaline cartilage
**articular cartilage**—on ends of epiphyses
**diaphysis**—shaft between epiphyses
**metaphysis**—expanded portion of bone between diaphysis and epiphysis
**periosteum**—strong membrane around bone (except articular cartilage) of dense irregular connective tissue
**compact bone**—forms diaphysis and epiphyseal surfaces
**spongy bone**—within epiphyses
  trabeculae—a structural lattice in spongy bone
**medullary cavity**—hollow chamber
**endosteum**—thin membrane of reticular connective tissue that lines the medullary cavity
**yellow marrow**—occupies medullary cavity and stores adipose tissue
**red marrow**—occupies spongy bone in some epiphyses and flat bones and produces blood cells

**FIGURE 12.2** Label the features associated with the microscopic structure of bone. 🄐 🄑

Trabeculae

Bone extracellular matrix

Lacuna (space)

Osteon

Lamella

Bone extracellular matrix

Central canal

Lacuna (occupied by osteocyte in living bone)

Canaliculi

**FIGURE 12.3** Micrograph of ground compact bone tissue (200×). **AP|R** (©Victor B. Eichler, Ph.D.)

7. Use the dissecting microscope to observe the compact bone and spongy bone of the sectioned specimens. Also, examine the marrow in the medullary cavity and the spaces within the spongy bone of the fresh specimen.

8. Complete Parts B and C of the laboratory report.

 **LEARN: ACTIVITY**

Examine a fresh chicken bone and a chicken bone that has been soaked for several days in vinegar or overnight in dilute hydrochloric acid. Wear disposable gloves for handling these bones. This acid treatment removes the inorganic salts from the bone extracellular matrix. Rinse the bones in water and note the texture and flexibility of each (fig. 12.4a). The bone becomes soft and flexible without the support of the inorganic salts with calcium.

Examine the specimen of chicken bone that has been exposed to high temperature (baked at 121°C [250°F] for 2 hours). This treatment removes the protein and other organic substances from the bone extracellular matrix (fig. 12.4b). The bone becomes brittle and fragile without the benefit of the collagen fibers. A living bone with a combination of the qualities of inorganic and organic substances possesses tensile strength.

(a)

(b)

**FIGURE 12.4** Results of fresh chicken bone demonstration: (a) soaked in vinegar; (b) baked in oven.
((a,b) ©J & J Photography)

Name _____

Date _____

Section _____

The Ⓐ corresponds to the indicated Learning Outcome(s) found at the beginning of the laboratory exercise.

# Bone Structure and Classification

## ♻ PART A ASSESSMENTS

Complete the following statements. (*Note*: Questions 1–6 pertain to bone classification by shape.)

1. A bone that is platelike is classified as a(an) _____ bone. Ⓐ4

2. The bones of the wrist are examples of _____ bones. Ⓐ4

3. The bone of the thigh is an example of a(an) _____ bone. Ⓐ4

4. Vertebrae are examples of _____ bones. Ⓐ4

5. The patella (kneecap) is a special type of short bone called a _____ bone. Ⓐ4

6. The bones of the skull that form a protective covering of the brain are examples of _____ bones. Ⓐ4

7. Distinguish between the epiphysis and the diaphysis of a long bone. Ⓐ1 _____

_____

8. Describe where cartilage is found on the surface of a long bone. Ⓐ1 _____

_____

9. Describe where the periosteum is found on the surface of a long bone. Ⓐ1 _____

_____

## ♻ PART B ASSESSMENTS

Complete the following:

1. Distinguish between the locations and tissues of the periosteum and those of the endosteum. Ⓐ1 _____

_____

_____

2. What structural differences did you note between compact bone and spongy bone? Ⓐ2 _____

_____

_____

3. How are these structural differences related to the locations and functions of these two types of bone? Ⓐ5 _____

_____

_____

4. From your observations, how does the marrow in the medullary cavity compare with the marrow in the spaces of the spongy bone? Ⓐ5 _____

_____

_____

Identify the structures indicated in figure 12.5.

Terms:
Compact bone
Diaphysis
Epiphysis (distal)
Epiphysis (proximal)
Medullary cavity
Spongy bone

Distal

1

2

3

Proximal

4

Area not sectioned

5

6

(a)                    (b)

**FIGURE 12.5** Identify the structures indicated in (*a*) the unsectioned long bone (fifth metatarsal) and (*b*) the partially sectioned long bone, using the terms provided.  A1  A2  (©J & J Photography)

# Organization of the Skeleton

## PURPOSE OF THE EXERCISE

To review the organization of the skeleton, the major bones of the skeleton, and the terms used to describe skeletal structures.

 ## LEARNING OUTCOMES AP|R

After completing this exercise, you should be able to

1. Distinguish between the axial skeleton and the appendicular skeleton.
2. Locate and label the major bones of the human skeleton.
3. Associate the terms used to describe skeletal structures and locate examples of such structures on the human skeleton.

The skeleton can be separated into two major portions: (1) the *axial skeleton,* which consists of the bones and cartilages of the head, neck, and trunk, and (2) the *appendicular skeleton,* which consists of the bones of the limbs and those that anchor the limbs to the axial skeleton. The bones that anchor the limbs include the pectoral and pelvic girdles. Men and women, although variation can exist, possess the same total bone number of 206.

 ## PRACTICE

### PROCEDURE—Organization of the Skeleton

1. Review the section entitled "Skeletal Organization" in chapter 7 of the textbook. (Pronunciations of the names for major skeletal structures are included within the narrative of chapter 7.)
2. As a review activity, label figure 13.1.

3. Examine the articulated human skeleton and locate the following parts. As you locate the following bones, note the number of each in the skeleton. Palpate as many of the corresponding bones in your skeleton as possible.

| | |
|---|---|
| **axial skeleton** | |
| skull | |
| cranium | (8) |
| face | (14) |
| middle ear bone | (6) |
| hyoid bone | (1) |
| vertebral column | |
| vertebra | (24) |
| sacrum | (1) |
| coccyx | (1) |
| thoracic cage | |
| rib | (24) |
| sternum | (1) |
| **appendicular skeleton** | |
| pectoral (shoulder) girdle | |
| scapula | (2) |
| clavicle | (2) |
| upper limbs | |
| humerus | (2) |
| radius | (2) |
| ulna | (2) |
| carpal | (16) |
| metacarpal | (10) |
| phalanx | (28) |
| pelvic girdle | |
| hip bone (coxal bone; pelvic bone; innominate bone) | (2) |
| lower limbs | |
| femur | (2) |
| tibia | (2) |
| fibula | (2) |
| patella | (2) |
| tarsal | (14) |
| metatarsal | (10) |
| phalanx | (28) |
| **Total** | **206 bones** |

1 _____

2 _____

3 _____

Hyoid bone

4 _____

5 _____

6 _____

7 _____

8 _____

9 _____

10 _____

11 _____

12 _____

13 _____

14 _____

15 _____

(a)

**FIGURE 13.1** Label the major bones of the skeleton: (*a*) anterior view; (*b*) posterior view. The axial portion is shown in orange; the appendicular portions are shown in yellow. **1** **2** **AP|R**

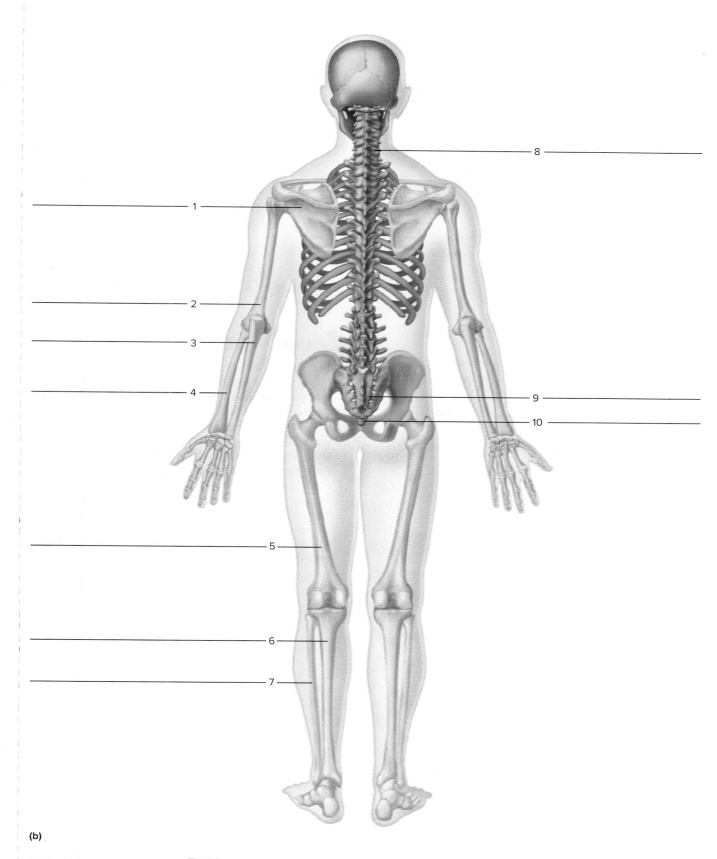

**(b)**

**FIGURE 13.1** *Continued.* **AP|R**

4. Study table 7.4 in chapter 7 of the textbook. Bone features (bone markings) can be grouped together in a category of *projections, articulations, depressions,* or *openings.* Within each category more specific examples occur. Each of the bones listed in this section represents only one example of a location in the human body. Locate an example of each of the following features on the example, bone from a disarticulated skeleton, noting the size, shape, and location in the human skeleton:

**Projections: sites for a tendon or ligament attachment**

**crest** (ridgelike)—hip bone

**epicondyle** (superior to condyle)—femur

**line (linea)** (slightly raised ridge)—femur

**process** (prominent)—vertebra

**protuberance** (outgrowth)—skull (occipital)

**ramus** (extension)—hip bone

**spine** (thornlike)—scapula

**trochanter** (large)—femur

**tubercle** (small, knoblike)—humerus

**tuberosity** (rough elevation)—radius

**Articulations: where bones connect at a joint**

**condyle** (rounded process)—skull (occipital)

**facet** (nearly flat)—vertebra

**head** (expanded end)—femur

**Depressions: recessed areas in bones**

**fossa** (shallow basin)—humerus

**fovea** (tiny pit)—femur

**Openings: open spaces in bones**

**fissure** (slit)—skull (orbit)

**foramen** (hole)—vertebra

**meatus** (tubelike passageway)—skull (temporal)

**sinus** (cavity)—skull (maxilla)

## ASSESS

### CRITICAL THINKING

Locate and name the largest foramen in the skull.

_____

_____

_____

Locate and name the largest foramen in the skeleton.

_____

_____

_____

5. Complete Parts A–D of Laboratory Report 13.

## LEARN: ACTIVITY

Images on radiographs (X-rays) are produced by allowing X-rays from an X-ray tube to pass through a body part and to expose photographic film positioned on the opposite side of the part. The image that appears on the film after it is developed reveals the presence of parts with different densities. Bone, for example, is very dense tissue and is a good absorber of X-rays. Thus, bone generally appears light on the film. Air-filled spaces, on the other hand, absorb almost no X-rays and appear as dark areas on the film. Liquids and soft tissues absorb intermediate quantities of X-rays, so they usually appear in various shades of gray.

Examine the available radiographs of skeletal structures by holding each film in front of a light source. Identify as many of the bones and features as you can.

The 🄰 corresponds to the indicated Learning Outcome(s) found at the beginning of the laboratory exercise.

# Organization of the Skeleton

## ♻ PART A ASSESSMENTS

Complete the following statements:

1. The two divisions of the skeleton are the _____ skeleton and the appendicular skeleton. 🄰

2. The extra bones that sometimes develop between the flat bones of the skull are called _____ bones. 🄂

3. The cranium and facial bones compose the _____. 🄰

4. The _____ bone supports the tongue. 🄂

5. The _____ at the inferior end of the sacrum is composed of four fused vertebrae. 🄂

6. Most ribs are attached anteriorly to the _____. 🄂

7. The thoracic cage is composed of _____ pairs of ribs. 🄂

8. The scapulae and clavicles together form the _____ girdle. 🄰

9. The humerus, radius, and _____ articulate to form the elbow joint. 🄂

10. The wrist is composed of eight bones called _____. 🄂

11. The hip bones are attached posteriorly to the _____. 🄂

12. The _____ covers the anterior surface of the knee. 🄂

13. The bones that articulate with the distal ends of the tibia and fibula are called _____. 🄂

14. All finger and toe bones are called _____. 🄂

## ♻ PART B ASSESSMENTS

Match the terms in column A with the definitions in column B. Place the letter of your choice in the space provided. 🄃

| Column A | Column B |
|---|---|
| a. Condyle | _____ 1. Small, nearly flat articular surface |
| b. Crest | |
| c. Facet | _____ 2. Shallow basin |
| d. Foramen | _____ 3. Rounded process |
| e. Fossa | _____ 4. Opening or hole |
| f. Line | |
| g. Tuberosity | _____ 5. Knoblike rough elevation |
| | _____ 6. Ridgelike projection |
| | _____ 7. Slightly raised ridge |

**PART C ASSESSMENTS**

Match the terms in column A with the definitions in column B. Place the letter of your choice in the space provided.

**Column A**

a. Fovea
b. Head
c. Meatus
d. Sinus
e. Spine
f. Trochanter
g. Tubercle

**Column B**

_____ 1. Tubelike passageway

_____ 2. Tiny pit or depression

_____ 3. Small, knoblike process

_____ 4. Thornlike projection

_____ 5. Rounded enlargement at end of bone

_____ 6. Air-filled cavity within bone

_____ 7. Relatively large process

 **PART D ASSESSMENTS**

1. Identify the bones indicated in figure 13.2.

**FIGURE 13.2** Identify the bones in this random arrangement, using the terms provided.  (©J & J Photography)

**Terms:**

| | |
|---|---|
| Clavicle | Rib |
| Femur | Sacrum |
| Fibula | Scapula |
| Hip bone | Sternum |
| Humerus | Tibia |
| Patella | Ulna |
| Radius | |

2. List any of the bones shown in figure 13.2 that are included as part of the axial skeleton.  _____

_____

# 14

# Skull

## MATERIALS NEEDED

Textbook
Human skull, articulated
Human skull, disarticulated (Beauchene)
Human skull, sagittal section
Colored pencils
Fetal skull

## PURPOSE OF THE EXERCISE

To examine the structure of the human skull and to identify the bones and major features of the skull.

##  LEARNING OUTCOMES AP|R

After completing this exercise, you should be able to

1. Distinguish between the cranium and the facial skeleton.
2. Locate and label the bones of the skull and their major features.
3. Locate and label the major sutures of the cranium.
4. Locate and label the sinuses of the skull.

**A** human skull consists of twenty-two bones that, except for the lower jaw, are firmly interlocked along sutures. Eight of these immovable bones make up the braincase, or cranium, and thirteen more immovable bones and the mandible form the facial skeleton.

##  PRACTICE

### PROCEDURE—Skull AP|R

1. Review the section entitled "Skull" in chapter 7 of the textbook.
2. As a review activity, label figures 14.1–14.5.
3. Study the bones with paranasal sinuses (fig. 14.6).

4. Examine the **cranial bones** of the articulated human skull and the sectioned skull. Also observe the corresponding disarticulated bones. Locate the following bones and features in the laboratory specimens and, at the same time, palpate as many of these bones and features in your skull as possible.

**frontal bone**      **(1)**
    supraorbital foramen
    frontal sinus

**parietal bone**      **(2)**
    sagittal suture
    coronal suture

**occipital bone**      **(1)**
    lambdoid suture
    foramen magnum
    occipital condyle

**temporal bone**      **(2)**
    squamous suture
    external acoustic meatus
    mandibular fossa
    mastoid process
    styloid process
    carotid canal
    jugular foramen
    internal acoustic meatus
    zygomatic process

**sphenoid bone**      **(1)**
    sella turcica
    greater and lesser wings
    sphenoidal sinus

**ethmoid bone**      **(1)**
    cribriform plate
    perpendicular plate
    superior nasal concha
    middle nasal concha
    ethmoidal sinus
    crista galli

5. Complete Parts A and B of Laboratory Report 14.

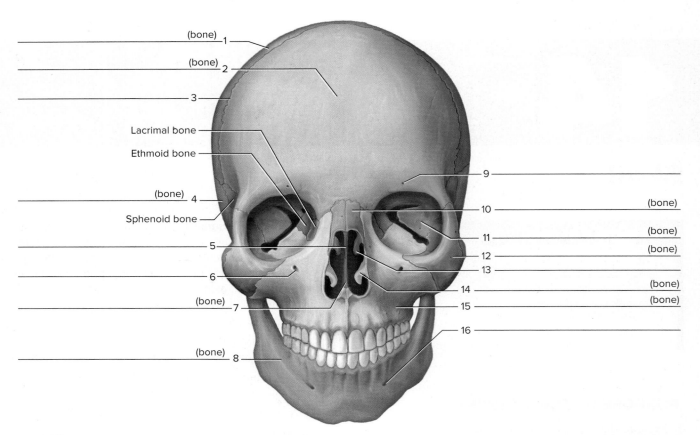

_____ (bone) 1

_____ (bone) 2

_____ 3

Lacrimal bone _____

Ethmoid bone _____

_____ (bone) 4

Sphenoid bone _____

_____ 5

_____ 6

_____ (bone) 7

_____ (bone) 8

9 _____

10 _____ (bone)

11 _____ (bone)

12 _____ (bone)

13 _____

14 _____ (bone)

15 _____ (bone)

16 _____

**FIGURE 14.1** Label the anterior bones and features of the skull. (If the line lacks the word *bone,* label the particular feature of that bone.) **A** **AP|R**

_____ (bone) 1

_____ 2

_____ 3

_____ (bone) 4

_____ (bone) 5

_____ 6

_____ 7

_____ 8

_____ 9

_____ 10

_____ 11

12 _____

13 _____ (bone)

14 _____ (bone)

Ethmoid bone _____

15 _____ (bone)

16 _____ (bone)

17 _____ (bone)

18 _____ (bone)

19 _____ (bone)

20 _____

**FIGURE 14.2** Label the lateral bones and features of the skull. **A** **A** **AP|R**

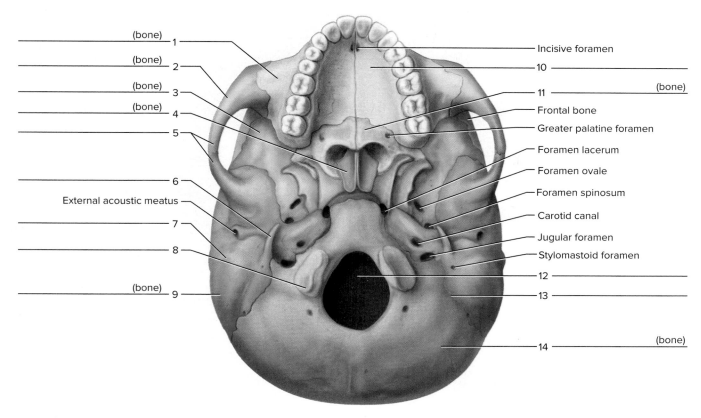

(bone) 1

(bone) 2

(bone) 3

(bone) 4

5

External acoustic meatus

6

7

8

(bone) 9

Incisive foramen

10

11 (bone)

Frontal bone

Greater palatine foramen

Foramen lacerum

Foramen ovale

Foramen spinosum

Carotid canal

Jugular foramen

Stylomastoid foramen

12

13

14 (bone)

**FIGURE 14.3** Label the inferior bones and features of the skull. 2 AP|R

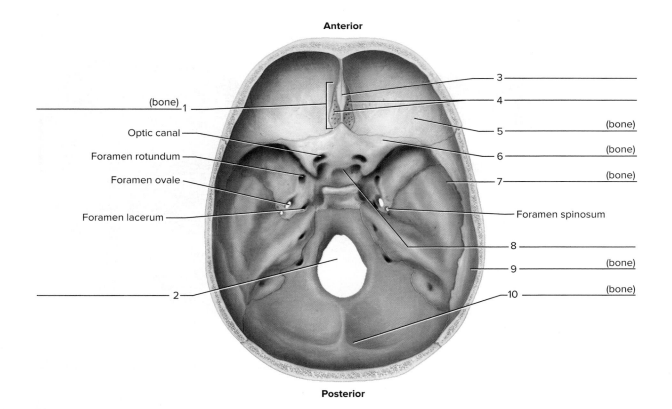

Anterior

(bone) 1

Optic canal

Foramen rotundum

Foramen ovale

Foramen lacerum

2

3

4

5 (bone)

6 (bone)

7 (bone)

Foramen spinosum

8

9 (bone)

10 (bone)

Posterior

**FIGURE 14.4** Label the bones and features of the floor of the cranial cavity as viewed from above. 2 AP|R

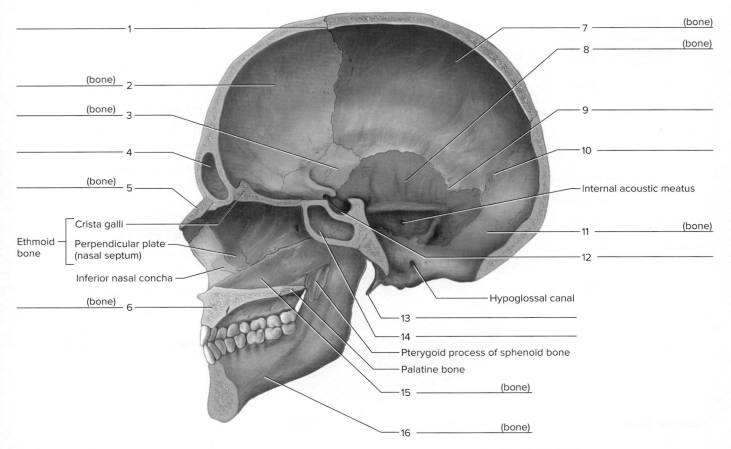

1 _____

2 _____ (bone)

3 _____ (bone)

4 _____

5 _____ (bone)

Ethmoid bone
- Crista galli _____
- Perpendicular plate (nasal septum) _____
- Inferior nasal concha _____

6 _____ (bone)

7 _____ (bone)

8 _____ (bone)

9 _____

10 _____

Internal acoustic meatus _____

11 _____ (bone)

12 _____

Hypoglossal canal _____

13 _____

14 _____

Pterygoid process of sphenoid bone _____

Palatine bone _____

15 _____ (bone)

16 _____ (bone)

**FIGURE 14.5**   Label the bones and features of the sagittal section of the skull.   2  3  4  AP|R

Frontal sinus

Ethmoidal air cells

Sphenoidal sinus

Maxillary sinus

**FIGURE 14.6**   Paranasal sinuses are located in the frontal bone, ethmoid bone, sphenoid bone, and both maxillary bones.

**6.** Examine the **facial bones** of the articulated and sectioned skulls and the corresponding disarticulated bones. Locate the following:

**maxilla** (2)
   maxillary sinus
   palatine process
   alveolar process
   alveolar arch

**palatine bone** (2)

**zygomatic bone** (2)
   temporal process
   zygomatic arch (formed by temporal and zygomatic processes)

**lacrimal bone** (2)

**nasal bone** (2)

**vomer** (1)

**inferior nasal concha** (2)

**mandible** (1)
   ramus
   mandibular condyle
   coronoid process
   alveolar border
   mandibular foramen
   mental foramen

**7.** Study the skull bones of a disarticulated skull (fig. 14.7).

**8.** Complete Part C of the laboratory report.

### LEARN: ACTIVITY

Use colored pencils to differentiate the bones illustrated in figure 14.2. Select a different color for each bone. This activity should help you locate various bones shown in a lateral view of a skull. You can check your work by referring to the corresponding figure in the textbook, presented in full color.

**9.** Study table 7.7 of chapter 7 in the textbook. Locate the following features of the human skull:

**carotid canal**
**foramen lacerum**
**foramen magnum**
**foramen ovale**
**foramen rotundum**
**foramen spinosum**
**greater palatine foramen**
**hypoglossal canal**
**incisive foramen**
**inferior orbital fissure**

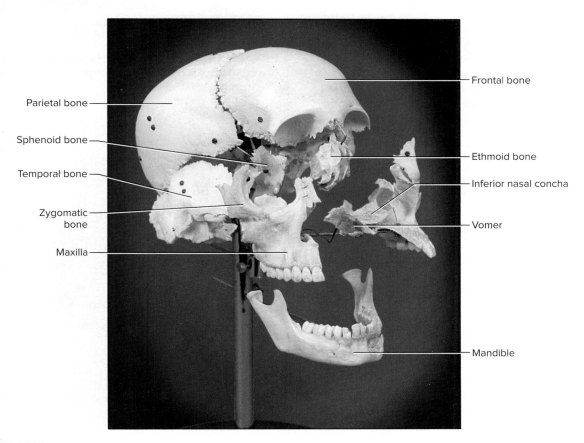

**FIGURE 14.7** Bones of a disarticulated skull. (©J & J Photography)

**infraorbital foramen**
**internal acoustic meatus**
**jugular foramen**
**mandibular foramen**
**mental foramen**
**optic canal**
**stylomastoid foramen**
**superior orbital fissure**
**supraorbital foramen**

**10.** Complete Parts D and E of the laboratory report.

 **ASSESS**

**CRITICAL THINKING**

Examine the inside of the cranium on a sectioned skull. What area appears to be the weakest area? Explain your answer.

_____

_____

_____

_____

 **LEARN: ACTIVITY**

Examine the fetal skeleton and skull (figs. 14.8 and 14.9). The skull is incompletely developed and the cranial bones are separated by fibrous membranes. These membranous areas are called as fontanels, or "soft spots." The fontanels close as the cranial bones grow together. The posterior and lateral fontanels usually close during the first year after birth, whereas the anterior fontanel may not close until the middle or end of the second year. What other features characterize the fetal skull?

_____

_____

_____

_____

_____

_____

**FIGURE 14.8** Anterior view of a fetal skeleton next to a metric scale. The gestational age of this skeleton is 7–8 months. (©J & J Photography)

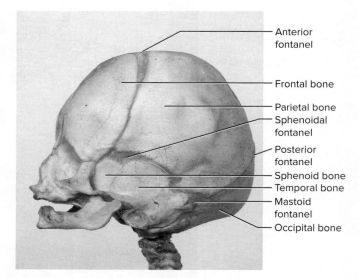

**FIGURE 14.9** Lateral view of a fetal skull. (©J & J Photography)

Name _____

Date _____

Section _____

The Ⓐ corresponds to the indicated Learning Outcome(s) found at the beginning of the laboratory exercise.

LABORATORY
REPORT

**14**

# Skull

## 🔄 PART A ASSESSMENTS

Complete the following statements:

**1.** Name six cranial bones that are visible on a lateral view of a skull. Ⓐ1 _____

_____

**2.** The _____ suture joins the frontal bone to the parietal bones. Ⓐ3

**3.** The parietal bones are firmly interlocked along the midline by the _____ suture. Ⓐ3

**4.** The _____ suture joins the parietal bones to the occipital bone. Ⓐ3

**5.** The temporal bones are joined to the parietal bones along the _____ sutures. Ⓐ3

**6.** Name the three cranial bones that contain sinuses. Ⓐ1 Ⓐ4 _____

**7.** Name a facial bone that contains a sinus. Ⓐ1 Ⓐ4 _____

## 🔄 PART B ASSESSMENTS

Match the bones in column A with the features in column B. Place the letter of your choice in the space provided. (Some answers are used more than once.) Ⓐ2

| Column A | Column B |
|----------|----------|
| **a.** Ethmoid bone | _____ **1.** Forms sagittal, coronal, squamous, and lambdoid sutures |
| **b.** Frontal bone | _____ **2.** Cribriform plate |
| **c.** Occipital bone | _____ **3.** Crista galli |
| **d.** Parietal bone | _____ **4.** External acoustic meatus |
| **e.** Sphenoid bone | _____ **5.** Foramen magnum |
| **f.** Temporal bone | _____ **6.** Mandibular fossa |
| | _____ **7.** Mastoid process |
| | _____ **8.** Middle nasal concha |
| | _____ **9.** Occipital condyle |
| | _____ **10.** Sella turcica |
| | _____ **11.** Styloid process |
| | _____ **12.** Supraorbital foramen |

## PART C ASSESSMENTS

Match the bones in column A with the characteristics in column B. Place the letter of your choice in the space provided. **2**

| Column A | Column B |
|---|---|
| a. Inferior nasal concha | _____ 1. Forms bridge of nose |
| b. Lacrimal bone | _____ 2. Only movable bone in the facial skeleton |
| c. Mandible | _____ 3. Contains coronoid process |
| d. Maxilla | _____ 4. Creates prominence of cheek inferior and lateral to the eye |
| e. Nasal bone | _____ 5. Contains sockets of upper teeth |
| f. Palatine bone | _____ 6. Forms inferior portion of nasal septum |
| g. Vomer | _____ 7. Forms anterior portion of zygomatic arch |
| h. Zygomatic bone | _____ 8. Scroll-shaped bone |
| | _____ 9. Forms anterior roof of mouth |
| | _____ 10. Contains mental foramen |
| | _____ 11. Forms posterior roof of mouth |
| | _____ 12. Small medial bone of each orbit |

## PART D ASSESSMENTS

Match the passageways in column A with the structures transmitted through them in column B. Place the letter of your choice in the space provided. **2**

| Column A | Column B |
|---|---|
| a. Foramen magnum | _____ 1. Maxillary division of trigeminal nerve |
| b. Foramen ovale | _____ 2. Nerve fibers of spinal cord |
| c. Foramen rotundum | _____ 3. Optic nerve |
| d. Incisive foramen | _____ 4. Vagus and accessory nerves |
| e. Internal acoustic meatus | _____ 5. Nasopalatine nerves |
| f. Jugular foramen | _____ 6. Mandibular division of trigeminal nerve |
| g. Optic canal | _____ 7. Vestibular and cochlear nerves |

## PART E ASSESSMENTS

Identify the numbered bones and features of the skulls indicated in figures 14.10–14.13. **2** **3**

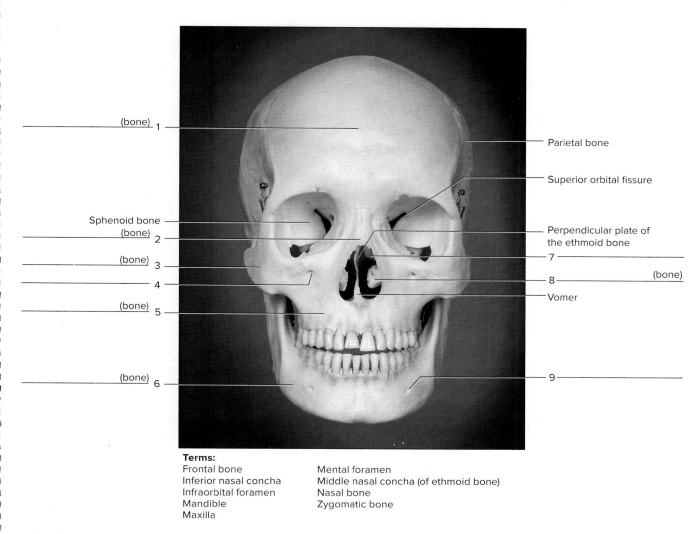

(bone) 1 ────

Parietal bone

Superior orbital fissure

Sphenoid bone ──── (bone) 2

Perpendicular plate of the ethmoid bone

(bone) 3 ────

7 ────

4 ────

8 ──── (bone)

Vomer

(bone) 5 ────

(bone) 6 ────

9 ────

**Terms:**

| | |
|---|---|
| Frontal bone | Mental foramen |
| Inferior nasal concha | Middle nasal concha (of ethmoid bone) |
| Infraorbital foramen | Nasal bone |
| Mandible | Zygomatic bone |
| Maxilla | |

**FIGURE 14.10**   Identify the bones and features indicated on this anterior view of the skull, using the terms provided.

(©J & J Photography)

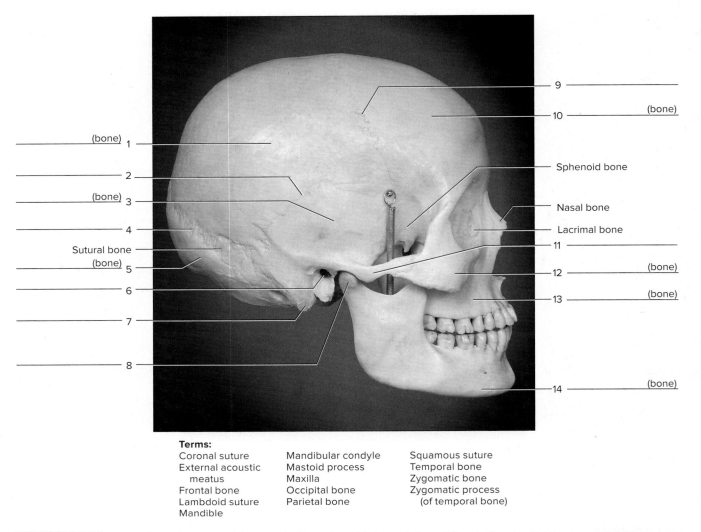

(bone) 1

2

(bone) 3

4

Sutural bone

(bone) 5

6

7

8

9

10 (bone)

Sphenoid bone

Nasal bone

Lacrimal bone

11

12 (bone)

13 (bone)

14 (bone)

**Terms:**

Coronal suture
External acoustic
   meatus
Frontal bone
Lambdoid suture
Mandible

Mandibular condyle
Mastoid process
Maxilla
Occipital bone
Parietal bone

Squamous suture
Temporal bone
Zygomatic bone
Zygomatic process
   (of temporal bone)

**FIGURE 14.11** Identify the bones and features indicated on this lateral view of the skull, using the terms provided.

(©J & J Photography)

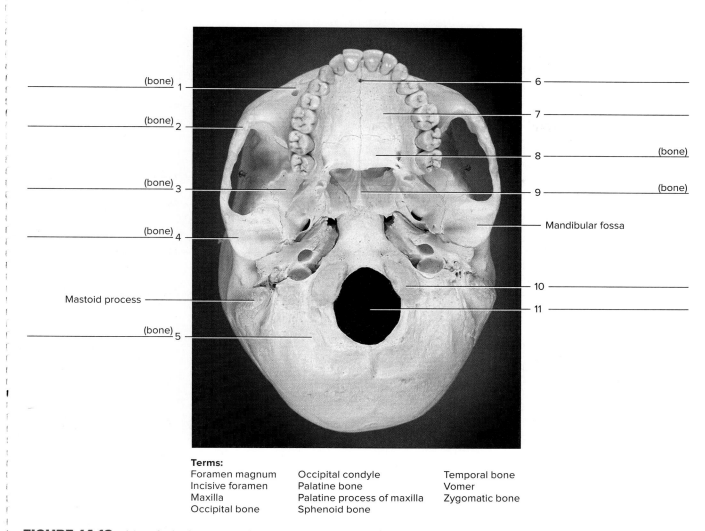

(bone) 1

(bone) 2

(bone) 3

(bone) 4

Mastoid process

(bone) 5

6

7

8 (bone)

9 (bone)

Mandibular fossa

10

11

**Terms:**

| | | |
|---|---|---|
| Foramen magnum | Occipital condyle | Temporal bone |
| Incisive foramen | Palatine bone | Vomer |
| Maxilla | Palatine process of maxilla | Zygomatic bone |
| Occipital bone | Sphenoid bone | |

**FIGURE 14.12**   Identify the bones and features indicated on this inferior view of the skull, using the terms provided.

(©J & J Photography)

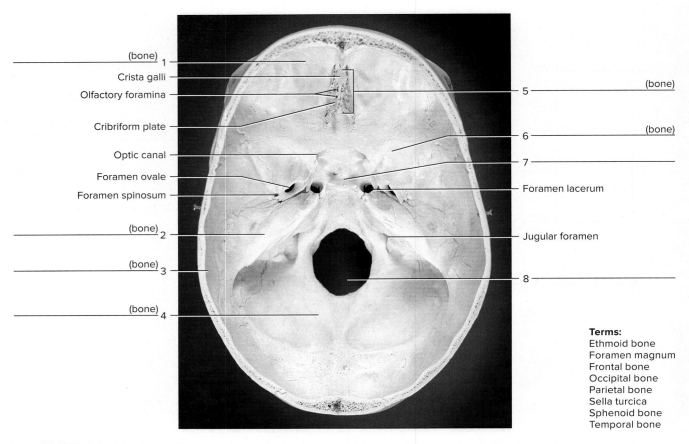

(bone) 1

Crista galli

Olfactory foramina

Cribriform plate

Optic canal

Foramen ovale

Foramen spinosum

(bone) 2

(bone) 3

(bone) 4

(bone) 5

(bone) 6

7

Foramen lacerum

Jugular foramen

8

**Terms:**
Ethmoid bone
Foramen magnum
Frontal bone
Occipital bone
Parietal bone
Sella turcica
Sphenoid bone
Temporal bone

**FIGURE 14.13** Identify the bones and features on this floor of the cranial cavity of a skull, using the terms provided. (©J & J Photography)

# 15

# Vertebral Column and Thoracic Cage

## MATERIALS NEEDED

Textbook
Human skeleton, articulated
Samples of cervical, thoracic, and lumbar vertebrae
Human skeleton, disarticulated

## PURPOSE OF THE EXERCISE

To examine the vertebral column and the thoracic cage of the human skeleton and to identify the bones and major features of these parts.

##  LEARNING OUTCOMES AP|R

After completing this exercise, you should be able to

1. Identify the major features of the vertebral column.
2. Locate the features of a vertebra.
3. Distinguish the cervical, thoracic, and lumbar vertebrae and locate the sacrum and coccyx.
4. Identify the structures and functions of the thoracic cage.
5. Distinguish between true and false ribs.

The *vertebral column,* consisting of twenty-six bones, extends from the skull to the pelvis and forms the vertical axis of the human skeleton. The vertebral column includes seven cervical vertebrae, twelve thoracic vertebrae, five lumbar vertebrae, one sacrum of five fused vertebrae, and one coccyx of usually four fused vertebrae. To help you to remember the number of cervical, thoracic, and lumbar vertebrae from superior to inferior, consider this saying: breakfast at 7, lunch at 12, and dinner at 5. These vertebrae are separated from one another by cartilaginous intervertebral discs and are held together by ligaments.

The *thoracic cage* surrounds the thoracic and upper abdominal cavities. It includes the ribs, the thoracic vertebrae, the sternum, and the costal cartilages. The thoracic cage protects organs in the thoracic and upper abdominal cavities, supports the pectoral girdle and arm, and aids breathing.

##  PRACTICE

### PROCEDURE A—Vertebral Column

1. Review the section entitled "Vertebral Column" in chapter 7 of the textbook.
2. As a review activity, label figures 15.1–15.4.
3. Examine the vertebral column of the human skeleton and locate the following bones and features. At the same time, locate as many of the corresponding bones and features in your skeleton as possible.

| | |
|---|---|
| **cervical vertebrae** | **(7)** |
|    atlas (C1) | |
|    axis (C2) | |
|    vertebra prominens (C7) | |
| **thoracic vertebrae** | **(12)** |
| **lumbar vertebrae** | **(5)** |
| **sacrum** | **(1)** |
| **coccyx** | **(1)** |
| **intervertebral discs** (fibrocartilage) | |
| **vertebral canal** (contains spinal cord) | |
| **cervical curvature** | |
| **thoracic curvature** | |
| **lumbar curvature** | |
| **sacral curvature** | |
| **intervertebral foramina** | |
|    (passageway for spinal nerves) | |

##  ASSESS

### CRITICAL THINKING

Note the four curvatures of the vertebral column. What functional advantages exist with curvatures for skeletal structure instead of a straight vertebral column?

_____

_____

_____

_____

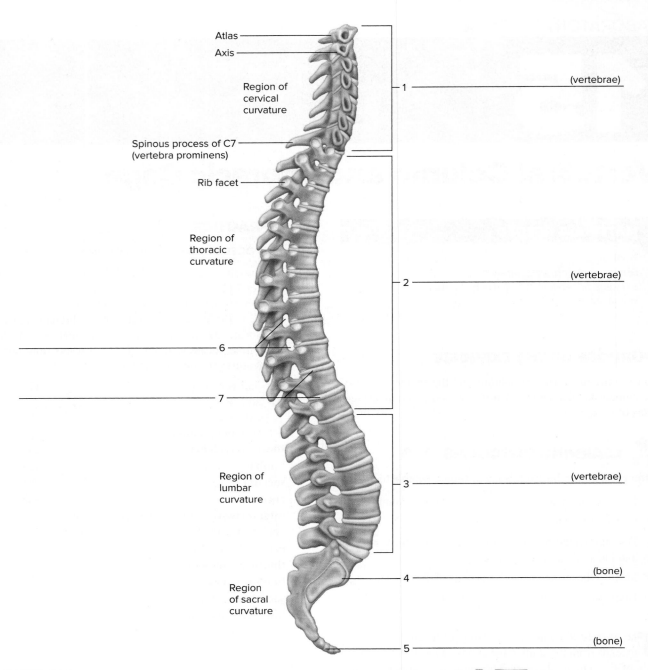

Atlas

Axis

Region of
cervical
curvature

Spinous process of C7
(vertebra prominens)

Rib facet

Region of
thoracic
curvature

Region of
lumbar
curvature

Region
of sacral
curvature

1 _____ (vertebrae)

2 _____ (vertebrae)

3 _____ (vertebrae)

4 _____ (bone)

5 _____ (bone)

6

7

**FIGURE 15.1**   Label the bones and features of the vertebral column (right lateral view).  A  AP|R

**(a)** Atlas (C1)

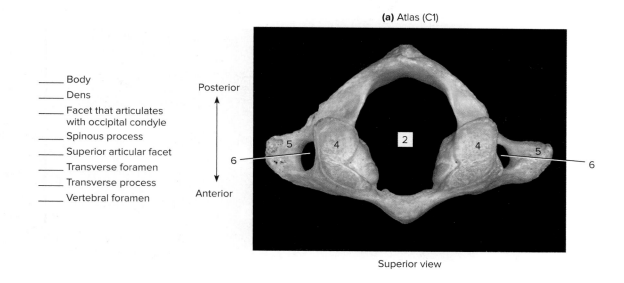

_____ Body
_____ Dens
_____ Facet that articulates with occipital condyle
_____ Spinous process
_____ Superior articular facet
_____ Transverse foramen
_____ Transverse process
_____ Vertebral foramen

Posterior
Anterior

Superior view

**(b)** Axis (C2)

Posterior
Anterior

Superior view

Right lateral view
Posterior ⟷ Anterior

**FIGURE 15.2**  Label the superior features of (*a*) the atlas and the superior and right lateral features of (*b*) the axis by placing the correct numbers in the spaces provided. (The broken arrow indicates a transverse foramen.)  **2** **3** **AP|R**
(©J & J Photography)

_____ Body
_____ Inferior vertebral notch
_____ Lamina
_____ Pedicle
_____ Spinous process
_____ Superior articular process
_____ Transverse foramen
_____ Transverse process
_____ Vertebral foramen

**(a)** Cervical vertebra

Posterior

Anterior

**(b)** Thoracic vertebra

**(c)** Lumbar vertebra

Posterior ⟷ Anterior

**FIGURE 15.3** Label the superior and right lateral features of the (a) cervical, (b) thoracic, and (c) lumbar vertebrae by placing the correct numbers in the spaces provided. (The broken arrow indicates a transverse foramen.) 2 3 AP R

(©J & J Photography)

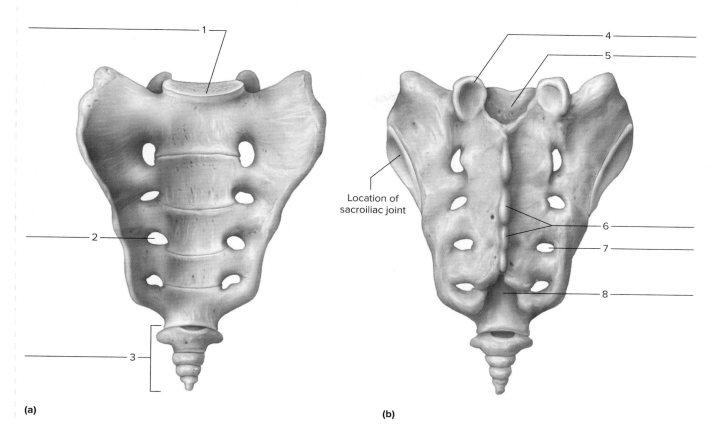

(a)

(b)

**FIGURE 15.4**    Label the coccyx and the features of the sacrum: (*a*) anterior view; (*b*) posterior view. [2] [3] [AP|R]

4. Compare the available samples of cervical, thoracic, and lumbar vertebrae by noting differences in size and shape and by locating the following features:

   **vertebral foramen**
   **body**
   **pedicles**
   **laminae**
   **spinous process**
   **transverse processes**
   **superior articular processes**
   **inferior articular processes**
   **inferior vertebral notch**
   **transverse foramina**
   **facets**
   **dens of axis**

5. Examine the sacrum and coccyx. Locate the following features:

   **sacrum**
      superior articular process
      posterior sacral foramen
      anterior sacral foramen
      sacral promontory
      sacral canal
      tubercles
      median sacral crest
      sacral hiatus

   **coccyx**

6. Complete Parts A and B of Laboratory Report 15.

## PRACTICE

### PROCEDURE B—Thoracic Cage

1. Review the section entitled "Thoracic Cage" in chapter 7 of the textbook.
2. As a review activity, label figure 15.5.
3. Study a typical rib with articulation sites with a thoracic vertebra (fig. 15.6).
4. Examine the thoracic cage of the human skeleton and locate the following bones and features:

   **rib**
      head
      tubercle
      neck
      shaft
      anterior (sternal) end
      facets
      true ribs (pairs 1–7)
      false ribs (pairs 8–12) (includes floating ribs) floating ribs (pairs 11–12)

   **costal cartilages** (hyaline cartilage)

   **sternum**
      jugular (suprasternal) notch
      clavicular notch
      manubrium
      sternal angle
      body
      xiphoid process

5. Complete Parts C and D of the laboratory report.

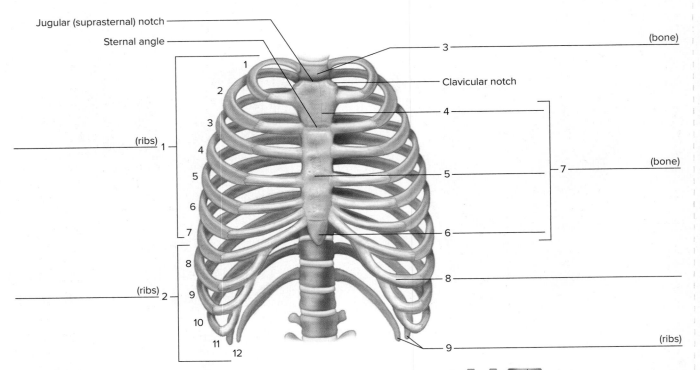

Jugular (suprasternal) notch

Sternal angle

(ribs) 1

(ribs) 2

1
2
3
4
5
6
7
8
9
10
11
12

3 (bone)

Clavicular notch

4

5

6

7 (bone)

8

9 (ribs)

**FIGURE 15.5** Label the bones and features of the thoracic cage (anterior view). 🄰 🄰 AP R

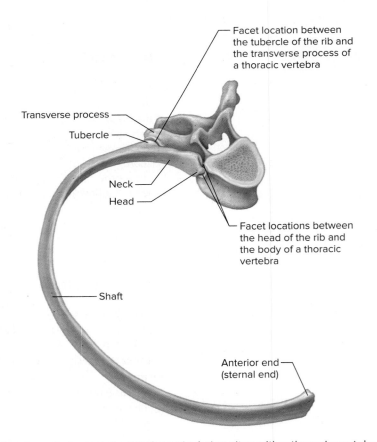

Facet location between the tubercle of the rib and the transverse process of a thoracic vertebra

Transverse process

Tubercle

Neck

Head

Facet locations between the head of the rib and the body of a thoracic vertebra

Shaft

Anterior end (sternal end)

**FIGURE 15.6** Superior view of a typical rib with the articulation sites with a thoracic vertebra.

Name _____

Date _____

Section _____

The  corresponds to the indicated Learning Outcome(s) found at the beginning of the laboratory exercise.

LABORATORY
REPORT

15

# Vertebral Column and Thoracic Cage

## PART A ASSESSMENTS

Complete the following statements:

1. The vertebral column encloses and protects the _____. 🄰

2. The number of separate bones in the vertebral column of an adult is _____. 🄰

3. The _____ of the vertebrae support the weight of the head and trunk. 🄰

4. The intervertebral foramina provide passageways for _____. 🄰

5. Transverse foramina of cervical vertebrae serve as passageways for _____ leading to the brain. 🄰

6. The first vertebra is also called the _____. 🄰

7. The second vertebra is also called the _____. 🄰

8. When the head is moved from side to side, the first vertebra pivots around the _____ of the second vertebra. 🄰

9. The _____ vertebrae have the largest and strongest bodies. 🄰

10. The typical number of vertebrae that fuse in the adult to form the sacrum is _____. 🄰

11. The upper, anterior margin of the sacrum that projects forward is called the _____. 🄰

12. An opening called the _____ exists at the tip of the sacral canal. 🄰

## PART B ASSESSMENTS

Based on your observations, compare typical cervical, thoracic, and lumbar vertebrae in relation to the characteristics indicated in the table. The table is partly completed. For your responses, consider characteristics such as size, shape, presence or absence, and unique features. 🄰 🄰

| Vertebra | Number | Size | Body | Spinous Process | Transverse Foramina |
|----------|--------|------|------|-----------------|---------------------|
| Cervical | 7 | | Smallest | C2 through C6 are forked (bifid) | |
| Thoracic | | Intermediate | | | |
| Lumbar | | | | | Absent |

## PART C ASSESSMENTS

Complete the following statements:

1. The last two pairs of ribs that have no cartilaginous attachments to the sternum are sometimes called

   _____ ribs.  **5**

2. There are _____ pairs of true ribs.  **5**

3. Costal cartilages are composed of _____ tissue.  **4**

4. The manubrium articulates with the _____ on its superior border.  **4**

5. List three general functions of the thoracic cage.  **4** _____

   _____

## PART D ASSESSMENTS

Identify the bones and features indicated in the radiograph of the neck in figure 15.7.  **1** **2**

**Terms:**
Atlas
Axis
Body
Intervertebral disc
Spinous process
Transverse process

Occipital bone

2 _____ (bone)

3 _____ (bone)

Mandible

4 _____

5 _____

6 _____

Body of seventh cervical vertebra

1 _____

Spinous process of C7 (vertebra prominens)

**FIGURE 15.7** Identify the bones and features indicated in this radiograph of the neck (lateral view), using the terms provided. **AP|R** (©Stockbyte/Punchstock RF)

# 16

# Pectoral Girdle and Upper Limb

## MATERIALS NEEDED

Textbook
Human skeleton, articulated
Human skeleton, disarticulated
Colored pencils

## PURPOSE OF THE EXERCISE

To examine the bones of the pectoral girdle and upper limb and to identify the major features of these bones.

 **LEARNING OUTCOMES** AP|R

After completing this exercise, you should be able to

1. Locate and identify the bones of the pectoral girdle and their major features.

2. Locate and identify the bones of the upper limb and their major features.

3. Recognize the anatomical terms pertaining to the pectoral girdle and upper limb.

The pectoral girdle (shoulder girdle) consists of two clavicles and two scapulae. These parts support the upper limbs and serve as attachments for various muscles that move these limbs.

Each upper limb includes a humerus, a radius, an ulna, eight carpals, five metacarpals, and fourteen phalanges. These bones form the framework of the arm, forearm, and hand. They also function as parts of levers when the limbs are moved.

 **PRACTICE**

### PROCEDURE A—Pectoral Girdle

1. Review the section entitled "Pectoral Girdle" in chapter 7 of the textbook.
2. As a review activity, label figures 16.1 and 16.2.

3. Examine the bones of the pectoral girdle and locate the following features. At the same time, locate as many of the corresponding surface bones and features of your own skeleton as possible.

**clavicle**
  medial (sternal) end
  lateral (acromial) end
**scapula**
  spine
  acromion process
  coracoid process
  glenoid cavity
  borders
    superior border
    medial (vertebral) border
    lateral (axillary) border
  fossae
    supraspinous fossa
    infraspinous fossa

 **ASSESS**

### CRITICAL THINKING

Why is the clavicle a bone that can easily fracture?

_____

_____

_____

_____

4. Complete Part A of Laboratory Report 16.

 **PRACTICE**

### PROCEDURE B—Upper Limb

1. Review the section entitled "Upper Limb" in chapter 7 of the textbook.
2. As a review activity, label figures 16.3–16.6.

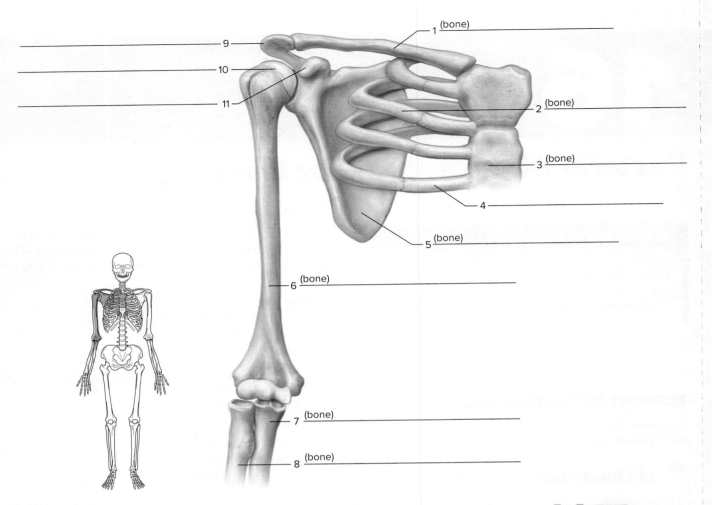

**FIGURE 16.1**    Label the bones and features of the right shoulder and upper limb (anterior view). Ⓐ Ⓐ **AP**|**R**

3. Examine the following bones and features of the upper limb:

**humerus**
  proximal features
    head
    greater tubercle
    lesser tubercle
    anatomical neck
    surgical neck
    intertubercular sulcus
  shaft
    deltoid tuberosity
  distal features
    capitulum
    trochlea
    medial epicondyle
      lateral epicondyle
      coronoid fossa
      olecranon fossa
**radius**
    head
    radial tuberosity
    styloid process
    ulnar notch
**ulna**
    trochlear notch (semilunar notch)
    radial notch
    olecranon process
    coronoid process
    styloid process
    head

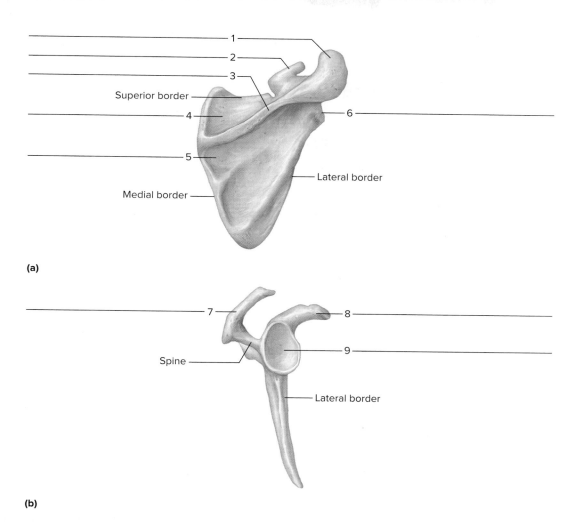

**FIGURE 16.2** Label (*a*) the posterior surface of the right scapula and (*b*) the lateral aspect of the right scapula. **A** **AP|R**

In figure (a):
- 1
- 2
- 3
- Superior border
- 4
- 5
- 6
- Lateral border
- Medial border
- (a)

In figure (b):
- 7
- 8
- Spine
- 9
- Lateral border
- (b)

**carpal bones**
    proximal row (listed lateral to medial)
        scaphoid
        lunate
        triquetrum
        pisiform
    distal row (listed medial to lateral)
        hamate
        capitate
        trapezoid
        trapezium
**metacarpal bones**
**phalanges**
    proximal phalanx
    middle phalanx
    distal phalanx

**4.** Complete Parts B–E of the laboratory report.

The following mnemonic device will help you learn the eight carpals:

**S**o **L**ong **T**op **P**art
**H**ere **C**omes **T**he **T**humb

The first letter of each word corresponds to the first letter of a carpal. This device arranges the carpals in order for the proximal, transverse row of four bones from lateral to medial, followed by the distal, transverse row from medial to lateral, which ends nearest the thumb. This arrangement assumes the hand is in the anatomical position.

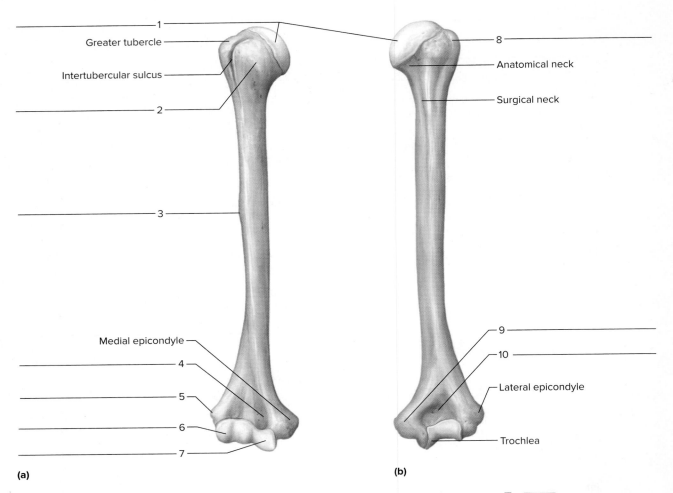

Greater tubercle

Intertubercular sulcus

1

2

3

Medial epicondyle

4

5

6

7

(a)

8

Anatomical neck

Surgical neck

9

10

Lateral epicondyle

Trochlea

(b)

**FIGURE 16.3** Label the (*a*) anterior surface and (*b*) posterior surface of the right humerus.  2  AP|R

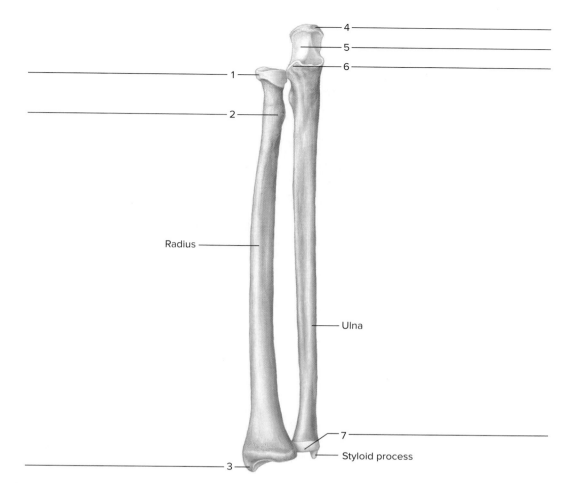

Radius

Ulna

Styloid process

**FIGURE 16.4** Label the major anterior features of the right radius and ulna (anterior view). 🄰 **AP|R**

2 _____ (bone)

3 _____

4 _____

5 _____ (bone)

6 _____ (bone)

**FIGURE 16.5** Label the bones and features of the right elbow, posterior view. 🄰

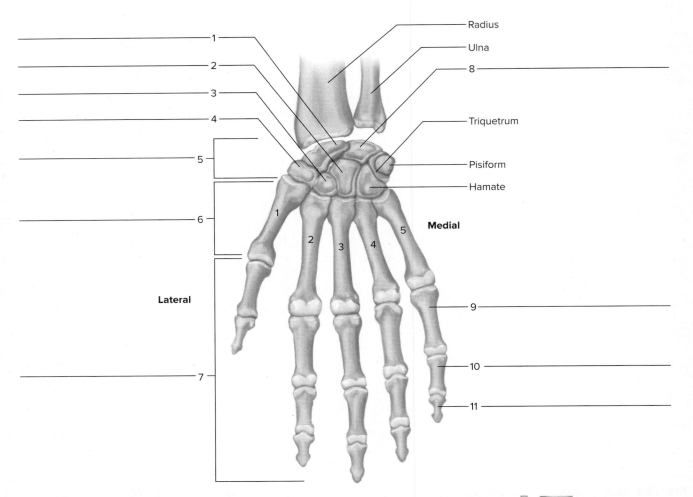

Radius

Ulna

1

2

3

4

5

6

7

8

Triquetrum

Pisiform

Hamate

Medial

Lateral

1

2

3

4

5

9

10

11

**FIGURE 16.6** Label the bones and groups of bones in this anterior view of the right hand.

**LEARN: ACTIVITY**

Use different colored pencils to distinguish the individual bones in figure 16.6.

Name _____

Date _____

Section _____

The   corresponds to the indicated Learning Outcome(s) found at the beginning of the laboratory exercise.

LABORATORY
REPORT

16

# Pectoral Girdle and Upper Limb

## PART A ASSESSMENTS

Complete the following statements:

**1.** The pectoral girdle is an incomplete ring because it is open in the back between the _____ . 🄰

**2.** The medial end of a clavicle articulates with the _____ of the sternum. 🄰

**3.** The lateral end of a clavicle articulates with the _____ process of the scapula. 🄰

**4.** The _____ is a bone that serves as a brace between the sternum and the scapula. 🄰

**5.** The _____ divides the posterior side of the scapula into unequal portions. 🄰

**6.** The lateral tip of the shoulder is the _____ process of the scapula. 🄰

**7.** Near the lateral end of the scapula, the _____ process curves anteriorly and inferiorly from the clavicle. 🄰

**8.** The glenoid cavity of the scapula articulates with the _____ of the humerus. 🄰

## PART B ASSESSMENTS

Match the bones in column A with the bones and features in column B. Place the letter of your choice in the space provided. 🄰

| Column A | Column B |
|---|---|
| **a.** Carpals | _____ **1.** Capitate |
| **b.** Humerus | _____ **2.** Coronoid fossa |
| **c.** Metacarpals | _____ **3.** Deltoid tuberosity |
| **d.** Phalanges | _____ **4.** Greater tubercle |
| **e.** Radius | _____ **5.** Five palmar bones |
| **f.** Ulna | _____ **6.** Fourteen bones in digits |
| | _____ **7.** Intertubercular sulcus |
| | _____ **8.** Lunate |
| | _____ **9.** Olecranon fossa |
| | _____ **10.** Radial tuberosity |
| | _____ **11.** Trapezium |
| | _____ **12.** Trochlear notch |

Identify the bones and features indicated in the radiographs of figures 16.7–16.9. 🅰 🅰

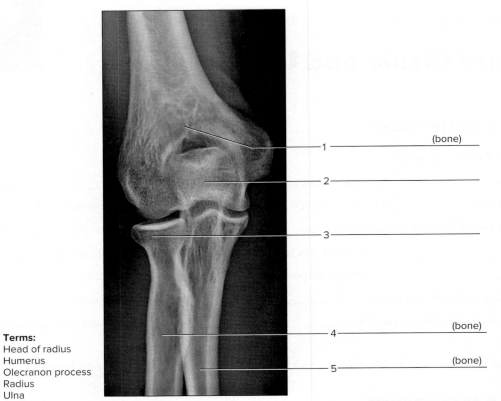

1 —————————————— (bone)

2 ——————————————

3 ——————————————

4 —————————————— (bone)

5 —————————————— (bone)

**Terms:**
Head of radius
Humerus
Olecranon process
Radius
Ulna

**FIGURE 16.7** Identify the bones and features indicated on this radiograph of the right elbow (anterior view), using the terms provided. **AP|R** (©Dale Butler)

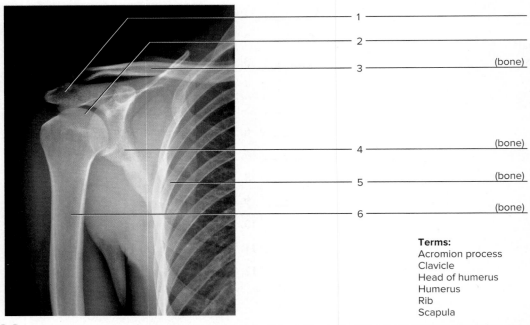

1 ——————————————

2 ——————————————

3 —————————————— (bone)

4 —————————————— (bone)

5 —————————————— (bone)

6 —————————————— (bone)

**Terms:**
Acromion process
Clavicle
Head of humerus
Humerus
Rib
Scapula

**FIGURE 16.8** Identify the bones and features indicated on this radiograph of the right shoulder (anterior view), using the terms provided. **AP|R** (©sanyanwuji/iStock/Getty Images RF)

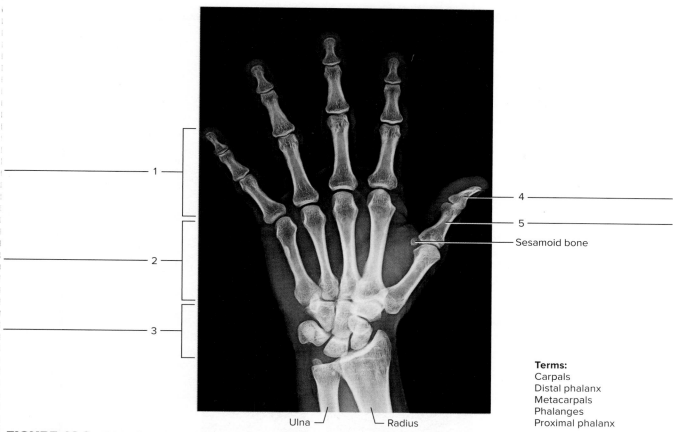

Terms:
Carpals
Distal phalanx
Metacarpals
Phalanges
Proximal phalanx

**FIGURE 16.9** Identify the bones indicated on this radiograph of the right hand (anterior view), using the terms provided. **AP|R** (©Dale Butler)

## PART D ASSESSMENTS

Complete the labeling of the bones of the hand in figure 16.10. **2**

| | |
|---|---|
| _____ | Capitate |
| _____ | Distal phalanges |
| _____ | Hamate |
| _____ | Lunate |
| _____ | Metacarpals |
| _____ | Middle phalanges |
| ___4____ | Pisiform |
| _____ | Proximal phalanges |
| _____ | Scaphoid |
| _____ | Trapezium |
| _____ | Trapezoid |
| ___3____ | Triquetrum |

**FIGURE 16.10** Label the bones numbered on this anterior view of the right hand by placing the correct numbers in the spaces provided. (©J & J Photography)

## PART E ASSESSMENTS

Locate the 21 anatomical terms pertaining to the "Pectoral Girdle and Upper Limb." **3**

| E | T | A | M | A | H | S | U | I | D | A | R |
|---|---|---|---|---|---|---|---|---|---|---|---|
| M | A | R | D | U | B | E | N | I | P | S | E |
| U | E | K | I | L | I | P | O | F | U | S | L |
| L | L | E | O | N | X | Z | H | C | Z | S | C |
| U | H | V | H | A | E | V | E | A | P | E | R |
| T | C | O | P | P | Q | F | B | P | H | C | E |
| I | O | O | A | S | S | O | F | I | A | O | B |
| P | R | R | C | D | A | E | H | T | L | R | U |
| A | T | G | S | H | A | F | T | A | A | P | T |
| C | O | R | A | C | O | I | D | T | N | G | K |
| H | U | M | E | R | U | S | T | E | X | J | M |
| P | N | E | C | K | A | L | U | P | A | C | S |

Find and circle words pertaining to Laboratory Exercise 16. The words may be horizontal, vertical, diagonal, or backward.

10 bones of the pectoral girdle and upper limb

11 features of bones of the pectoral girdle and upper limb

# 17

# Pelvic Girdle and Lower Limb

## MATERIALS NEEDED

Textbook
Human skeleton, articulated
Human skeleton, disarticulated
Male and female pelves
Colored pencils

## PURPOSE OF THE EXERCISE

To examine the bones of the pelvic girdle and lower limb and to identify the major features of these bones.

## LEARNING OUTCOMES AP|R

After completing this exercise, you should be able to

(1) Locate and identify the bones of the pelvic girdle and their major features.

(2) Locate and identify the bones of the lower limb and their major features.

(3) Recognize the anatomical terms pertaining to the pelvic girdle and lower limb.

The pelvic girdle includes two hip bones that articulate with each other anteriorly at the pubic symphysis and posteriorly with the sacrum. Together, the pelvic girdle, sacrum, and coccyx constitute the pelvis. The pelvis, in turn, provides support for the trunk of the body and provides attachments for the lower limbs.

The bones of the lower limb form the framework of the thigh, leg, and foot. Each limb includes a femur, a patella, a tibia, a fibula, seven tarsals, five metatarsals, and fourteen phalanges.

## PRACTICE

### PROCEDURE A—Pelvic Girdle

1. Review the section entitled "Pelvic Girdle" in chapter 7 of the textbook.
2. As a review activity, label figures 17.1 and 17.2.

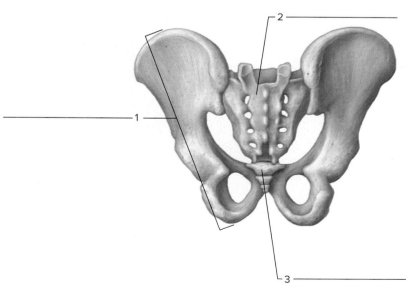

**FIGURE 17.1** Label the bones of the pelvis (posterior view).

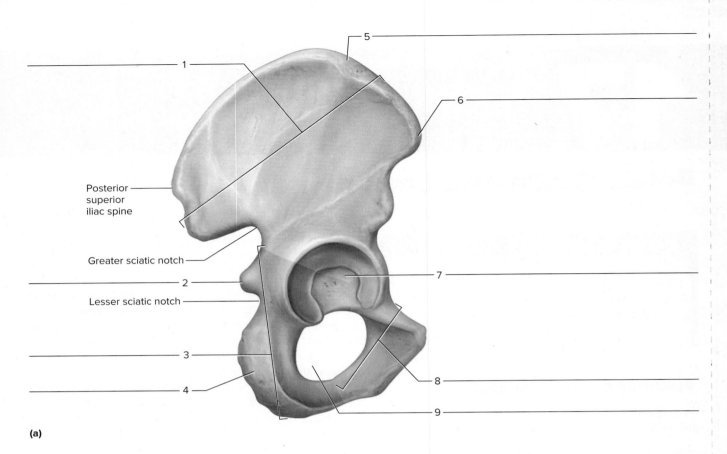

Posterior
superior
iliac spine

Greater sciatic notch

Lesser sciatic notch

**(a)**

Posterior superior
iliac spine

**(b)**

**FIGURE 17.2**   Label (*a*) the lateral and (*b*) the medial features of the right hip bone.   🅰 **AP|R**

**3.** Examine the bones of the pelvic girdle and locate the following:

**hip bone (coxal bone; pelvic bone; innominate bone)**

acetabulum
ilium
iliac crest
iliac fossa
sacroiliac joint
anterior superior iliac spine
posterior superior iliac spine
greater sciatic notch
lesser sciatic notch
ischium
ischial tuberosity
ischial spine
pubis
pubic symphysis
pubic arch
obturator foramen

**4.** Complete Part A of Laboratory Report 17.

**ASSESS**

**CRITICAL THINKING**

Examine the male and female pelves and study figure 17.3. Look for major differences between them. Note especially the flare of the iliac bones, the angle of the pubic arch, the distance between the ischial spines and ischial tuberosities, and the curve and width of the sacrum. In what ways are the differences you observed related to the function of the female pelvis as a birth canal?

_____

_____

_____

_____

(a)

(b)

**FIGURE 17.3** The female pelvis is usually wider in all diameters and roomier than that of the male pelvis. (*a*) Female pelvis. (*b*) Male pelvis. **AP|R**

### PROCEDURE B—Lower Limb

1. Review the section entitled "Lower Limb" in chapter 7 of the textbook.
2. As a review activity, label figures 17.4–17.7.
3. Examine the bones of the lower limb and locate each of the following:

**femur**
    proximal features
      head
      fovea capitis
      neck
      greater trochanter
      lesser trochanter
    shaft
      gluteal tuberosity
      linea aspera
    distal features
      lateral epicondyle
      medial epicondyle
      lateral condyle
      medial condyle

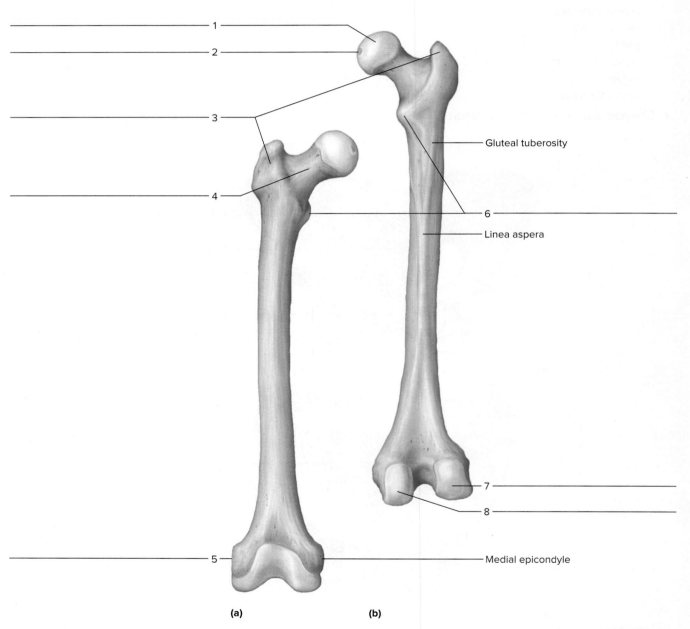

(a)        (b)

**FIGURE 17.4**   Label the features of (*a*) the anterior surface and (*b*) the posterior surface of the right femur.

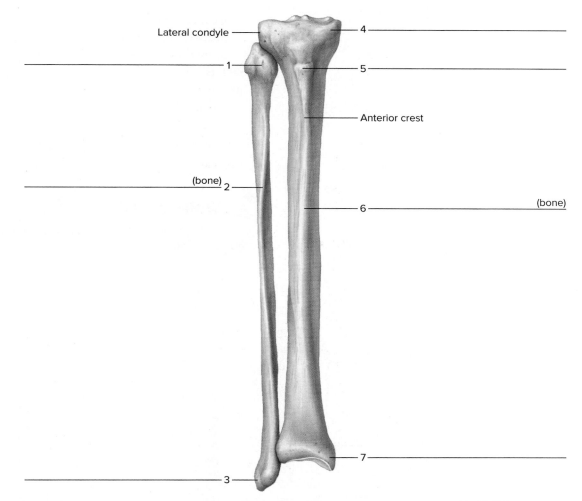

Lateral condyle

Anterior crest

(bone) 2

(bone)

**FIGURE 17.5**  Label the bones and features of the right tibia and fibula in this anterior view.  2  **AP|R**

2 (bone)

Lateral epicondyle

Lateral condyle

(bone)

(bone)

**FIGURE 17.6**  Label the bones and features of the right knee, posterior view.  2  **AP|R**

**patella**
**tibia**
    medial condyle
    lateral condyle
    tibial tuberosity
    anterior crest (border)
    medial malleolus
**fibula**
    head
    lateral malleolus
**tarsal bones**
    talus
    calcaneus
    navicular
    cuboid

    lateral cuneiform
    intermediate cuneiform
    medial cuneiform
**metatarsal bones**
**phalanges**
    proximal phalanx
    middle phalanx
    distal phalanx

**4.** Complete Parts B–E of the laboratory report.

### LEARN: ACTIVITY

Use different colored pencils to distinguish the individual bones in figure 17.7.

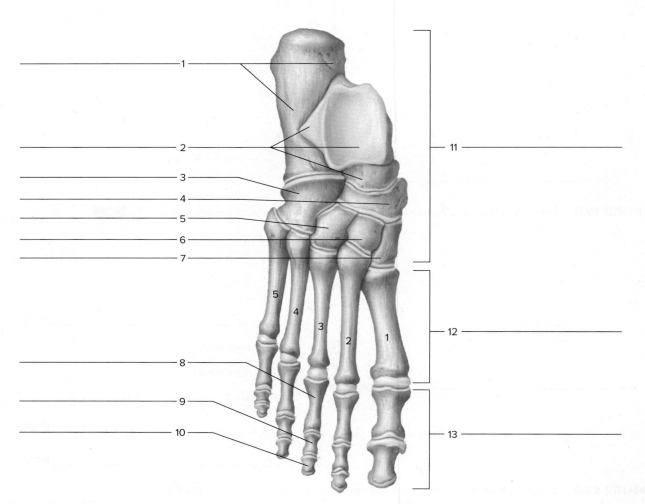

**FIGURE 17.7**　　Label the bones of the superior surface of the right foot. **2** **AP|R**

The Ⓐ corresponds to the indicated Learning Outcome(s) found at the beginning of the laboratory exercise.

# Pelvic Girdle and Lower Limb

## 🔄 PART A ASSESSMENTS

Complete the following statements:

1. The pelvic girdle consists of two _____ . Ⓐ

2. The head of the femur articulates with the _____ of the hip bone. Ⓐ

3. The _____ is the largest portion of the hip bone. Ⓐ

4. The distance between the _____ represents the shortest diameter of the pelvic outlet. Ⓐ

5. The pubic bones come together anteriorly to form a cartilaginous joint called the _____ . Ⓐ

6. The _____ is the portion of the ilium that causes the prominence of the hip. Ⓐ

7. When a person sits, the _____ of the ischium supports the weight of the body. Ⓐ

8. The angle formed by the pubic bones below the pubic symphysis is called the _____ . Ⓐ

9. The _____ is the largest foramen in the skeleton. Ⓐ

10. The ilium joins the sacrum at the _____ joint. Ⓐ

## 🔄 PART B ASSESSMENTS

Match the bones in column A with the features in column B. Place the letter of your choice in the space provided. ②

| Column A | Column B |
|---|---|
| a. Femur | _____ 1. Middle phalanx |
| b. Fibula | _____ 2. Lesser trochanter |
| c. Metatarsals | _____ 3. Medial malleolus |
| d. Patella | _____ 4. Fovea capitis |
| e. Phalanges | _____ 5. Calcaneus |
| f. Tarsals | _____ 6. Lateral cuneiform |
| g. Tibia | _____ 7. Tibial tuberosity |
| | _____ 8. Talus |
| | _____ 9. Linea aspera |
| | _____ 10. Lateral malleolus |
| | _____ 11. Sesamoid bone |
| | _____ 12. Five bones that form the instep |

Identify the bones and features indicated in the radiographs of figures 17.8–17.10. 🅐 🄰

**Terms:**
Head of femur
Ilium
Obturator foramen
Pubic symphysis
Pubis
Sacrum

3 _____

4 _____ (bone)

1 _____

2 _____

5 _____

6 _____

**FIGURE 17.8** Identify the bones and features indicated on this radiograph of the anterior view of the pelvic region, using the terms provided. **AP|R** (©Dale Butler)

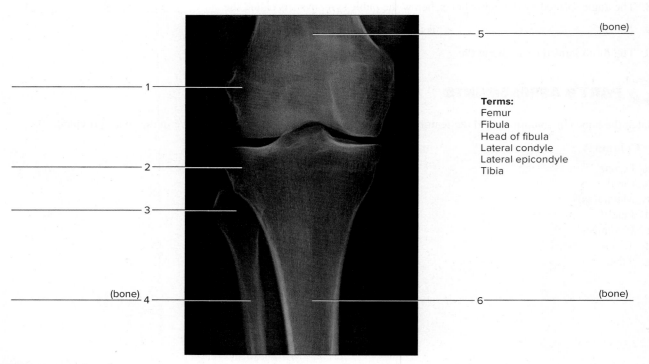

5 _____ (bone)

1 _____

**Terms:**
Femur
Fibula
Head of fibula
Lateral condyle
Lateral epicondyle
Tibia

2 _____

3 _____

(bone) 4 _____

6 _____ (bone)

**FIGURE 17.9** Identify the bones and features indicated in this radiograph of the right knee (anterior view), using the terms provided. **AP|R** (©Dale Butler)

**Terms:**
Calcaneus
Distal phalanx
Metatarsal
Proximal phalanx
Talus
Tibia

Navicular ———

Medial cuneiform ———

——— 1

——— 2

——— 3

————— 4

————— 5

————— 6

**FIGURE 17.10** Identify the bones indicated on this radiograph of the right foot (medial view), using the terms provided.
(©Dale Butler)

 **PART D ASSESSMENTS**

Identify the bones of the foot in figure 17.11. 🄰

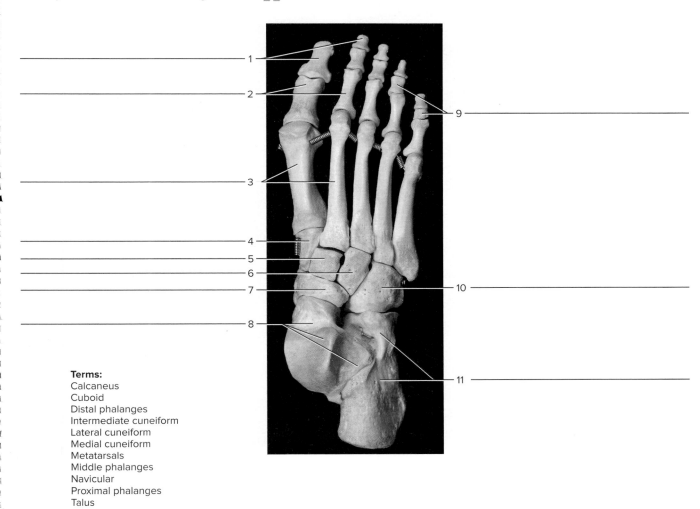

——— 1

——— 2

——— 3

——— 4

——— 5

——— 6

——— 7

——— 8

————— 9

————— 10

————— 11

**Terms:**
Calcaneus
Cuboid
Distal phalanges
Intermediate cuneiform
Lateral cuneiform
Medial cuneiform
Metatarsals
Middle phalanges
Navicular
Proximal phalanges
Talus

**FIGURE 17.11** Identify the bones indicated on this superior view of the right foot, using the terms provided.
(©J & J Photography)

Locate the 22 anatomical terms pertaining to the "Pelvic Girdle and Lower Limb." 

| P | L | A | S | R | A | T | A | T | E | M | B |
|---|---|---|---|---|---|---|---|---|---|---|---|
| A | C | E | T | A | B | U | L | U | M | Q | H |
| T | E | L | Y | D | N | O | C | B | T | S | C |
| E | A | T | A | L | U | S | W | E | O | U | U |
| L | E | D | B | N | K | H | C | R | A | L | N |
| L | N | E | C | K | A | E | V | O | F | O | E |
| A | I | L | I | U | M | D | P | S | E | E | I |
| T | L | X | R | S | I | W | D | I | M | L | F |
| Z | I | U | I | O | W | H | A | T | U | L | O |
| H | V | B | B | M | Y | W | E | Y | R | A | R |
| I | U | U | I | I | S | C | H | I | U | M | M |
| P | C | H | Y | A | F | O | R | A | M | E | N |

Find and circle words pertaining to Laboratory Exercise 17. The words may be horizontal, vertical, diagonal, or backward.

9 bones of the pelvic girdle and lower limb

13 features of the pelvic girdle and lower limb

# 18

# Joint Structure and Movements

## MATERIALS NEEDED

Textbook
Human skull
Human skeleton, articulated
Models of synovial joints (shoulder, elbow, hip, and knee)
Fresh animal joint (knee joint preferred)
Radiographs of major joints

## ⚠ SAFETY

■ Wear disposable gloves when handling the fresh animal joint.
■ Wash your hands before leaving the laboratory.

## PURPOSE OF THE EXERCISE

To examine examples of the three types of joints, to identify the major features of these joints, and to review the types of movements produced at synovial joints.

##  LEARNING OUTCOMES AP|R

After completing this exercise, you should be able to

1 Identify key structural and functional characteristics of fibrous, cartilaginous, and synovial joints.

2 Locate examples of each type of joint.

3 Locate examples of each of the six types of synovial joints.

4 Demonstrate the types of movements that occur at synovial joints.

5 Examine the structure of the shoulder, elbow, hip, and knee joints.

Joints are junctions between bones. Although they vary considerably in structure, they can be classified according to the type of tissue that binds the bones together. Thus, the three groups of *structural joints* can be identified as

*fibrous joints, cartilaginous joints,* and *synovial joints.* Fibrous joints are filled with dense fibrous connective tissue, cartilaginous joints are filled with a type of cartilage, and synovial joints contain synovial fluid inside a joint cavity.

*Functional joints* can also be classified by the degree of movement allowed: *synarthroses* are immovable, *amphiarthroses* allow slight movement, and *diarthroses* allow free movement. Movements occurring at freely movable synovial joints are due to the contractions of skeletal muscles. In each case, the type of movement depends on the type of joint involved and the way in which the muscles are attached to the bones on either side of the joint.

##  PRACTICE

### PROCEDURE A—Types of Joints AP|R

1. Review the sections entitled "Fibrous Joints," "Cartilaginous Joints," and "Synovial Joints" in chapter 8 of the textbook.

2. Examine the human skull and articulated skeleton to locate examples of the following types of joints:

**fibrous joints**

suture (synarthrotic joint)—skull
gomphosis (synarthrotic joint)—tooth sockets
syndesmosis (amphiarthrotic joint)—tibiofibular joint

**cartilaginous joints**

synchondrosis (synarthrotic joint)—epiphyseal plates
symphysis (amphiarthrotic joint)—intervertebral discs; pubic symphysis

3. Complete Part A of Laboratory Report 18.

4. Locate examples of the following types of synovial joints (diarthrotic joints) in the skeleton and in your own body. Experiment with each joint to experience its range of movements.  3 4

**synovial joints**

ball-and-socket joint—hip; shoulder
condylar (ellipsoidal) joint—radiocarpal; metacarpophalangeal

plane (gliding) joint—intercarpal; intertarsal
hinge joint—elbow; knee; interphalangeal
pivot joint—radioulnar at elbow; dens at atlas
saddle joint—base of thumb with trapezium

5. Examine models of the shoulder, elbow, hip, and knee joints. Locate the major knee joint structures of figure 18.1 that are visible on the knee joint model.

6. Complete Parts B and C of the laboratory report.

 **LEARN: ACTIVITY**

Examine a longitudinal section of a fresh synovial animal knee joint. Locate the dense connective tissue that forms the joint capsule and the hyaline cartilage that forms the articular cartilage on the ends of the bones. Locate the synovial membrane on the inner surface of the joint capsule. Does the joint have any semilunar cartilages (menisci)?

_____

What is the function of such cartilages?

_____

_____

**FIGURE 18.1** Anterior view of right bent knee (patella removed). AP R

Femur
Lateral condyle
Lateral meniscus
Lateral condyle
Fibular collateral ligament
Fibula
Tibia

Posterior cruciate ligament
Medial condyle
Anterior cruciate ligament
Medial meniscus
Medial condyle
Tibial collateral ligament
Patellar ligament (cut)

 **PRACTICE**

**PROCEDURE B—Joint Movements**

1. Review the section entitled "Types of Joint Movements" in chapter 8 of the textbook and study figures 18.2 and 18.3.

2. Most joints and body parts are extended and/or adducted when the body is positioned in anatomical position. Skeletal muscle action involves the more movable end (*insertion*) being pulled toward the more stationary end (*origin*). In the limbs, the origin is usually proximal to the insertion; in the trunk, the origin is usually medial to the insertion. Use these concepts as reference points as you move joints. Move various parts of your body to demonstrate the synovial joint movements described in tables 18.1 and 18.2. A

3. Have your laboratory partner do some of the movements and see if you can correctly identify the movements made. A

 **ASSESS**

**CRITICAL THINKING**

Describe a body position that can exist when all major body parts are flexed.

_____

_____

4. Complete Parts D and E of the laboratory report.

**LEARN: ACTIVITY**

Study the available radiographs of joints by holding the films in front of a light source. Identify the type of joint and the bones incorporated in the joint. Also identify other major visible features.

**TABLE 18.1 Angular and Rotational Movements**

| Movement | Description |
|---|---|
| Flexion | Decrease of an angle (usually in the sagittal plane) |
| Extension | Increase of an angle (usually in the sagittal plane) |
| Hyperextension | Extension beyond anatomical position |
| Abduction | Movement of a body part away from the midline (usually in the frontal plane) |
| Adduction | Movement of a body part toward the midline (usually in the frontal plane) |
| Circumduction | Circular movement (combines flexion, abduction, extension, and adduction) |
| Rotation | Movement of part around its long axis |
| Medial (internal) | Inward rotation |
| Lateral (external) | Outward rotation |

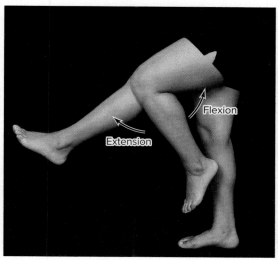

(a) Flexion and extension of the knee joint

(b) Flexion, extension, and hyper-extension of the hand at the wrist joint

(c) Abduction and adduction of the hip joint

(d) Circumduction of the shoulder joint

(e) Medial and lateral rotation of the arm at the shoulder joint

**FIGURE 18.2** Examples of angular and rotational movements of synovial joints (*a–e*). **AP|R** ((*a–e*) ©J & J Photography)

**TABLE 18.2 Special Movements (pertain to specific joints)**

| Movement | Description |
|---|---|
| Supination | Rotation of forearm and palm of hand anteriorly or upward |
| Pronation | Rotation of forearm and palm of hand posteriorly or downward |
| Elevation | Movement of body part upward |
| Depression | Movement of body part downward |
| Dorsiflexion | Movement of ankle joint so dorsum (superior) of foot becomes closer to anterior surface of leg (as standing on heels) |
| Plantar flexion | Movement of ankle joint so the plantar surface of foot becomes closer to the posterior surface of leg (as standing on toes) |
| Inversion (called supination in some health professions) | Medial movement of sole of foot at ankle joint |
| Eversion (called pronation in some health professions) | Lateral movement of sole of foot at ankle joint |
| Protraction | Anterior movement in the transverse plane |
| Retraction | Posterior movement in the transverse plane |

(a) Supination and pronation of the forearm and hand involving movement at the radioulnar joint **AP|R**

(b) Elevation and depression of the shoulder (scapula) **AP|R**

(c) Dorsiflexion and plantar flexion of the foot at the ankle joint **AP|R**

(d) Inversion and eversion of the right foot at the ankle joint (The left foot is unchanged for comparison.) **AP|R**

(e) Protraction and retraction of the head

**FIGURE 18.3**  Special movements of synovial joints (a–e).  ((a–e) ©J & J Photography)

The ▲ corresponds to the indicated Learning Outcome(s) found at the beginning of the laboratory exercise.

# Joint Structure and Movements

## 🔁 PART A ASSESSMENTS

Match the terms in column A with the descriptions in column B. Place the letter of your choice in the space provided. ▲1 ▲2

| Column A | Column B |
|---|---|
| **a.** Gomphosis | _____ **1.** Immovable joint between flat bones of the skull united by a thin layer of dense connective tissue |
| **b.** Suture | |
| **c.** Symphysis | _____ **2.** Fibrocartilage pad fills the slightly movable joint |
| **d.** Synchondrosis | _____ **3.** Temporary joint in which bones are united by bands of hyaline cartilage |
| **e.** Syndesmosis | |
| | _____ **4.** Slightly movable joint in which bones are united by interosseous membrane |
| | _____ **5.** Joint formed by union of tooth root in bony socket |

## 🔁 PART B ASSESSMENTS

Identify the types of structural and functional joints numbered in figure 18.4. ▲1 ▲2

| Structural Classification | Functional Classification |
|---|---|
| **1.** _____ | _____ |
| **2.** _____ | _____ |
| **3.** _____ | _____ |
| **4.** _____ | _____ |
| **5.** _____ | _____ |
| **6.** _____ | _____ |
| **7.** _____ | _____ |
| **8.** _____ | _____ |
| **9.** _____ | _____ |

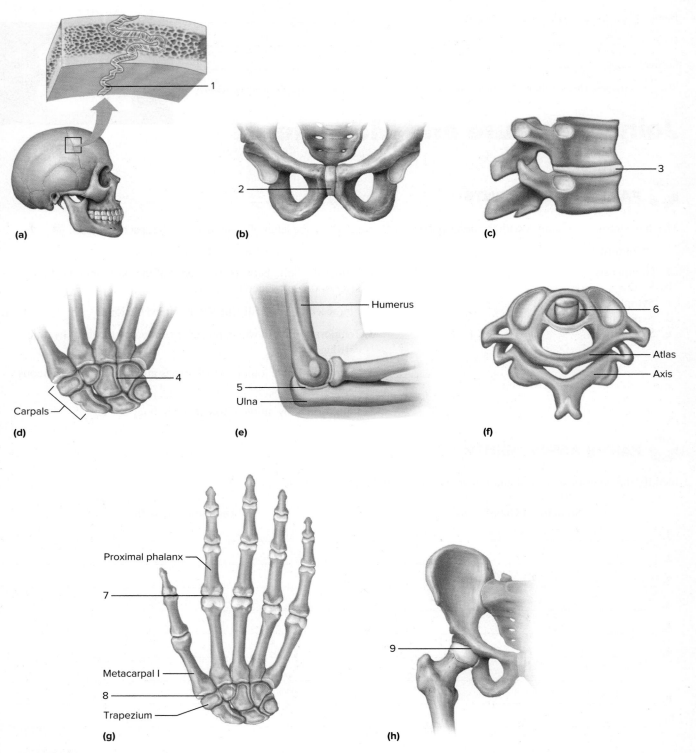

(a)

(b)

(c)

1

2

3

(d)

Carpals

4

(e)

Humerus

5

Ulna

(f)

6

Atlas

Axis

(g)

Proximal phalanx

7

Metacarpal I

8

Trapezium

(h)

9

**FIGURE 18.4** Identify the types of structural and functional joints numbered in these illustrations (a–h).

**PART C ASSESSMENTS**

Match the types of synovial joints in column A with the examples in column B. Place the letter of your choice in the space provided. **3**

| Column A | Column B |
|---|---|
| **a.** Ball-and-socket | _____ **1.** Hip joint |
| **b.** Condylar (ellipsoidal) | _____ **2.** Metacarpal–phalanx |
| **c.** Hinge | _____ **3.** Proximal radius–ulna |
| **d.** Pivot | _____ **4.** Humerus–ulna of the elbow joint |
| **e.** Plane (gliding) | _____ **5.** Phalanx–phalanx |
| **f.** Saddle | _____ **6.** Shoulder joint |
| | _____ **7.** Knee joint |
| | _____ **8.** Carpal–metacarpal of the thumb |
| | _____ **9.** Carpal–carpal |
| | _____ **10.** Tarsal–tarsal |

 **PART D ASSESSMENTS**

Complete the missing components of the following table: **4** **5**

| Name of Joint | Type of Joint | Bones Included | Types of Movement Possible |
|---|---|---|---|
| Shoulder joint | | Humerus, scapula | |
| Elbow joint | Hinge, plane, pivot | | Flexion and extension between humerus and ulna; twisting between radius and humerus; rotation between head of radius and ulna |
| Hip joint | | | Movements in all planes and rotation |
| Knee joint | Hinge (modified), condylar, plane | | |

**PART E ASSESSMENTS**

Identify the types of joint movements numbered in figure 18.5. **4**

1. _____ (of head)
2. _____ (of shoulder)
3. _____ (of shoulder)
4. _____ (of hand at radioulnar joint)
5. _____ (of hand at radioulnar joint)
6. _____ (of arm at shoulder)
7. _____ (of arm at shoulder)
8. _____ (of wrist)
9. _____ (of wrist)
10. _____ (of thigh at hip)
11. _____ (of thigh at hip)
12. _____ (of lower limb at hip)
13. _____ (of chin/mandible)
14. _____ (of chin/mandible)
15. _____ (of vertebral column/trunk)
16. _____ (of vertebral column/trunk)
17. _____ (of head and neck)
18. _____ (of head and neck)
19. _____ (of arm at shoulder)
20. _____ (of arm at shoulder)
21. _____ (of elbow)
22. _____ (of elbow)
23. _____ (of thigh at hip)
24. _____ (of thigh at hip)
25. _____ (of knee)
26. _____ (of knee)
27. _____ (of foot at ankle)
28. _____ (of foot at ankle)

**FIGURE 18.5** Identify each of the types of movements numbered and illustrated: (*a*) anterior view; (*b*) lateral view of head; (*c*) lateral view.

# 19

# Skeletal Muscle Structure and Function

## PURPOSE OF THE EXERCISE

To review the structure of a skeletal muscle.

 ## LEARNING OUTCOMES APR

After completing this exercise, you should be able to

1. Locate the major structures of a skeletal muscle fiber
   on a microscope slide of skeletal muscle tissue and on a
   model.

2. Describe how connective tissue is associated with
   muscle tissue within a skeletal muscle.

3. Distinguish between the origin and insertion of a
   muscle.

4. Describe and demonstrate the general actions of
   agonists, synergists, and antagonists.

 A skeletal muscle represents an organ of the muscular system and is composed of several types of tissues. These tissues include skeletal muscle tissue, nervous tissue, blood, and various connective tissues.

Each skeletal muscle is encased within and permeated by connective tissue sheaths. The connective tissue often projects beyond the end of a muscle, providing an attachment to other muscles or to bones as a tendon. The connective tissues also extend into the structure of a muscle and separate it into compartments. The outer layer, called *epimysium,* extends deeper into the muscle as *perimysium,* containing bundles of cells (fascicles), and farther inward extends around each muscle cell (fiber) as a thin *endomysium.* The connective tissues provide support and reinforcement during muscular contractions and allow portions of a muscle to contract somewhat independently. Some of the collagen fibers are continuous with the tendon, periosteum, and inward extensions as bone fibers, allowing a strong structural continuity.

Muscles are named according to location, size, action, shape, attachments, or number of origins, or the direction of the fibers. Examples of how muscles are named include these: gluteus maximus (location and size); adductor longus (action and size); sternocleidomastoid (attachments); serratus anterior (shape and location); biceps brachii (two origins and location); and orbicularis oculi (direction of fibers and location).

 ## PRACTICE

### PROCEDURE—Skeletal Muscle Structure and Function

1. Study the section entitled "Skeletal Muscle Tissue" in chapter 5 of the textbook.

2. Examine the microscopic structure of a longitudinal section of skeletal muscle by observing a prepared microscope slide of this tissue. Use figure 19.1 of skeletal muscle tissue to locate the following features:

   **skeletal muscle fiber** (cell)
   **nuclei**
   **striations** (alternating light and dark)

3. Examine the microscopic structure of a cross section of a skeletal muscle by observing a prepared microscope

slide of this tissue. Use figure 19.2 of skeletal muscle tissue to locate the following features:

**connective tissue**
    perimysium
    endomysium

**fascicle**

**muscle cell**
    nucleus
    myofibrils

4. Review the section entitled "Structure of a Skeletal Muscle" in chapter 9 of the textbook.
5. As a review activity, label figure 19.3 and study figure 19.4.
6. Examine the human torso model and locate examples of fascia, tendons, and aponeuroses. An origin tendon is attached to a less movable end of the muscle, whereas the insertion tendon is attached to a more movable end of the muscle. Sheets of connective tissue, called aponeuroses, also serve for some muscle attachments. Locate examples of tendons in your body.
7. Complete Part A of Laboratory Report 19.

 **LEARN: ACTIVITY**

Examine the fresh round beefsteak. It represents a cross section through the beef thigh muscles. Note the white lines of connective tissue that separate the individual skeletal muscles. Also note how the connective tissue extends into the structure of a muscle and separates it into small compartments of muscle tissue. Locate the epimysium and the perimysium of an individual muscle.

**FIGURE 19.1** Micrograph of a longitudinal section of skeletal muscle fibers (cells) (400×). **AP|R** (©McGraw-Hill Education/Al Telser, photographer)

**FIGURE 19.2** Micrograph of a cross section of a fascicle and associated connective tissues (800×). (©Ed Reschke/Photolibrary/ Getty Images)

Muscle

↓

Fascicles

↓

Muscle fibers (muscle cells)

↓

Myofibrils

↓

Thick and thin filaments (myofilaments)

Bone

3

Fascia (located between adjacent muscles)

Epimysium

4

Endomysium

1

Axon of motor neuron

Blood vessel

Muscle fiber (single cell)

Sarcolemma

2

5

Myofibrils

Filaments (actin and myosin)

**FIGURE 19.3** Label the structures of a skeletal muscle. A A AP R

Myofibrils

Cisternae of
sarcoplasmic reticulum — Triad

Transverse tubule

Nucleus

Sarcoplasmic
reticulum

Openings into
transverse tubules

Thick and thin
filaments

Nucleus

Mitochondria

Sarcoplasm

Sarcolemma

**FIGURE 19.4** Structures of a skeletal muscle fiber.

8. Examine the model of the skeletal muscle fiber and locate the following:

**sarcolemma**
**sarcoplasm**
**myofibril**
    thick (myosin) filament
    thin (actin, tropomyosin, and troponin) filament
**sarcomere** (functional contractile unit)
    A band (dark)
    I band (light)
    H zone
    M line
    Z line (disc)
**sarcoplasmic reticulum**
    cisternae
**transverse (T) tubules**

9. Complete Part B of the laboratory report.

10. Review the section entitled "Skeletal Muscle Actions" in chapter 9 of the textbook.

11. Locate the biceps brachii and its origins and insertions in the human torso model and in your body.

12. Make various movements with your upper limb at the shoulder and elbow. For each movement, determine the location of the muscles that function as *agonists* and as *antagonists*. An agonist causes an action, whereas an antagonist works against the action; however, the role of the muscle is dependent on the particular movement involved. Antagonistic muscle pairs pull from opposite sides of the same body region. *Synergistic* muscles often supplement the contraction force of an agonist, or by acting to stabilize nearby joints. **The role of a muscle as an agonist, an antagonist, or a synergist depends upon the movement under consideration, as their roles can change.** These voluntary movements of complex actions are controlled by the nervous system (see table 19.1).

13. Complete Part C of the laboratory report.

**TABLE 19.1 Various Roles of Muscles**

| Functional Category | Description |
| --- | --- |
| Agonist* | Muscle responsible for the action (movement) |
| Antagonist | Muscle responsible for action in the opposite direction of an agonist or for resistance to an agonist |
| Synergist | Muscle that assists an agonist, often by supplementing the contraction force |

*The term *agonist* is often used interchangeably with *prime mover*. When the term prime mover is used, it usually refers to the agonist that provides most of the force of the movement of the action involved.

Name _____

Date _____

Section _____

The  corresponds to the indicated Learning Outcome(s) found at the beginning of the laboratory exercise.

# Skeletal Muscle Structure and Function

## PART A ASSESSMENTS

Match the terms in column A with the definitions in column B. Place the letter of your choice in the space provided. Ⓐ Ⓐ

**Column A**

a. Endomysium
b. Epimysium
c. Fascia
d. Fascicle
e. Myosin
f. Perimysium
g. Sarcolemma
h. Sarcomere
i. Sarcoplasm
j. Sarcoplasmic reticulum
k. Tendon
l. Transverse (T) tubule

**Column B**

_____ 1. Membranous channel extending inward from muscle fiber membrane

_____ 2. Cytoplasm of a muscle fiber

_____ 3. Connective tissue located between adjacent muscles

_____ 4. Layer of connective tissue that separates a muscle into small bundles called fascicles

_____ 5. Cell membrane of a muscle fiber

_____ 6. Layer of connective tissue that surrounds a skeletal muscle

_____ 7. Unit of alternating light and dark striations between Z lines

_____ 8. Layer of connective tissue that surrounds an individual muscle fiber

_____ 9. Cellular organelle in muscle fiber corresponding to the endoplasmic reticulum

_____ 10. Cordlike part that attaches a muscle to a bone

_____ 11. Protein found within thick filament

_____ 12. Small bundle of muscle fibers within a muscle

## PART B ASSESSMENTS

Provide the labels for the electron micrograph in figure 19.5. Ⓐ

1. _____

2. _____

3. _____

4. _____

**Terms:**
A band (dark)
I band (light)
Sarcomere
Z line

**FIGURE 19.5**    Identify the bands and lines of the striations of a transmission electron micrograph of relaxed sarcomeres (16,000×), using the terms provided. **AP|R**  (©Biology Pics/Science Source)

## PART C ASSESSMENTS

Complete the following statements:

1. The _____ of a muscle is typically attached to a more immovable or fixed part. **3**

2. The _____ of a muscle is typically attached to a more movable part on the other side of the joint. **3**

3. When the forearm is extended at the elbow joint, the _____ muscle acts as the agonist. **4**

4. The forearm is flexed at the elbow when the _____ muscle contracts. **4**

5. A muscle responsible for an action is called a(an) _____ . **4**

6. Assisting muscles of an agonist are called as_____ . **4**

7. Antagonists are muscles that resist the actions of _____ and cause movement in the opposite direction. **4**

8. An agonist that provides the most force of a specific action is sometimes referred to as a _____ . **4**

**150**

# 20

# Muscles of the Head and Neck

## MATERIALS NEEDED

Textbook
Human torso model with musculature
Human skull
Human skeleton, articulated
Long rubber bands

## PURPOSE OF THE EXERCISE

To review the locations, actions, origins, and insertions of the muscles of the head and neck.

##  LEARNING OUTCOMES AP|R

After completing this exercise, you should be able to

1. Locate and identify the muscles of facial expression, the muscles of mastication, and the muscles that move the head.

2. Describe and demonstrate the action of each of these muscles.

3. Locate the origin and insertion of each of these muscles in a human skeleton and the musculature of the human torso model.

The skeletal muscles of the face and head include the muscles of facial expression, which lie just beneath the skin; the muscles of mastication, attached to the mandible; and the muscles that move the head, located in the neck.

##  PRACTICE

### PROCEDURE—Muscles of the Head and Neck

1. Review the sections entitled "Muscles of Facial Expression," "Muscles of Mastication," and "Muscles That Move the Head and Vertebral Column" in chapter 9 of the textbook.

2. As a review activity, label figures 20.1–20.3.

3. Locate the following muscles in the human torso model and in your body whenever possible:

**muscles of facial expression**
  epicranius
    frontalis
    occipitalis
  orbicularis oculi
  orbicularis oris
  buccinator
  zygomaticus major
  zygomaticus minor
  platysma

**muscles of mastication**
  masseter
  temporalis
  medial pterygoid
  lateral pterygoid

**muscles that move the head and neck**
  sternocleidomastoid
  splenius capitis
  semispinalis capitis
  scalene muscles (scalenes)

4. Demonstrate the action of these muscles in your body.

5. Locate the origins and insertions of these muscles in the human skull and skeleton.

6. Complete Parts A–C of Laboratory Report 20.

##  LEARN: LAB IN MOTION

A long rubber band can be used to simulate muscle locations, origins, insertions, and actions on the human torso model, the skeleton, or a laboratory partner. Hold one end of the rubber band firmly on the origin location of a muscle; then slightly stretch the rubber band and hold the other end on the insertion site. Allow the insertion end to slowly move toward the origin end to simulate the contraction and action of the muscle. 1 2 3

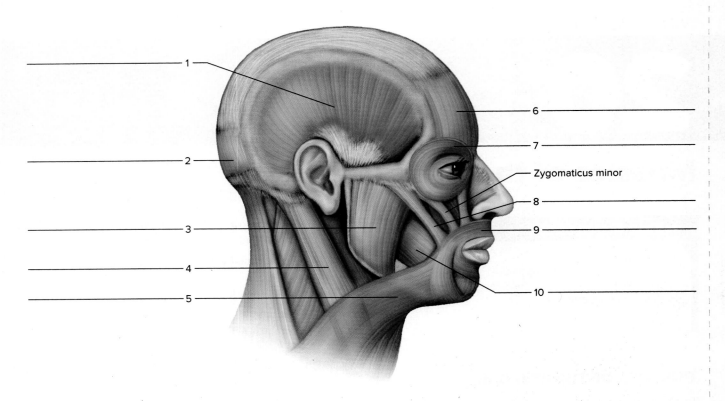

**FIGURE 20.1**   Label the muscles of expression and mastication.  Ⓐ **AP|R**

Zygomaticus minor

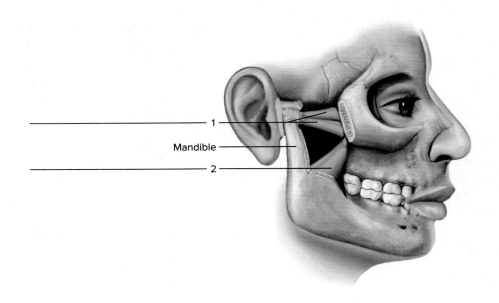

Mandible

**FIGURE 20.2**   Label these deep muscles of mastication.  Ⓐ **AP|R**

Splenius capitis (cut)

Scalenes

1

2

**FIGURE 20.3** Label these deep muscles of the posterior neck (trapezius removed). A AP R

# Notes

The Ⓐ corresponds to the indicated Learning Outcome(s) found at the beginning of the laboratory exercise.

# Muscles of the Head and Neck

## PART A ASSESSMENTS

Complete the following statements:

1. When the _____ contracts, the corner of the mouth is elevated. Ⓐ**2**

2. The _____ acts to compress the wall of the cheeks, as when air is blown out of the mouth. Ⓐ**2**

3. The _____ causes the lips to close and pucker during kissing, whistling, and speaking. Ⓐ**2**

4. The _____ and platysma help to depress the mandible. Ⓐ**2**

5. The _____ acts to elevate and retract the mandible. Ⓐ**2**

6. The _____ pterygoid can elevate the mandible and pull it sideways. Ⓐ**2**

7. The _____ pterygoid can protract the mandible and pull the jaw sideways. Ⓐ**2**

8. The _____ can close the eye, as in blinking. Ⓐ**2**

9. The _____ together pull the head and neck forward and down toward the chest. Ⓐ**2**

10. The _____ can rotate head and laterally flex and extend neck. Ⓐ**2**

11. The muscle used to pout and to express horror by pulling the lower lip downward is the _____ . Ⓐ**2**

12. The muscle used to smile and laugh is the _____ . Ⓐ**2**

## PART B ASSESSMENTS

Name the muscle indicated by the following combinations of origin and insertion. Ⓐ**3**

| Origin | Insertion | Muscle |
|--------|-----------|--------|
| 1. Occipital bone | Skin around eye | _____ |
| 2. Zygomatic bone | Corner of mouth skin and muscle | _____ |
| 3. Zygomatic arch | Lateral surface of mandible | _____ |
| 4. Sphenoid bone | Anterior surface of mandibular condyle | _____ |
| 5. Manubrium of sternum and clavicle | Mastoid process of temporal bone | _____ |
| 6. Alveolar processes of mandible and maxilla | Orbicularis oris | _____ |
| 7. Fascia in upper chest | Mandible and skin below mouth | _____ |
| 8. Temporal bone | Coronoid process of mandible | _____ |
| 9. Processes of lower cervical and upper thoracic vertebrae | Occipital bone | _____ |
| 10. Transverse processes of cervical vertebrae | Surfaces of ribs 1–2 | _____ |

**ASSESS**

### CRITICAL THINKING

Identify the muscles of various facial expressions in the photographs of figure 20.4. **1** **2**

1. _____

2. _____

3. _____

4. _____

5. _____

(a)

(b)

(c)

**Terms:**
Epicranius/frontalis
Orbicularis oculi
Orbicularis oris
Platysma
Zygomaticus major

**FIGURE 20.4**   Identify the muscles of facial expression being contracted in each of these photographs (*a–c*), using the terms provided. ((*a–c*) ©J & J Photography)

LABORATORY EXERCISE

# 21

# Muscles of the Chest, Shoulder, and Upper Limb

## MATERIALS NEEDED

Textbook
Human torso model
Human skeleton, articulated
Muscular models of the upper limb
Long rubber bands

## PURPOSE OF THE EXERCISE

To review the locations, actions, origins, and insertions of the muscles in the chest, shoulder, and upper limb.

 **LEARNING OUTCOMES** AP|R

After completing this exercise, you should be able to

① Locate and identify the muscles of the chest, shoulder, and upper limb.

② Describe and demonstrate the action of each of these muscles.

③ Locate the origin and insertion of each of these muscles in a human skeleton and on muscular models.

The muscles of the chest and shoulder are responsible for moving the scapula and arm, whereas those within the arm and forearm move joints in the elbow and hand.

 **PRACTICE**

**PROCEDURE—Muscles of the Chest, Shoulder, and Upper Limb**

1. Review the sections entitled "Muscles That Move the Pectoral Girdle," "Muscles That Move the Arm," "Muscles That Move the Forearm," and "Muscles That Move the Hand" in chapter 9 of the textbook.

2. As a review activity, label figures 21.1–21.4.

3. Locate the following muscles in the human torso model and models of the upper limb. Also locate in your body as many of the muscles as you can.

**muscles that move the pectoral girdle**
  anterior muscles
    pectoralis minor
    serratus anterior
  posterior muscles
    trapezius
    rhomboid major
    rhomboid minor
    levator scapulae

**muscles that move the arm**
  origins on axial skeleton
    pectoralis major
    latissimus dorsi
  origins on scapula
    deltoid
    teres major
    coracobrachialis
  rotator cuff (SITS) muscles (origins also on scapula)
    supraspinatus
    infraspinatus
    teres minor
    subscapularis

**muscles that move the forearm**
  muscle bellies in arm
    biceps brachii
    brachialis
    triceps brachii
  muscle bellies in forearm
    brachioradialis
    supinator
    pronator teres
    pronator quadratus

**muscles that move the hand**
  anterior flexor muscles
    flexor carpi radialis
    flexor carpi ulnaris
    palmaris longus
    flexor digitorum profundus
    flexor digitorum superficialis

posterior extensor muscles
  extensor carpi radialis longus
  extensor carpi radialis brevis
  extensor carpi ulnaris
  extensor digitorum

**FIGURE 21.1**  Label the muscles of the posterior shoulder. Superficial muscles are illustrated on the left side and deep muscles on the right side.  **A** **AP|R**

**FIGURE 21.2**  Label the muscles of the anterior chest. Superficial muscles are illustrated on the left side and deep muscles on the right side.  **A** **AP|R**

(a)

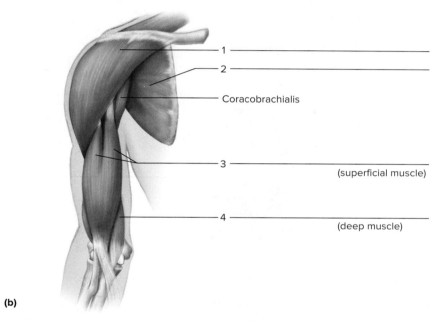

1 _____
2 _____
Coracobrachialis
3 _____
(superficial muscle)
4 _____
(deep muscle)

(b)

**FIGURE 21.3**  Label (*a*) the muscles of the posterior shoulder and arm and (*b*) the muscles of the anterior shoulder and arm, with the rib cage removed.  🅰 **AP|R**

4. Demonstrate the action of these muscles in your body.
5. Locate the origins and insertions of these muscles in the human skeleton.
6. Complete Parts A, B, and C of Laboratory Report 21.

**LEARN: LAB IN MOTION**

A long rubber band can be used to simulate muscle locations, origins, insertions, and actions on muscular models, the skeleton, or a laboratory partner. Hold one end of the rubber band firmly on the origin location of a muscle; then slightly stretch the rubber band and hold the other end on the insertion site. Allow the insertion end to slowly move toward the origin end to simulate the contraction and action of the muscle. 🅐 🅐 🅐

Biceps brachii

Brachialis

1

2

Extensor carpi radialis longus

3

4

Flexor carpi ulnaris

Pronator quadratus

Retinaculum

**(a)**

Triceps brachii

Brachioradialis

3

1

Extensor carpi radialis brevis

2

4

Retinaculum

**(b)**

**FIGURE 21.4**   Label (*a*) the muscles of the anterior forearm and (*b*) the muscles of the posterior forearm.

Name _____

Date _____

Section _____

The  corresponds to the indicated Learning Outcome(s) found at the beginning of the laboratory exercise.

LABORATORY REPORT

# 21

# Muscles of the Chest, Shoulder, and Upper Limb

## PART A ASSESSMENTS

Match the muscles in column A with the actions in column B. Place the letter of your choice in the space provided.

**Column A**

a. Brachialis
b. Coracobrachialis
c. Deltoid
d. Extensor carpi ulnaris
e. Flexor carpi ulnaris
f. Infraspinatus
g. Pectoralis major
h. Pectoralis minor
i. Rhomboid major
j. Serratus anterior
k. Teres major
l. Triceps brachii

**Column B**

_____ 1. Abducts arm and flexes and extends shoulder

_____ 2. Flexes shoulder, adducts and rotates arm medially

_____ 3. Flexes wrist and adducts hand

_____ 4. Elevates and retracts scapula

_____ 5. Depresses and protracts scapula

_____ 6. Used to thrust shoulder anteriorly (protraction), as when pushing something

_____ 7. Flexes elbow

_____ 8. Flexes shoulder and adducts arm

_____ 9. Extends elbow

_____ 10. Extends shoulder and adducts and rotates arm medially

_____ 11. Extends wrist and adducts hand

_____ 12. Rotates arm laterally

## PART B ASSESSMENTS

Name the muscle indicated by the following combinations of origin and insertion. **3**

| Origin | Insertion | Muscle |
|---|---|---|
| 1. Spinous processes of upper thoracic vertebrae | Medial border of scapula | _____ |
| 2. Anterior surfaces of upper ribs | Medial border of scapula | _____ |
| 3. Anterior surfaces of ribs 3–5 | Coracoid process of scapula | _____ |
| 4. Coracoid process of scapula | Shaft of humerus | _____ |
| 5. Lateral border of scapula | Intertubercular sulcus of humerus | _____ |
| 6. Anterior surface of scapula | Lesser tubercle of humerus | _____ |
| 7. Lateral border of scapula | Greater tubercle of humerus | _____ |
| 8. Anterior shaft of humerus | Coronoid process of ulna | _____ |
| 9. Medial epicondyle of humerus and coronoid process of ulna | Lateral surface of proximal radius | _____ |
| 10. Distal lateral end of humerus | Lateral surface of radius above styloid process | _____ |
| 11. Medial epicondyle of humerus | Base of second and third metacarpals | _____ |
| 12. Medial epicondyle of humerus | Fascia of palm | _____ |

## PART C ASSESSMENTS

## ASSESS

### CRITICAL THINKING

Identify the muscles indicated in figure 21.5. **1**

1. _____
2. _____
3. _____
4. _____
5. _____
6. _____
7. _____
8. _____
9. _____
10. _____
11. _____

12. _____
13. _____
14. _____
15. _____
16. _____
17. _____
18. _____
19. _____
20. _____
21. _____

**Terms:**
Biceps brachii
Deltoid
External oblique
Pectoralis major
Rectus abdominis
Serratus anterior
Sternocleidomastoid
Trapezius

(a)

**Terms:**
Biceps brachii
Deltoid
Infraspinatus
Latissimus dorsi
Trapezius
Triceps brachii

(b)

**FIGURE 21.5** Identify the muscles that appear as body surface features in these photographs (*a, b,* and *c*), using the terms provided. ((*a–c*) ©J & J Photography)

**Terms:**
Biceps brachii
Brachioradialis
Deltoid
Pectoralis major
Serratus anterior
Trapezius
Triceps brachii

(c)

**FIGURE 21.5** *Continued.*

# Muscles of the Deep Back, Abdominal Wall, and Pelvic Floor

## PURPOSE OF THE EXERCISE

To review the actions, origins, and insertions of the muscles of the deep back, abdominal wall, and pelvic floor.

## LEARNING OUTCOMES AP|R

After completing this exercise, you should be able to

1. Locate and identify the muscles of the deep back, abdominal wall, and pelvic floor.
2. Describe the action of each of these muscles.
3. Locate the origin and insertion of each of these muscles in a human skeleton or on muscular models.

The deep muscles of the back extend the vertebral column. Because the muscles have numerous origins, insertions, and subgroups, the muscles overlap each other. The deep back muscles can extend the spine when contracting as a group but also help to maintain posture and normal spine curvatures.

The anterior and lateral walls of the abdomen contain broad, flattened muscles arranged in layers. These muscles connect the rib cage and vertebral column to the pelvic girdle. The abdominal wall muscles compress the abdominal visceral organs, help maintain posture, assist in forceful exhalation, and contribute to trunk flexion and waist rotation.

The muscles of the pelvic floor are arranged in two muscular sheets: (1) a pelvic diaphragm that spans the outlet of the pelvic cavity and (2) a urogenital diaphragm that fills the space within the pubic arch. The pelvic floor is penetrated by the urethra, vagina, and anus in a female; thus, pelvic floor muscles are important in obstetrics.

## PRACTICE

### PROCEDURE A—Muscles of the Deep Back and Abdominal Wall

1. Review the sections entitled "Muscles That Move the Head and Vertebral Column" and "Muscles of the Abdominal Wall" in chapter 9 of the textbook.
2. As a review activity, label figures 22.1 and 22.2.
3. Locate the following muscles in the human torso model:

   **erector spinae group**
      iliocostalis (lateral group)
      longissimus (intermediate group)
      spinalis (medial group)
   **external oblique**
   **internal oblique**
   **transversus abdominis**
   **rectus abdominis**

4. Demonstrate the actions of these muscles in your body.
5. Locate the origin and insertion of each of these muscles in the human skeleton.
6. Complete Part A of Laboratory Report 22.

## ASSESS

### CRITICAL THINKING

List the muscles from superficial to deep for an appendectomy incision.

_____

_____

_____

_____

Splenius capitis (cut)

Semispinalis capitis (cut)

Splenius capitis

Semispinalis capitis

1

2

3

**Terms:**
Iliocostalis (lateral group)
Longissimus (intermediate group)
Spinalis (medial group)

Quadratus lumborum

**FIGURE 22.1** Label the three deep back muscle groups of the erector spinae group, using the terms provided. (*Note:* The three erector spinae muscle groups are not illustrated on both sides of the body.) A AP|R

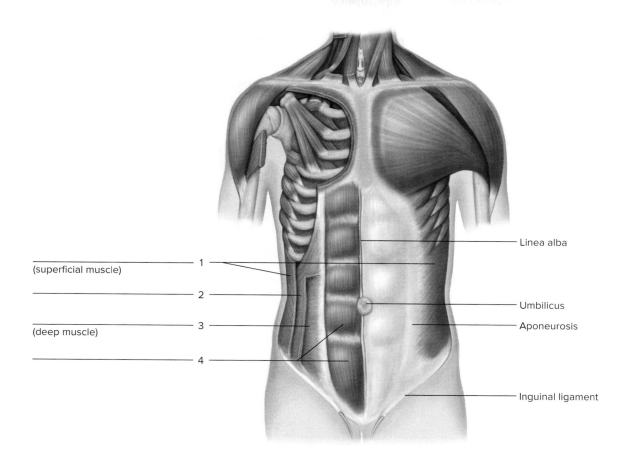

(superficial muscle) ____ 1

____ 2

(deep muscle) ____ 3

____ 4

Linea alba

Umbilicus

Aponeurosis

Inguinal ligament

**FIGURE 22.2** Label the muscles of the abdominal wall. 🅐 **AP|R**

 **PRACTICE**

**PROCEDURE B—Muscles of the Pelvic Floor**

1. Review the section entitled "Muscles of the Pelvic Floor and Perineum" in chapter 9 of the textbook.
2. As a review activity, label figures 22.3 and 22.4.
3. Study figure 22.5 as it relates to the pelvic muscles indicated in figures 22.3 and 22.4.
4. Locate the following muscles in the models of the male and female pelves:

**pelvic floor (diaphragm)**
   levator ani
   coccygeus

**urogenital diaphragm**
   deep transversus perinei
   external urethral sphincter

**other perineal muscles**
   bulbospongiosus
   ischiocavernosus
   external anal sphincter
   superficial transversus perinei

5. Locate the origin and insertion of each of these muscles in the human skeleton. 🄲
6. Complete Part B of the laboratory report.

**FIGURE 22.3** Label the muscles of the male pelvic floor and perineum. 🅐

Penis
Scrotum
1
2
3
4
External anal sphincter
Anus
Gluteus maximus

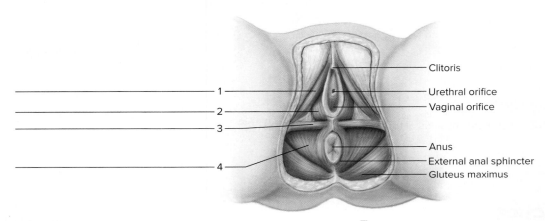

**FIGURE 22.4** Label the muscles of the female pelvic floor and perineum. 🅐

Clitoris
Urethral orifice
Vaginal orifice
1
2
3
4
Anus
External anal sphincter
Gluteus maximus

**FIGURE 22.5** Internal view of the female pelvis.

Pelvic diaphragm
Coccygeus
Levator ani
Urogenital diaphragm
Coccyx
Rectum
Vagina
Urethra
Pubic symphysis

Name _____

Date _____

Section _____

LABORATORY
REPORT

The Ⓐ corresponds to the indicated Learning Outcome(s) found at the beginning of the laboratory exercise.

# Muscles of the Deep Back, Abdominal Wall, and Pelvic Floor

## PART A ASSESSMENTS

Complete the following statements:

1. A band of tough connective tissue in the midline of the anterior abdominal wall called the _____ serves as a muscle attachment. Ⓐ1

2. The _____ muscle spans from the ribs and the xiphoid process of the sternum to the pubic bones. Ⓐ3

3. The _____ forms the third layer (deepest layer) of the abdominal wall muscles. Ⓐ1

4. The action of the external oblique muscle is to compress the abdomen and _____. Ⓐ2

5. The action of the rectus abdominis is to compress the abdomen and _____. Ⓐ2

6. The iliocostalis, longissimus, and spinalis muscles together form the _____. Ⓐ1

## PART B ASSESSMENTS

Complete the following statements:

1. The levator ani and coccygeus together form the _____. Ⓐ1

2. The action of the external anal sphincter is to _____. Ⓐ2

3. The action of the coccygeus is to _____
   _____. Ⓐ2

4. The _____ surrounds the base of the penis. Ⓐ1

5. In females, the bulbospongiosus acts to constrict the _____. Ⓐ2

6. The ischiocavernosus extends from the penis or clitoris of the pubic arch to the _____. Ⓐ1

7. In the female, the _____ muscles are separated by the urethra, vagina, and rectum. Ⓐ1

8. The action of the deep transversus perinei is to _____. Ⓐ2

9. The coccygeus extends from the coccyx and sacrum to the _____. Ⓐ3

10. The _____ assists in closing the urethra. Ⓐ2

# Notes

# 23

# Muscles of the Hip and Lower Limb

## MATERIALS NEEDED

Textbook
Human torso model with musculature
Human skeleton, articulated
Muscular models of the lower limb
Long rubber bands

## PURPOSE OF THE EXERCISE

To review the actions, origins, and insertions of the muscles that move the thigh, leg, and foot.

##  LEARNING OUTCOMES AP|R

After completing this exercise, you should be able to

1. Locate and identify the muscles that move the thigh, leg, and foot.

2. Describe and demonstrate the actions of each of these muscles.

3. Locate the origin and insertion of each of these muscles in a human skeleton and on muscular models.

The muscles that move the thigh are attached to the femur and to some part of the pelvic girdle. Those attached anteriorly primarily act to flex the thigh at the hip, whereas those attached posteriorly act to extend, abduct, or rotate the thigh.

The muscles that move the leg connect the tibia or fibula to the femur or to the pelvic girdle. They function to flex or extend the leg at the knee. Other muscles, located in the leg, act to move the foot.

##  PRACTICE

### PROCEDURE—Muscles of the Hip and Lower Limb

1. Review the sections entitled "Muscles That Move the Thigh," "Muscles That Move the Leg," and "Muscles That Move the Foot" in chapter 9 of the textbook.

2. As a review activity, label figures 23.1–23.6.

3. Locate the following muscles in the human torso model and in the lower limb models. Also locate as many of them as possible in your body.

**muscles that move the thigh**
anterior hip muscles
iliopsoas group
psoas major
iliacus
posterior and lateral hip muscles
gluteus maximus
gluteus medius
gluteus minimus
tensor fasciae latae
medial adductor muscles
pectineus
adductor longus
adductor magnus
adductor brevis
gracilis

**muscles that move the leg**
anterior thigh muscles
sartorius
quadriceps femoris group
rectus femoris
vastus lateralis
vastus medialis
vastus intermedius
posterior thigh muscles
hamstring group
biceps femoris
semitendinosus
semimembranosus

**muscles that move the foot**
anterior leg muscles
tibialis anterior
fibularis (peroneus) tertius
extensor digitorum longus

posterior leg muscles

    gastrocnemius

    soleus

    flexor digitorum longus

    tibialis posterior

lateral leg muscles

    fibularis (peroneus) longus

    fibularis (peroneus) brevis

**4.** Demonstrate the action of each of these muscles in your body.

**5.** Locate the origin and insertion of each of these muscles in the human skeleton.

**6.** Complete Parts A, B, and C of Laboratory Report 23.

## LEARN: LAB IN MOTION

A long rubber band can be used to simulate muscle locations, origins, insertions, and actions on muscular models, the skeleton, or a laboratory partner. Hold one end of the rubber band firmly on the origin location of a muscle; then slightly stretch the rubber band and hold the other end on the insertion site. Allow the insertion end to slowly move toward the origin end to simulate the contraction and action of the muscle. [1] [2] [3]

**FIGURE 23.1** Label the muscles of the anterior right hip and thigh. [1] AP|R

**FIGURE 23.2** Label the muscles of the lateral right hip and thigh. Ⓐ

Labels visible on figure:
- 1
- 2
- 3
- 4
- Sartorius
- 5
- 6
- Iliotibial tract (band)

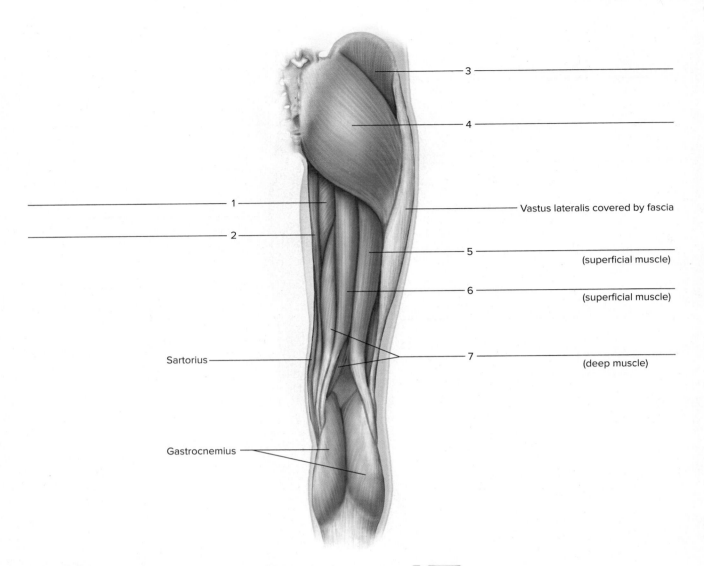

1 _____

2 _____

3 _____

4 _____

Vastus lateralis covered by fascia

5 _____
(superficial muscle)

6 _____
(superficial muscle)

Sartorius _____

7 _____
(deep muscle)

Gastrocnemius _____

**FIGURE 23.3**  Label the muscles of the posterior right hip and thigh.

Patellar ligament

Gastrocnemius

1

2

3

Soleus

Fibularis (peroneus) brevis

Tibia

Extensor retinacula

**FIGURE 23.4**    Label the muscles of the anterior right leg.  A  AP|R

**FIGURE 23.5** Label the muscles of the lateral right leg.

Iliotibial tract (band)

1

2

3

4

5

6

Fibularis (peroneus) tertius

Retinaculum

(superficial muscle)

1

(deep muscle)

2

Fibularis (peroneus) longus

Calcaneal tendon

Flexor digitorum longus

Retinaculum

**FIGURE 23.6** Label the muscles of the posterior right leg. A AP|R

# Notes

Name _____

Date _____

Section _____

The A corresponds to the indicated Learning Outcome(s) found at the beginning of the laboratory exercise.

LABORATORY
REPORT

23

# Muscles of the Hip and Lower Limb

## PART A ASSESSMENTS

Match the muscles in column A with the actions in column B. Place the letter of your choice in the space provided. A2

| Column A | Column B |
|---|---|
| **a.** Biceps femoris | _____ **1.** Adducts thigh and flexes knee |
| **b.** Fibularis (peroneus) longus | _____ **2.** Plantar flexion and eversion of foot |
| **c.** Fibularis (peroneus) tertius | _____ **3.** Flexes hip |
| **d.** Gluteus medius | _____ **4.** Abducts and rotates thigh laterally; flexes knee and hip |
| **e.** Gracilis | _____ **5.** Dorsiflexion and eversion of foot |
| **f.** Psoas major and iliacus | _____ **6.** Abducts thigh and rotates it medially |
| **g.** Quadriceps femoris group | _____ **7.** Plantar flexion and inversion of foot |
| **h.** Sartorius | _____ **8.** Flexes knee and laterally rotates leg |
| **i.** Tibialis anterior | _____ **9.** Extends knee |
| **j.** Tibialis posterior | _____ **10.** Dorsiflexion and inversion of foot |

## PART B ASSESSMENTS

Name the muscle indicated by the following combinations of origin and insertion. A3

| Origin | Insertion | Muscle |
|---|---|---|
| **1.** Lateral surface of ilium | Greater trochanter of femur | _____ |
| **2.** Ischial tuberosity | Posterior surface of femur | _____ |
| **3.** Anterior superior iliac spine | Medial surface of proximal tibia | _____ |
| **4.** Lateral and medial condyles of femur | Posterior surface of calcaneus | _____ |
| **5.** Anterior iliac crest | Fascia (iliotibial tract) of the thigh | _____ |
| **6.** Greater trochanter and posterior surface of femur | Patella and to tibial tuberosity | _____ |
| **7.** Ischial tuberosity | Medial surface of proximal tibia | _____ |
| **8.** Medial surface of femur | Patella and to tibial tuberosity | _____ |
| **9.** Posterior surface of tibia | Distal phalanges of four lateral toes | _____ |
| **10.** Lateral condyle and lateral surface of tibia | Medial cuneiform and first metatarsal | _____ |

**PART C ASSESSMENTS**

 **ASSESS**

**CRITICAL THINKING**

Identify the muscles indicated in figure 23.7.

1. _____
2. _____
3. _____
4. _____
5. _____
6. _____
7. _____

**(a)** Left thigh, anterior view

**(b)** Right lower limb, medial view

**Terms:**

Gastrocnemius     Tibialis anterior
Rectus femoris     Vastus lateralis
Sartorius          Vastus medialis
Soleus

**FIGURE 23.7**   Identify the muscles that appear as lower limb surface features in these photographs (*a* and *b*), using the terms provided. ((*a* and *b*) ©J & J Photography)

# 24

# Surface Anatomy

## MATERIALS NEEDED

Textbook
Small, round stickers
Colored pencils (black and red)

## PURPOSE OF THE EXERCISE

To examine the surface features of the human body and the terms used to describe them.

## LEARNING OUTCOMES

After completing this exercise, you should be able to

1. Locate and identify major surface features of the human body.

2. Arrange surface features by body region.

3. Distinguish between surface features that are bony landmarks and soft tissue.

**E**xternal landmarks, called *surface anatomy*, located on the human body provide an opportunity to examine surface features and to help locate other internal structures. This exercise will focus on bony landmarks and superficial soft tissue structures, primarily related to the skeletal and muscular systems. The technique used for the examination of surface features is called *palpation*, and is accomplished by touching with the hands or fingers. This enables us to determine the location, size, and texture of the underlying anatomical structure. Some of the surface features can be visible without the need of palpation, especially on

body builders and very thin individuals. If the individual involved has considerable subcutaneous adipose tissue, additional pressure might be necessary to palpate some of the structures.

These surface features will help us to locate additional surface features and deeper structures during the study of the organ systems. Surface features will help us to determine the pulse locations of superficial arteries and for proper assessments of injection sites; descriptions of pain and injury sites; insertion of examination tubes; and heart, lung, and bowel sounds heard with a stethoscope. For those who choose careers in emergency medical services, nursing, medicine, physical education, physical therapy, chiropractic, and massage therapy, surface anatomy is especially valuable.

## PRACTICE

### PROCEDURE—Surface Anatomy

1. Review textbook figures on body regions, human organism reference plates, the skeletal system, and the muscular system. These figures are located in chapters 1, 7, 8, and 9.

2. Use the textbook figures and the other figures provided to complete figures 24.1 through 24.8. Many features have already been labeled. Features in **boldface** are *bony features*. Other labeled features (not boldface) are *soft tissue features*. Bony features are a part of a bone and the skeletal system. Soft tissue features are composed of tissue other than bone, such as a muscle, tendon, ligament, cartilage, or skin. The missing features are listed by each figure.

Terms:
**Hyoid bone**
**Mandible**
Sternocleidomastoid
Trapezius (upper)

Frontalis

Zygomatic arch

1

2

Laryngeal prominence ("Adam's apple") of thyroid cartilage

(a)

Zygomatic arch

External acoustic meatus

Temporomandibular joint

3

Trapezius (upper)

(b)

External occipital protuberance

Occipital ridge

Mastoid process

Vertebra prominens (C7)

4

Acromion process

Deltoid (posterior)

Medial border of scapula

(c)

**FIGURE 24.1**  Label the surface features of (*a*) the anterior head and neck, (*b*) the lateral head and neck, and (*c*) the posterior head and neck, using the terms provided. (**Boldface** indicates bony features; not boldface indicates soft tissue features.)  **3**  A ((*a*) ©McGraw-Hill Education/Joe DeGrandis, photographer; (*b,c*) ©Eric Wise)

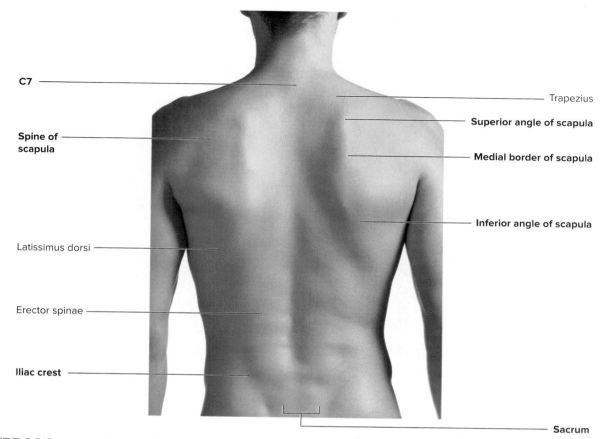

C7

Spine of
scapula

Latissimus dorsi

Erector spinae

Iliac crest

Trapezius

**Superior angle of scapula**

**Medial border of scapula**

**Inferior angle of scapula**

**Sacrum**

**FIGURE 24.2**   Surface features of the posterior shoulder and torso. (**Boldface** indicates bony features; not boldface indicates soft tissue features.) (©momcilog/E+/Getty Images RF)

**Jugular (suprasternal) notch**

**Clavicle**

**Acromioclavicular joint**

1

Anterior axillary fold

Serratus anterior

(a)

2

**Sternoclavicular joint**

Supraclavicular fossa

Upper trapezius

**Acromion process**

Deltoid

Pectoralis major

3

Terms:
Biceps brachii
Serratus anterior
Sternocleidomastoid
**Sternum**

**FIGURE 24.3**   Label the anterior surface features of (*a*) upper torso and (*b*) lower torso, using the terms provided. (**Boldface** indicates bony features; not boldface indicates soft tissue features.)  ((*a*) ©McGraw-Hill Education/Joe DeGrandis, photographer; (*b*) ©Brasoveanu Bogdan/Alamy Stock Photo RF)

Axilla

Sternum

Xiphoid process

4

Rectus abdominis

Linea alba

External oblique

Umbilicus

Iliac crest

**FIGURE 24.3**  (b)
· Continued.

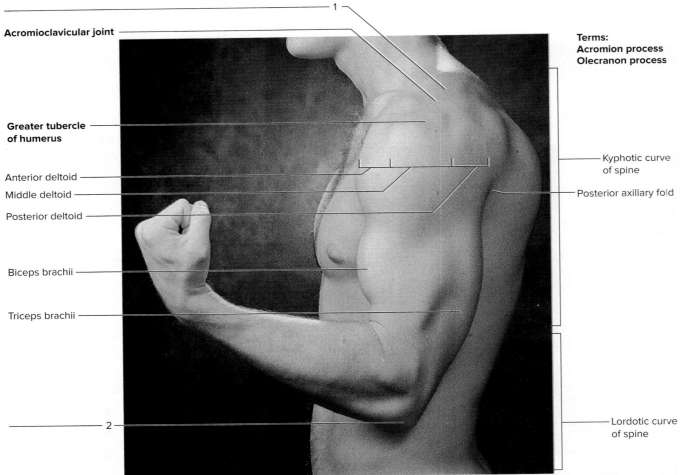

1

Acromioclavicular joint

**Terms:**
**Acromion process**
**Olecranon process**

**Greater tubercle**
**of humerus**

Kyphotic curve
of spine

Anterior deltoid

Middle deltoid

Posterior axillary fold

Posterior deltoid

Biceps brachii

Triceps brachii

2

Lordotic curve
of spine

**FIGURE 24.4**  Label the surface features of the lateral shoulder and upper limb, using the terms provided. (**Boldface**
indicates bony features; not boldface indicates soft tissue features.)  A  (©McGraw-Hill Education/Joe DeGrandis, photographer)

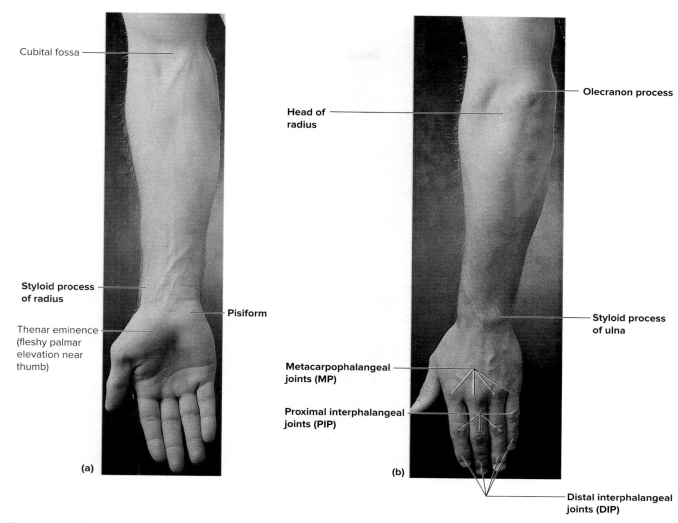

Cubital fossa

Olecranon process

Head of radius

Styloid process of radius

Thenar eminence (fleshy palmar elevation near thumb)

Pisiform

Styloid process of ulna

Metacarpophalangeal joints (MP)

Proximal interphalangeal joints (PIP)

Distal interphalangeal joints (DIP)

(a)

(b)

**FIGURE 24.5** Surface features of the upper limb (*a*) anterior view and (*b*) posterior view. (**Boldface** indicates bony features; not boldface indicates soft tissue features.) ((*a,b*) ©McGraw-Hill Education/Joe DeGrandis, photographer)

Terms:
Lateral malleolus
Medial malleolus
Patella

Anterior superior iliac spine

Greater trochanter of femur

Vastus lateralis

Vastus medialis

2

**Head of fibula**

**Tibial tuberosity**

Tibialis anterior

Gastrocnemius

**Tibia**

1

3

**FIGURE 24.6** Label the surface features of the anterior view of the body, using the terms provided. (**Boldface** indicates bony features; not boldface indicates soft tissue features.) **A** 1 (©Eric Wise)

Transverse arch

**Head of metatarsal I**

Medial longitudinal arch

**FIGURE 24.7** Surface features of the plantar surface of the foot. (**Boldface** indicates bony features; not boldface indicates soft tissue features.) (©McGraw-Hill Education/Joe DeGrandis, photographer)

Popliteal fossa

Gastrocnemius

Calcaneal tendon

**Terms:**
Calcaneal tendon
Gastrocnemius
**Lateral malleolus**
**Patella**

1

**Calcaneus**

**(a)**

Iliotibial tract

2

**Patella**

**Head of fibula**

4

3

Soleus

**Medial malleolus**

**Lateral malleolus**

**(b)**

**(c)**

**FIGURE 24.8**   Label the surface features of the (*a*) posterior lower limb, (*b*) lateral lower limb, and (*c*) medial lower limb, using the terms provided. (**Boldface** indicates bony features; not boldface indicates soft tissue features.)   **3**   ((*a–c*) ©McGraw-Hill Education/Joe DeGrandis, photographer)

3. Complete Parts A and B of Laboratory Report 24.
4. Use figures 24.1 through 24.8 to determine which items listed in table 24.1 are bony features and which items are soft tissue features. If the feature listed is a bony surface feature, place an "X" in the second column; if the feature listed is a soft tissue feature, place an "X" in the third column. **3**
5. Complete Part C of the laboratory report.

 **LEARN: LAB IN MOTION**

Select fifteen surface features presented in the laboratory exercise figures. Using small, round stickers, write the name of a surface feature on each sticker. Working with a laboratory partner, locate the selected surface features on yourself and on your partner. Use the stickers to accurately "label" your partner. **1**

■ Were you able to find all the features on both of you?
_____

■ Was it easier to find the surface features on yourself or your partner? _____

**TABLE 24.1** Representative Bony and Soft Tissue Surface Features (*Note*: After indicating your "X" in the appropriate column, use the extra space in columns 2 and 3 to add personal descriptions of the nature of the bony feature or the soft tissue feature palpation.)

| Bony and Soft Tissues | Bony Features | Soft Tissue Features |
|---|---|---|
| Biceps brachii | | |
| Calcaneal tendon | | |
| Deltoid | | |
| Gastrocnemius | | |
| Greater trochanter | | |
| Head of fibula | | |
| Iliac crest | | |
| Iliotibial tract | | |
| Lateral malleolus | | |
| Mandible | | |
| Mastoid process | | |
| Medial border of scapula | | |
| Rectus abdominis | | |
| Sacrum | | |
| Serratus anterior | | |
| Sternocleidomastoid | | |
| Sternum | | |
| Thenar eminence | | |
| Tibialis anterior | | |
| Zygomatic arch | | |

Name _____

Date _____

Section _____

The  corresponds to the indicated Learning Outcome(s) found at the beginning of the laboratory exercise.

LABORATORY
REPORT

24

# Surface Anatomy

## ⟳ PART A ASSESSMENTS

Match the body regions in column A with the surface features in column B. Place the letter of your choice in the space provided. ⒜ ⒝

### Column A

a. Head
b. Trunk (torso, including shoulder, and hip)
c. Upper limb
d. Lower limb

### Column B

_____ **1.** Umbilicus

_____ **2.** Medial malleolus

_____ **3.** Iliac crest

_____ **4.** Transverse arch

_____ **5.** Spine of scapula

_____ **6.** External occipital protuberance

_____ **7.** Metacarpophalangeal joints

_____ **8.** Sternum

_____ **9.** Olecranon process

_____ **10.** Zygomatic arch

_____ **11.** Mastoid process

_____ **12.** Thenar eminence

_____ **13.** Popliteal fossa

_____ **14.** Cubital fossa

Label figure 24.9 with the surface features provided.

Terms:
**Acromion process**
Biceps brachii
**Clavicle**
**Greater trochanter**
**Head of fibula**
**Iliac crest**
Pectoralis major
Rectus abdominis

Thenar eminence
**Tibia**
**Tibial tuberosity**
Trapezius (upper)
Vastus lateralis
**Xiphoid process**
**Zygomatic arch**

(a)

**FIGURE 24.9** Label the surface features of the (a) anterior view of the body and (b) posterior view of the body, using the terms provided. (**Boldface** indicates bony features; not boldface indicates soft tissue features.) **A** ((a,b) ©Eric Wise)

7
8
9
10
11
12

**Ischial tuberosity**

13

1
2
3
4
5
6

14
15

(b)

**FIGURE 24.9** *Continued.*

**Terms:**
**C7**
Calcaneal tendon
**Distal interphalangeal joint (DIP)**
Erector spinae
**External occipital protuberance**
**Iliac crest**
**Iliotibial tract**
**Inferior angle of scapula**

**Mastoid process**
**Medial border of scapula**
**Olecranon process**
Popliteal fossa
**Sacrum**
Soleus
**Styloid process of ulna**

Using table 24.1, indicate the locations of the surface features with an "X" on figure 24.10. Use a black "X" for the bony features and a red "X" for the soft tissue features.

**Terms from table 24.1:**
Biceps brachii
Calcaneal tendon
Deltoid
Gastrocnemius
Greater trochanter
Head of fibula
Iliac crest
Iliotibial tract
Lateral malleolus
Mandible
Mastoid process
Medial border of scapula
Rectus abdominis
Sacrum
Serratus anterior
Sternocleidomastoid
Sternum
Thenar eminence
Tibialis anterior
Zygomatic arch

(a)                                                                 (b)

**FIGURE 24.10**   Indicate bony surface features with a black "X" and soft tissue surface features with a red "X," using the results from table 24.1, on the (*a*) anterior and (*b*) posterior diagrams of the body.  **3**

# 25

# Nervous Tissue and Nerves

## MATERIALS NEEDED

Textbook
Compound light microscope
Prepared microscope slides of the following:
    Spinal cord (smear)
    Posterior (dorsal) root ganglion (section)
    Neuroglia (astrocytes)
    Peripheral nerve (cross section and longitudinal
       section)
Neuron model
Prepared microscope slide of Purkinje cells from
    cerebellum

## PURPOSE OF THE EXERCISE

To review the characteristics of nervous tissue and to observe neurons, neuroglia, and various features of a nerve.

 **LEARNING OUTCOMES** AP|R

After completing this exercise, you should be able to

(1) Describe the location and characteristics of nervous tissue.

(2) Distinguish structural and functional characteristics of neurons and neuroglia.

(3) Identify and sketch the major structures of a neuron and a nerve.

**N**ervous tissue, which occurs in the brain, spinal cord, and nerves, contains neurons and neuroglia. The neurons are the basic structural and functional units of the nervous system involved in making decisions, detecting stimuli, and conducting impulses. The neuroglia perform various supportive and protective functions for neurons.

 **PRACTICE**

**PROCEDURE—Nervous Tissue and Nerves** AP|R

1. Review the sections entitled "General Functions of the Nervous System" and "Classification of Cells of the Nervous System" in chapter 10 of the textbook.
2. As a review activity, label figures 25.1 and 25.2.
3. Complete Parts A and B of Laboratory Report 25.
4. Obtain a prepared microscope slide of a spinal cord smear. Using low-power magnification, search the slide and locate the relatively large, deeply stained cell bodies of motor neurons (multipolar neurons).
5. Observe a single, multipolar motor neuron, using high-power magnification, and note the following features:

**cell body**
    nucleus
    nucleolus
    chromatophilic substance (Nissl bodies)
    neurofibrils (threadlike structures extending into the
      nerve fibers)
**dendrites**
**axon (nerve fiber)**

    Compare the slide to the neuron model and to figure 25.3. You also may note small, darkly stained nuclei of neuroglia around the motor neuron.
6. Sketch and label a motor (efferent) neuron in the space provided in Part C of the laboratory report.
7. Obtain a prepared microscope slide of a posterior (dorsal) root ganglion. Search the slide and locate a cluster of sensory neuron cell bodies. You also may note bundles of nerve fibers passing among groups of neuron cell bodies (fig. 25.4).
8. Sketch and label a sensory (afferent) neuron cell body in the space provided in Part C of the laboratory report.

Chromatophilic substance (Nissl bodies)

Neurofibrils

Axon hillock

Direction
of
impulse

Nucleus

Axon

Myelin (cut)

5

6

1

2

3

4

**FIGURE 25.1**   Label this diagram of a multipolar motor neuron.  A  AP|R

Neurofibril

**FIGURE 25.2** Label the features of the myelinated axon (nerve fiber). A AP|R

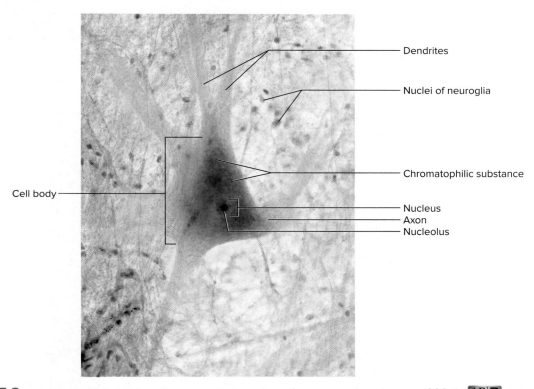

Dendrites

Nuclei of neuroglia

Cell body

Chromatophilic substance

Nucleus
Axon
Nucleolus

**FIGURE 25.3** Micrograph of a multipolar neuron and neuroglia from a spinal cord smear (600×). AP|R (©McGraw-Hill Education/Al Telser, photographer)

**FIGURE 25.4** Micrograph of a posterior (dorsal) root ganglion containing cell bodies of sensory neurons (100×). **AP|R** (©Ed Reschke)

Sensory neuron cell bodies

Nerve fibers (axons)

**FIGURE 25.5** Micrograph of astrocytes (1,000×). (©Ed Reschke)

Blood vessel

Astrocyte

9. Obtain a prepared microscope slide of neuroglia. Search the slide and locate some darkly-stained astrocytes with numerous long, slender processes (fig. 25.5).
10. Sketch neuroglia in the space provided in Part C of the laboratory report.
11. Obtain a prepared microscope slide of a nerve. Locate the cross section of the nerve and note the many round nerve fibers inside. Also note the dense layer of connective tissue (perineurium) that encircles a fascicle of nerve fibers and holds them together in a bundle. The individual nerve fibers are surrounded by a layer of more delicate connective tissue (endoneurium) (fig. 25.6).
12. Using high-power magnification, observe a single nerve fiber and note the following features:

    **axon**
    **myelin sheath around the axon of Schwann cell**
    (most of the myelin may have been dissolved and lost during the slide preparation)
    **neurilemma of Schwann cell**

Connective tissue (perineurium)

Connective tissue (endoneurium)

Axon (nerve fiber)

Myelin sheath of Schwann cell

Neurilemma of Schwann cell

**FIGURE 25.6** Cross section of a bundle of axons within a nerve (400×). (©Ed Reschke)

13. Sketch and label a nerve fiber with Schwann cell (cross section) in the space provided in Part D of the laboratory report.

14. Locate the longitudinal section of the nerve on the slide (fig. 25.7). Note the following:

    **axon**
    **myelin sheath of Schwann cell**
    **neurilemma of Schwann cell**
    **node of Ranvier**

15. Sketch and label a nerve fiber with Schwann cell (longitudinal section) in the space provided in Part D of the laboratory report.

 **LEARN: ACTIVITY**

Obtain a prepared microscope slide of Purkinje cells (fig. 25.8). To locate these neurons, search the slide for large, flask-shaped cell bodies. Each cell body has one or two large, thick dendrites that give rise to extensive branching networks of dendrites. These large cells are located in a particular region of the brain (cerebellar cortex).

Node of Ranvier
Axon (nerve fiber)
Neurilemma of Schwann cell
Myelin sheath of Schwann cell

**FIGURE 25.7**   Longitudinal section of a nerve (650×).
**AP|R**  (©Ed Reschke)

Dendrites with extensive branching

Cell body

Axon

**FIGURE 25.8**   Large multipolar Purkinje cell from cerebellum of the brain (400×).  (©David Becker/Science Photo Library/Getty Images)

# Notes

The **A** corresponds to the indicated Learning Outcome(s) found at the beginning of the laboratory exercise.

# Nervous Tissue and Nerves

## PART A ASSESSMENTS

Match the terms in column A with the descriptions in column B. Place the letter of your choice in the space provided. **A₁ A₂**

| Column A | Column B |
|---|---|
| **a.** Astrocyte | _____ **1.** Sheath of Schwann cell containing cytoplasm and nucleus that encloses myelin |
| **b.** Axon | |
| **c.** Chromatophilic substance (Nissl bodies) | _____ **2.** Corresponds to rough endoplasmic reticulum in other cells |
| **d.** Collateral | _____ **3.** Network of fine threads extending into nerve fiber |
| **e.** Dendrite | _____ **4.** Substance of Schwann cell composed of lipoprotein that insulates axons and increases impulse speed |
| **f.** Myelin sheath | |
| **g.** Neurilemma | _____ **5.** Neuron process with many branches that conducts an impulse toward the cell body |
| **h.** Neurofibrils | |
| | _____ **6.** Branch of an axon |
| | _____ **7.** Star-shaped neuroglia between neurons and blood vessels |
| | _____ **8.** Nerve fiber arising from a slight elevation of the cell body that conducts an impulse away from the cell body |

## PART B ASSESSMENTS

Match the terms in column A with the descriptions in column B. Place the letter of your choice in the space provided. **A₁ A₂**

| Column A | Column B |
|---|---|
| **a.** Effector | _____ **1.** Transmits impulse from sensory to motor neuron within central nervous system |
| **b.** Ependymal cell | |
| **c.** Ganglion | _____ **2.** Transmits impulse out of the brain or spinal cord to effectors (muscles and glands) |
| **d.** Interneuron | |
| **e.** Microglia | _____ **3.** Transmits impulse into brain or spinal cord from receptors |
| **f.** Motor (efferent) neuron | _____ **4.** Myelin-forming neuroglia in brain and spinal cord |
| **g.** Oligodendrocyte | _____ **5.** Phagocytic neuroglia |
| **h.** Sensory (afferent) neuron | _____ **6.** Structure capable of responding to motor impulse |
| | _____ **7.** Specialized mass of neuron cell bodies outside the brain or spinal cord |
| | _____ **8.** Cells that cover the inside spaces of the brain ventricles and help regulate cerebrospinal fluid |

In the space that follows, sketch the indicated cells. Label any of the cellular structures observed, and indicate the magnification of each sketch. 2 3

| | |
|---|---|
| Motor neuron (_____X) | Sensory neuron cell body (_____X) |

Neuroglia (_____X)

In the space that follows, sketch the indicated view of a nerve fiber (axon). Label any structures observed, and indicate the magnification. 2 3

| | |
|---|---|
| Nerve fiber cross section with Schwann cell (_____X) | Nerve fiber longitudinal section with Schwann cell (_____X) |

# 26

# Brain and Cranial Nerves

## MATERIALS NEEDED

Textbook
Dissectible model of the human brain
Preserved human brain
Anatomical charts of the human brain

## PURPOSE OF THE EXERCISE

To review the structural and functional characteristics of the human brain and cranial nerves.

 ## LEARNING OUTCOMES AP|R

After completing this exercise, you should be able to

1. Identify the major external and internal structures in the human brain.

2. Locate the major functional regions of the brain.

3. Identify the twelve pairs of cranial nerves.

4. Differentiate the functions of the cranial nerves.

The brain, the largest and most complex part of the nervous system, contains nerve centers associated with sensory functions and is responsible for sensations and perceptions. It issues motor commands to skeletal muscles and carries on higher mental activities. It also functions to coordinate muscular movements, and it contains centers and neural pathways necessary for the regulation of internal organs.

Twelve pairs of cranial nerves arise from the inferior surface of the brain and are designated by number and name. The cranial nerves are part of the PNS; most arise from the brainstem region of the brain. Although most of these nerves conduct both sensory and motor impulses, some contain only sensory fibers associated with special sense organs. Others are primarily composed of motor fibers and are involved with the activities of muscles and glands.

 ## PRACTICE

### PROCEDURE A—Human Brain

1. Review the section entitled "Brain" in chapter 11 of the textbook.
2. As a review activity, label figures 26.1 and 26.2.
3. Complete Part A of Laboratory Report 26.
4. Observe the anatomical charts, dissectible model, and preserved specimen of the human brain. Locate each of the following features:

**cerebrum**

**cerebral hemispheres**

**corpus callosum**

**gyri (convolutions)**

**sulci**

    central sulcus

    lateral sulcus

**fissures**

    longitudinal fissure

    transverse fissure

**lobes**

    frontal lobe

    parietal lobe

    temporal lobe

    occipital lobe

    insula (insular lobe)

**cerebral cortex**

**basal nuclei** (basal ganglia, a widely used clinical term)

    caudate nucleus

    putamen

    globus pallidus

**ventricles**
lateral ventricles
third ventricle
fourth ventricle
choroid plexuses
cerebral aqueduct

**diencephalon**
thalamus
hypothalamus
optic chiasma
mammillary bodies
pineal gland

**cerebellum**
lateral (right and left) hemispheres
falx cerebelli
vermis
cerebellar cortex
arbor vitae

**brainstem**
midbrain (mesencephalon)
cerebral aqueduct
cerebral peduncles
corpora quadrigemina
pons
medulla oblongata

5. Locate the labeled areas in figure 26.3 that represent the following functional regions of the cerebrum:

**motor area for voluntary muscle control**
**motor speech area (Broca's area)**
**cutaneous sensory area**
**auditory area**
**visual area**
**sensory speech area (Wernicke's area)**

6. Complete Part B of the laboratory report.

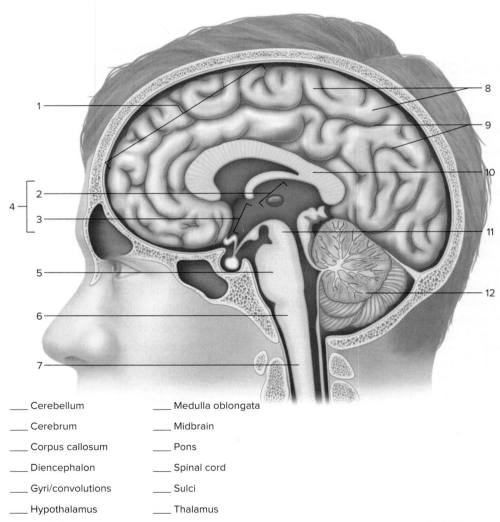

___ Cerebellum     ___ Medulla oblongata

___ Cerebrum     ___ Midbrain

___ Corpus callosum     ___ Pons

___ Diencephalon     ___ Spinal cord

___ Gyri/convolutions     ___ Sulci

___ Hypothalamus     ___ Thalamus

**FIGURE 26.1**   Label this diagram by placing the correct numbers in the spaces provided.

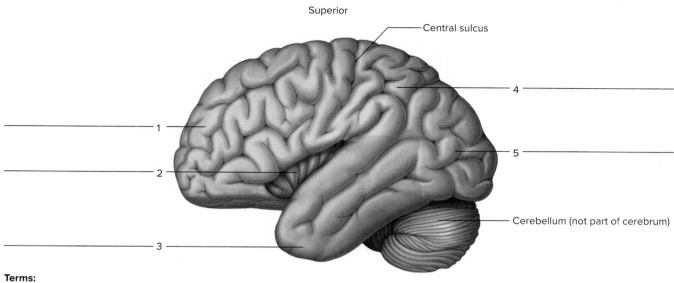

Anterior ← → Posterior

Superior

Central sulcus

4

1

2

5

3

Cerebellum (not part of cerebrum)

Inferior

**Terms:**
Frontal lobe
Insula (deep lobe)
Occipital lobe
Parietal lobe
Temporal lobe

**FIGURE 26.2**    Label the five lobes of the left cerebral hemisphere, using the terms provided. Lobe 3 is retracted to expose the deep lobe 2 in this figure. **1** **AP|R**

1

4

5

2

6

3

Cerebellum

Brainstem

**Terms:**
Auditory area
Motor speech area (Broca's area)
Cutaneous sensory area
Motor area for voluntary muscle control
Visual area
Sensory speech area (Wernicke's area)

**FIGURE 26.3**    Label the functional areas of the cerebrum, using the terms provided. (*Note:* These areas are not visible as distinct parts of the brain.) **2**

## PRACTICE

### PROCEDURE B—Cranial Nerves

1. Review the section entitled "Cranial Nerves" in chapter 11 of the textbook.
2. As a review activity, label figure 26.4. The cranial nerves arise from the base of the brain and can be viewed from the inferior surface.
3. Observe the model and preserved specimen of the human brain, and locate as many of the following cranial nerves as possible as you differentiate some of their associated functions:

**olfactory nerves (I)**—smell

**optic nerves (II)**—vision

**oculomotor nerves (III)**—pupil constriction and opening of eyelid

**trochlear nerves (IV)**—stimulation of superior oblique eye muscles

**trigeminal nerves (V)**—sensation from face and teeth and mastication movements

**abducens nerves (VI)**—lateral eye movements

**facial nerves (VII)**—salivation, tear secretions, and taste

**vestibulocochlear nerves (VIII)**—hearing and balance (equilibrium)

**glossopharyngeal nerves (IX)**—control of swallowing movements

**vagus nerves (X)**—regulation of many visceral organs, including the heart rate

**accessory nerves (XI)**—control of neck and shoulder muscles

**hypoglossal nerves (XII)**—control of tongue movements

The following mnemonic device will help you learn the twelve pairs of cranial nerves in the proper order:

**Ol**d **Op**ie **oc**casionally **t**ries **trig**onometry, **an**d **f**eels **very glo**omy, **vagu**e, **an**d **hypo**active.[1]

4. Complete Part C of the laboratory report.

---

1. From *HAPS-EDucator*, Winter 2002. An official publication of the Human Anatomy & Physiology Society (HAPS).

**FIGURE 26.4**   Provide the names of the cranial nerves in this inferior view.   A  AP|R

Name _____

Date _____

Section _____

The ⚞A⚟ corresponds to the indicated Learning Outcome(s) found at the beginning of the laboratory exercise.

LABORATORY
REPORT

**26**

# Brain and Cranial Nerves

## 🔄 PART A ASSESSMENTS

Match the terms in column A with the descriptions in column B. Place the letter of your choice in the space provided. ⚞A⚟

### Column A

a. Central sulcus
b. Cerebral cortex
c. Corpus callosum
d. Falx cerebelli
e. Gyrus (convolution)
f. Hypothalamus
g. Insula
h. Medulla oblongata
i. Midbrain
j. Optic chiasma
k. Pineal gland
l. Pons

### Column B

_____ 1. Structure formed by the crossing-over of the optic nerves

_____ 2. Part of diencephalon that forms lower walls and floor of third ventricle

_____ 3. Cone-shaped gland attached to upper posterior portion of diencephalon

_____ 4. Connects cerebral hemispheres

_____ 5. Ridge on surface of cerebrum

_____ 6. Separates frontal and parietal lobes

_____ 7. Part of brainstem between diencephalon and pons

_____ 8. Rounded bulge on underside of brainstem

_____ 9. Part of brainstem continuous with the spinal cord

_____ 10. Layer of dura mater that separates cerebellar hemispheres

_____ 11. Cerebral lobe located deep within lateral sulcus

_____ 12. Thin layer of gray matter on surface of cerebrum

## 🔄 PART B ASSESSMENTS

Identify the features indicated in the sagittal (median) section of the right half of the human brain in figure 26.5. ⚞A⚟

1. _____    6. _____

2. _____    7. _____

3. _____    8. _____

4. _____    9. _____

5. _____    10. _____

Anterior

Posterior

1

4 — 2
3

5

8 — 6

7

9

**Terms:**
Brainstem
Cerebellum
Cerebrum
Corpus callosum
Diencephalon
Hypothalamus
Medulla oblongata
Midbrain
Pons
Thalamus

Fornix

10

Fourth ventricle

Spinal cord

**FIGURE 26.5** Identify the features on this sagittal (median) section of the right half of the human brain, using the terms provided. AP|R (©Martin M. Rotker/Science Source)

 **PART C ASSESSMENTS**

Match the cranial nerves in column A with the associated functions in column B. Place the letter of your choice in the space provided. A

| Column A | Column B |
|---|---|
| **a.** Abducens | _____ **1.** Regulates heart rate |
| **b.** Accessory | _____ **2.** Equilibrium and hearing |
| **c.** Facial | _____ **3.** Stimulates superior oblique muscle of eye |
| **d.** Glossopharyngeal | _____ **4.** Sensory impulses from teeth and face |
| **e.** Hypoglossal | _____ **5.** Pupil constriction and eyelid opening |
| **f.** Oculomotor | _____ **6.** Smell |
| **g.** Olfactory | _____ **7.** Controls neck and shoulder movements |
| **h.** Optic | _____ **8.** Controls tongue movements |
| **i.** Trigeminal | _____ **9.** Vision |
| **j.** Trochlear | _____ **10.** Stimulates lateral rectus muscle of eye |
| **k.** Vagus | _____ **11.** Taste, salivation, and secretion of tears |
| **l.** Vestibulocochlear | _____ **12.** Swallowing |

# 27

# Dissection of the Sheep Brain

## MATERIALS NEEDED

Dissectible model of human brain
Preserved sheep brain
Dissecting tray
Dissection instruments
Long knife
Disposable gloves
Frontal sections of sheep brains

## ⚠ SAFETY

- Wear disposable gloves and safety goggles when handling the sheep brains.
- Save or dispose of the brains as instructed.
- Wash your hands before leaving the laboratory.

## PURPOSE OF THE EXERCISE

To observe the major features of the sheep brain and to compare these features with those of the human brain.

##  LEARNING OUTCOMES AP|R

After completing this exercise, you should be able to

1. Examine the major structures of the sheep brain.
2. Summarize several differences and similarities between the sheep brain and the human brain.
3. Locate the most developed cranial nerves of the sheep brain.

**M**ammalian brains have many features in common. Human brains may not be available, so sheep brains often are dissected as an aid to understanding mammalian brain structure. However, the adaptations of the sheep differ from the adaptations of the human, so comparisons of their structural features may not be precise. The sheep is a quadruped; therefore, the spinal cord is horizontal, unlike the vertical orientation in a bipedal human. Preserved sheep brains have a different appearance and are firmer than those that are removed directly from the cranial cavity; this is caused by the preservatives used.

##  PRACTICE

### PROCEDURE—Dissection of the Sheep Brain

1. Obtain a preserved sheep brain and rinse it thoroughly in water to remove as much of the preserving fluid as possible.
2. Examine the surface of the brain for the presence of meninges. (The outermost layers of these membranes may have been mostly lost during removal of the brain from the cranial cavity.) Locate the following:

   **dura mater**—the thick, opaque outer layer
   **arachnoid mater**—the weblike, semitransparent middle layer spanning the fissures of the brain
   **pia mater**—the thin, delicate, innermost layer that adheres to the gyri and sulci of the surface of the brain

3. Remove any remaining dura mater by pulling it gently from the surface of the brain.
4. Position the brain with its ventral surface down in the dissecting tray. Study figure 27.1, and locate the following structures on the specimen:

   **cerebral hemispheres**
   **gyri (convolutions)**
   **sulci**
   **longitudinal fissure**
   **frontal lobe**
   **parietal lobe**
   **temporal lobe**
   **occipital lobe**
   **cerebellum**
   **medulla oblongata**
   **spinal cord**

5. Gently separate the cerebral hemispheres along the longitudinal fissure and expose the transverse band of white fibers within the fissure that connects the hemispheres. This band is the *corpus callosum*.
6. Bend the cerebellum and medulla oblongata slightly downward and away from the cerebrum (fig. 27.2). This will expose the *pineal gland* in the upper midline and the *corpora quadrigemina*, which consists of four

**FIGURE 27.1** Dorsal surface of the sheep brain. AP|R

Labels in Figure 27.1:
- Sulcus
- Gyrus (convolution)
- Cerebral hemisphere
- Cerebellum
- Medulla oblongata
- Frontal lobe
- Parietal lobe
- Temporal lobe
- Longitudinal fissure
- Occipital lobe
- Spinal cord

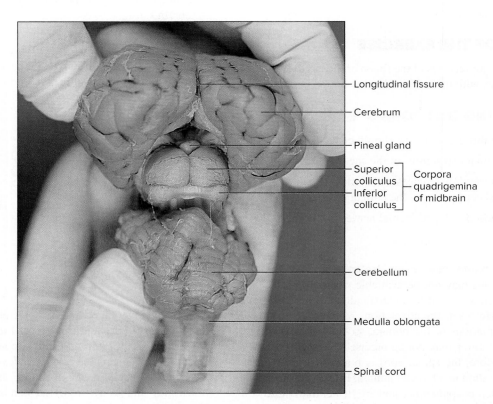

**FIGURE 27.2** Gently bend the cerebellum and medulla oblongata away from the cerebrum to expose the pineal gland of the diencephalon and the corpora quadrigemina of the midbrain. (©J & J Photography)

Labels in Figure 27.2:
- Longitudinal fissure
- Cerebrum
- Pineal gland
- Superior colliculus
- Inferior colliculus
- Corpora quadrigemina of midbrain
- Cerebellum
- Medulla oblongata
- Spinal cord

rounded structures, called *colliculi*, associated with the midbrain.

7. Position the brain with its ventral surface upward. Study figures 27.3 and 27.4, and locate the following structures on the specimen:

**longitudinal fissure**
**olfactory bulbs**
**optic nerves**

**optic chiasma**
**optic tract**
**mammillary body** (single in sheep)
**infundibulum (pituitary stalk)**
**midbrain**
**pons**

8. Although some of the cranial nerves may be missing or are quite small and difficult to find, locate as many of

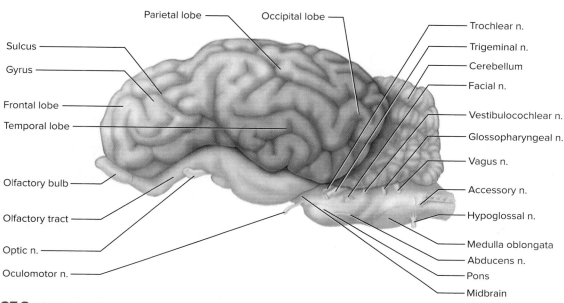

**FIGURE 27.3**  Lateral surface of the sheep brain. (The "n." stands for nerve.)

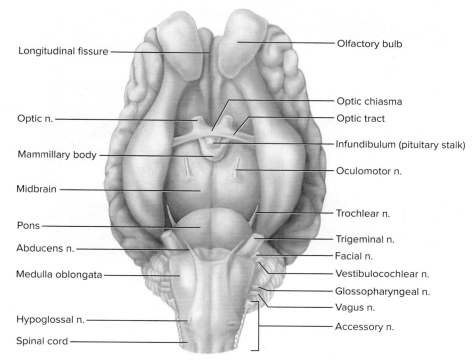

**FIGURE 27.4**  Ventral surface of the sheep brain. **AP|R**

the following as possible, using figures 27.3 and 27.4 as references:

**oculomotor nerves**
**trochlear nerves**
**trigeminal nerves**
**abducens nerves**
**facial nerves**
**vestibulocochlear nerves**
**glossopharyngeal nerves**
**vagus nerves**
**accessory nerves**
**hypoglossal nerves**

9. Using a long, sharp knife, cut the sheep brain along the midline to produce a median section. Study figures 27.2 and 27.5, and locate the following structures on the specimen:

**cerebrum**
**olfactory bulb**
**corpus callosum**
**cerebellum**
    white matter
    gray matter
**lateral ventricle** (one in each cerebral hemisphere)
**third ventricle** (within diencephalon)
**fourth ventricle** (between brainstem and cerebellum)

**diencephalon**
    optic chiasma
    infundibulum
    pituitary gland (this structure may be missing)
    mammillary body (single in sheep)
    thalamus
    hypothalamus
    pineal gland
**midbrain**
    corpora quadrigemina
        superior colliculi
        inferior colliculi

**pons**
**medulla oblongata**

### LEARN: ACTIVITY

Observe a frontal section from a sheep brain. Note the longitudinal fissure, gray matter, white matter, corpus callosum, lateral ventricles, third ventricle, and thalamus.

10. Dispose of the sheep brain as directed by the laboratory instructor.
11. Complete Parts A, B, and C of Laboratory Report 27.

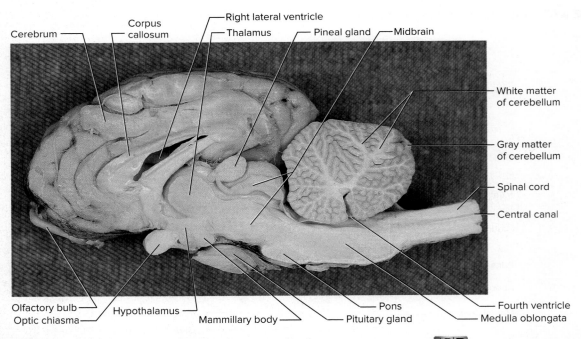

**FIGURE 27.5** Sagittal (median) section of the right half of the sheep brain dissection. **AP|R** (©J & J Photography)

Name _____

Date _____

Section _____

The Ⓐ corresponds to the indicated Learning Outcome(s) found at the beginning of the laboratory exercise.

# Dissection of the Sheep Brain

## ♻ PART A ASSESSMENTS

Answer the following questions:

**1.** Describe the location of any meninges observed to be associated with the sheep brain. Ⓐ _____

_____

_____

**2.** Compare the relative sizes of the sheep and human cerebral hemispheres. ②  _____

_____

_____

**3.** How do the gyri and sulci of the sheep cerebrum compare with the human cerebrum in numbers? ②  _____

_____

_____

**4.** What is the significance of the differences you noted in your answers for questions 2 and 3? ②  _____

_____

_____

**5.** What difference did you note in the structures of the sheep cerebellum and the human cerebellum? ②  _____

_____

_____

**6.** How do the sizes of the olfactory bulbs of the sheep brain compare with those of the human brain? ②  _____

_____

_____

**7.** Based on their relative sizes, which of the cranial nerves seems to be most highly developed in the sheep brain? ③  _____

_____

_____

**8.** What is the significance of the observations you noted in your answers for questions 6 and 7? ②  _____

_____

_____

  **ASSESS**

### CRITICAL THINKING

Prepare a list of at least six features to illustrate ways in which the brains of sheep and humans are similar.  **2**

1. _____

2. _____

3. _____

4. _____

5. _____

6. _____

   Interpret the significance of these similarities.  **2** _____

   _____

   _____

   _____


  **PART C ASSESSMENTS**

Identify the features indicated in the sagittal (median) section of the sheep brain in figure 27.6.  **1**

| _____ Cerebellum | _____ Hypothalamus | _____ Pineal gland |
| _____ Cerebrum | _____ Medulla oblongata | _____ Pons |
| _____ Corpus callosum | _____ Midbrain | _____ Thalamus |

**FIGURE 27.6** Label the features of this sagittal (median) section of the sheep brain by placing the correct numbers in the spaces provided. (©J & J Photography)

# 28

# Spinal Cord and Meninges

## MATERIALS NEEDED

Textbook
Compound light microscope
Prepared microscope slide of a spinal cord cross
    section with spinal nerve roots
Spinal cord model with meninges
Preserved spinal cord with meninges intact

## PURPOSE OF THE EXERCISE

To review the characteristics of the spinal cord and the meninges and to observe the major features of these structures.

##  LEARNING OUTCOMESS AP|R

After completing this exercise, you should be able to

1. Identify the major features and functions of the spinal cord.

2. Locate the ascending and descending tracts of the spinal cord.

3. Arrange the layers of the meninges and describe the structure of each.

The spinal cord is a column of nerve fibers that extends down through the vertebral canal. Together with the brain, it makes up the central nervous system.

Neurons within the spinal cord provide a two-way communication system between the brain and body parts outside the central nervous system. The cord also contains the processing centers for spinal reflexes.

Three membranes, or meninges, surround the entire CNS. The meninges consist of layers of membranes located between the bones of the skull and vertebral column and the soft tissues of the central nervous system. They include the dura mater, the arachnoid mater, and the pia mater.

##  PRACTICE

### PROCEDURE A—Structure of the Spinal Cord

1. Review the section entitled "Spinal Cord" in chapter 11 of the textbook.
2. As a review activity, label figures 28.1 and 28.2.
3. Complete Part A of Laboratory Report 28.
4. Study figure 28.3 and complete Part B of the laboratory report.
5. Obtain a prepared microscope slide of a spinal cord cross section. Use the low power of the microscope to locate the following features:

**posterior median sulcus**
**anterior median fissure**
**central canal**
**gray matter**
    gray commissure
    posterior (dorsal) horn
    lateral horn
    anterior (ventral) horn
**white matter**
    posterior (dorsal) funiculus (column)
    lateral funiculus (column)
    anterior (ventral) funiculus (column)
**roots of spinal nerve**
    posterior (dorsal) roots
    posterior (dorsal) root ganglia
    anterior (ventral) roots

6. Observe the model of the spinal cord, and locate the features listed in step 5.
7. Complete Part C of the laboratory report.

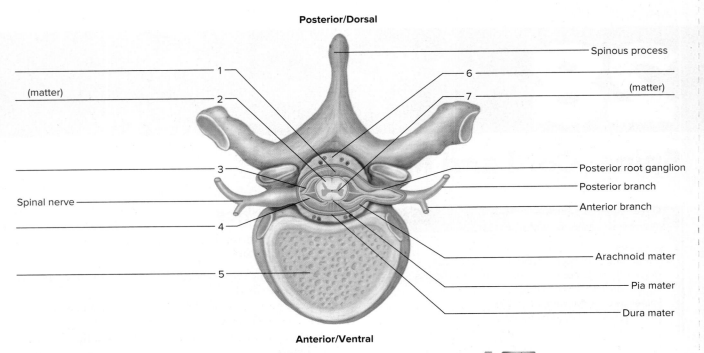

**Posterior/Dorsal**

Spinous process

1

(matter)

2

(matter)

6

7

3

Posterior root ganglion

Posterior branch

Spinal nerve

Anterior branch

4

Arachnoid mater

5

Pia mater

Dura mater

**Anterior/Ventral**

**FIGURE 28.1**   Label the features of the spinal cord and the surrounding structures. Ⓐ AP｜R

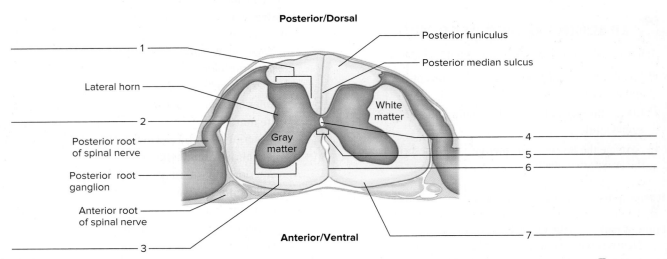

**Posterior/Dorsal**

Posterior funiculus

1

Posterior median sulcus

Lateral horn

White matter

2

Gray matter

4

Posterior root
of spinal nerve

5

6

Posterior root
ganglion

Anterior root
of spinal nerve

**Anterior/Ventral**

7

3

**FIGURE 28.2**   Label this cross section of the spinal cord, including the features of the white and gray matter. Ⓐ

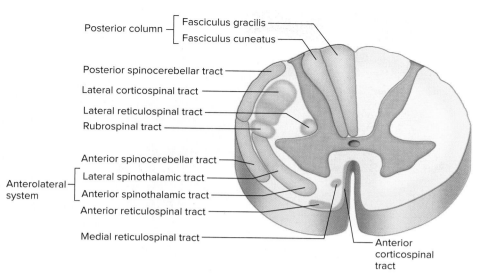

Posterior column ⎯ Fasciculus gracilis
Fasciculus cuneatus

Posterior spinocerebellar tract

Lateral corticospinal tract

Lateral reticulospinal tract

Rubrospinal tract

Anterolateral system ⎯ Anterior spinocerebellar tract
Lateral spinothalamic tract
Anterior spinothalamic tract
Anterior reticulospinal tract

Medial reticulospinal tract

Anterior corticospinal tract

**FIGURE 28.3** Major ascending (sensory) and descending (motor) tracts within a cross section of the spinal cord. Ascending tracts are in pink, descending tracts in light brown, and are shown only on one side. This pattern varies with the level of the spinal cord. This pattern is representative of the midcervical region. (*Note*: These tracts are not visible as individually stained structures on microscope slides.)

 **PRACTICE**

### PROCEDURE B—Meninges

1. Review the section entitled "Meninges" in chapter 11 of the textbook.
2. Study figure 28.4, showing the meninges associated with the brain and spinal cord.
3. Observe the spinal cord model with meninges and locate the following features:

 **dura mater**

 **arachnoid mater**

 **subarachnoid space**

 **pia mater**

4. As a review activity, label figure 28.5.
5. Complete Part D of the laboratory report.

 **LEARN: ACTIVITY**

Observe the preserved section of spinal cord with the meninges intact. Note the appearance of the central gray matter and the surrounding white matter. Note the heavy covering of dura mater, firmly attached to the cord on each side by a set of ligaments (denticulate ligaments) originating in the pia mater. The intermediate layer of meninges, the arachnoid mater, is devoid of blood vessels, but in a live human being, the space beneath this layer contains cerebrospinal fluid. The pia mater, closely attached to the surface of the spinal cord, contains many blood vessels.

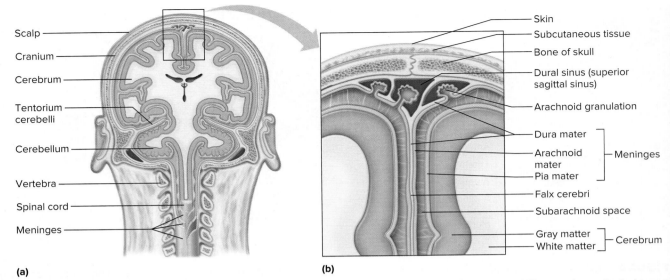

**Scalp**

**Cranium**

**Cerebrum**

**Tentorium cerebelli**

**Cerebellum**

**Vertebra**

**Spinal cord**

**Meninges**

(a)

**Skin**

**Subcutaneous tissue**

**Bone of skull**

**Dural sinus (superior sagittal sinus)**

**Arachnoid granulation**

**Dura mater**

**Arachnoid mater**

**Pia mater**

Meninges

**Falx cerebri**

**Subarachnoid space**

**Gray matter**

**White matter**

Cerebrum

(b)

**FIGURE 28.4** Meninges. (*a*) Membranes called meninges enclose the brain and spinal cord. (*b*) The meninges include three layers: dura mater, arachnoid mater, and pia mater. **AP|R**

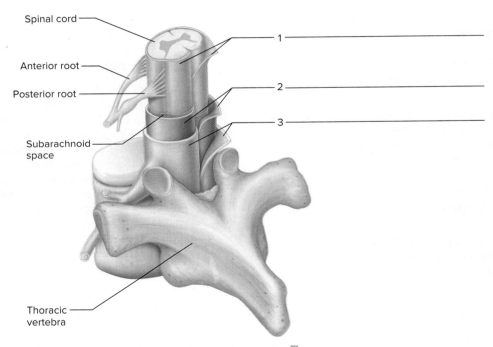

**Spinal cord**

**Anterior root**

**Posterior root**

**Subarachnoid space**

1

2

3

**Thoracic vertebra**

**FIGURE 28.5** Label the meninges associated with the spinal cord. **A**

The A corresponds to the indicated Learning Outcome(s) found at the beginning of the laboratory exercise.

# Spinal Cord and Meninges

## ♻ PART A ASSESSMENTS

Complete the following statements:

1. The spinal cord gives rise to 31 pairs of _____ . A

2. The bulge in the spinal cord that gives off nerves to the upper limbs is called the _____ . A

3. The bulge in the spinal cord that gives off nerves to the lower limbs is called the _____ . A

4. The _____ is a groove that extends the length of the spinal cord posteriorly. A

5. In a spinal cord cross section, the posterior _____ of the gray matter resemble the upper wings of a butterfly. A

6. The cell bodies of motor neurons are found in the _____ horns of the spinal cord. A

7. The _____ connects the gray matter on the left and right sides of the spinal cord. A

8. The _____ in the gray commissure of the spinal cord contains cerebrospinal fluid and is continuous with the ventricles of the brain. A

9. The white matter of the spinal cord is divided into anterior, lateral, and posterior _____ . A

10. The longitudinal bundles of nerve fibers within the spinal cord constitute major nerve pathways called _____ . A

## ♻ PART B ASSESSMENTS

Match the nerve tracts in column A with the descriptions in column B. Place the letter of your choice in the space provided. A

| Column A | Column B |
|---|---|
| **a.** Corticospinal <br> **b.** Fasciculus gracilis <br> **c.** Lateral spinothalamic <br> **d.** Posterior spinocerebellar <br> **e.** Reticulospinal | _____ **1.** Ascending tract to the brain to interpret touch, pressure, and body movements |
| | _____ **2.** Descending tract whose fibers conduct motor impulses to sweat glands and muscles to control tone |
| | _____ **3.** Descending tract whose fibers conduct motor impulses to skeletal muscles |
| | _____ **4.** Ascending tract to the cerebellum necessary for coordination of skeletal muscles |
| | _____ **5.** Ascending tract to the brain to give rise to sensations of temperature and pain |

Identify the features indicated in the spinal cord cross section of figure 28.6.

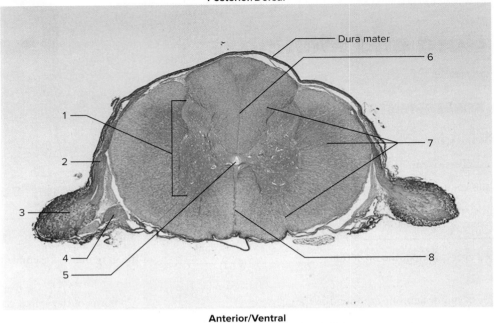

**Posterior/Dorsal**

Dura mater

**Anterior/Ventral**

_____ Anterior median fissure       _____ Gray matter                    _____ Posterior root of spinal nerve

_____ Anterior root of spinal nerve  _____ Posterior median sulcus        _____ White matter

_____ Central canal                  _____ Posterior root ganglion

**FIGURE 28.6**    Micrograph of a spinal cord cross section with spinal nerve roots (7.5×). Label the features by placing the correct numbers in the spaces provided.  (©Ed Reschke)

**PART D ASSESSMENTS**

Match the terms in column A with the descriptions in column B. Place the letter of your choice in the space provided.

| Column A | Column B |
|---|---|
| **a.** Arachnoid mater | _____ **1.** Band of pia mater that anchors dura mater to cord |
| **b.** Denticulate ligament | _____ **2.** Channel through which venous blood flows |
| **c.** Dural sinus | |
| **d.** Dura mater | _____ **3.** Outermost layer of meninges |
| **e.** Epidural space | _____ **4.** Follows irregular contours of spinal cord surface |
| **f.** Pia mater | |
| **g.** Subarachnoid space | _____ **5.** Contains cerebrospinal fluid |
| | _____ **6.** Thin, weblike middle membrane |
| | _____ **7.** Separates dura mater from bone of vertebra |

# 29

# Reflex Arc and Reflexes

## MATERIALS NEEDED

Textbook
Rubber percussion hammer

## PURPOSE OF THE EXERCISE

To review the characteristics of reflex arcs and reflex behavior and to demonstrate some of the reflexes that occur in the human body.

 ## LEARNING OUTCOMES AP|R

After completing this exercise, you should be able to

1. Describe the components of a reflex arc.
2. Demonstrate and record stretch reflexes that occur in humans.
3. Analyze the components and patterns of stretch reflexes.

A reflex arc represents the simplest type of neural pathway found in the nervous system. This pathway begins with a receptor at the dendrite end of a sensory (afferent) neuron. The sensory neuron leads into the central nervous system and may communicate with one or more interneurons. Some of these interneurons, in turn, communicate with motor (efferent) neurons, whose axons (nerve fibers) lead outward to effectors.

Thus, when a sensory receptor is stimulated by a change occurring inside or outside the body, impulses may pass through a reflex arc, and, as a result, effectors may respond. Such an automatic, subconscious response is called a *reflex*.

A *stretch reflex* involves a single synapse (monosynaptic) between a sensory and a motor neuron within the gray matter of the spinal cord. Examples of stretch reflexes include the patellar, calcaneal, biceps, triceps, and plantar reflexes. Other, more complex *withdrawal reflexes* involve interneurons in combination with sensory and motor neurons; thus, they are polysynaptic. Examples of withdrawal reflexes include responses to touching hot objects or stepping on sharp objects.

Reflexes demonstrated in this laboratory exercise are stretch reflexes. When a muscle is stretched by a tap over its tendon, stretch receptors (proprioceptors) called *muscle spindles* are stretched within the muscle, which initiates an impulse over a reflex arc. A sensory neuron conducts an impulse from the muscle spindle into the gray matter of the spinal cord, where it synapses with a motor neuron, which conducts the impulse to the effector muscle. The stretched muscle responds by contracting to resist or reverse further stretching. These stretch reflexes are important to maintaining proper posture, balance, and movements. Observations of many of these reflexes in clinical tests on patients may indicate damage to a level of the spinal cord or peripheral nerves of the particular reflex arc.

 ## PRACTICE

### PROCEDURE—Reflex Arc and Reflexes

1. Review the sections entitled "Reflex Arcs" in chapter 11 of the textbook and "Proprioception" in chapter 12 of the textbook.
2. Study figure 29.1.
3. Complete Part A of Laboratory Report 29.
4. Work with a laboratory partner to demonstrate each of the reflexes listed (see fig. 29.2). *It is important that the muscles involved in the reflexes be totally relaxed to observe proper responses.* If a person is trying too hard to experience the reflex or is trying to suppress the reflex, assign a multitasking activity while the stimulus with the rubber hammer occurs. For example, assign a physical task with the upper limbs along with a complex mental activity during the patellar (knee-jerk) demonstration. After each demonstration, record your observations in the table provided in Part B of the laboratory report.
   a. *Patellar (knee-jerk) reflex.* Have your laboratory partner sit on a table (or sturdy chair) with legs relaxed and hanging freely over the edge without touching the floor. Gently strike your partner's patellar ligament (just below the patella) with the blunt side of a rubber percussion hammer (fig. 29.2a). The reflex involves the femoral nerve and the spinal cord. The normal response is a moderate extension of the leg at the knee joint.

**FIGURE 29.1** Diagram of a stretch (monosynaptic) reflex arc. The patellar reflex represents a specific example of a stretch reflex. AP|R

**b.** *Calcaneal (ankle-jerk) reflex.* Have your partner kneel on a chair with back toward you and with feet slightly dorsiflexed over the edge and relaxed. Gently strike the calcaneal tendon (just above its insertion on the calcaneus) with the blunt side of the rubber hammer (fig. 29.2*b*). The reflex involves the tibial nerve and the spinal cord. The normal response is plantar flexion of the foot.

**c.** *Biceps (biceps-jerk) reflex.* Have your partner place a bare arm bent about 90° at the elbow on the table. Press your thumb on the inside of the elbow over the tendon of the biceps brachii, and gently strike your thumb with the rubber hammer (fig. 29.2*c*). The reflex involves the musculocutaneous nerve and the spinal cord. Watch the biceps brachii for a response. The response might be a slight twitch of the muscle or flexion of the forearm at the elbow joint.

**d.** *Triceps (triceps-jerk) reflex.* Have your partner lie supine with an upper limb bent about 90° across the abdomen. Gently strike the tendon of the triceps brachii near its insertion just proximal to the olecranon process at the tip of the elbow (fig. 29.2*d*). The reflex involves the radial nerve and the spinal cord. Watch the triceps brachii for a response. The response might be a slight twitch of the muscle or extension of the forearm at the elbow joint.

**e.** *Plantar reflex.* Have your partner remove a shoe and sock and lie supine with the lateral surface of the foot resting on the table. Draw the metal tip of the rubber hammer, applying firm pressure, over the sole from the heel to the base of the large toe (fig. 29.2*e*). The normal response is flexion (curling) of the toes and plantar flexion of the foot. If the toes spread apart and dorsiflexion of the great toe occurs, the reflex is the abnormal *Babinski reflex* response (normal in infants until the nerve fibers have complete myelinization). If the Babinski reflex occurs later in life, it may indicate damage to the corticospinal tract of the CNS.

**5.** Complete Part B of the laboratory report.

**(a)** Patellar (knee-jerk) reflex

**(b)** Calcaneal (ankle-jerk) reflex

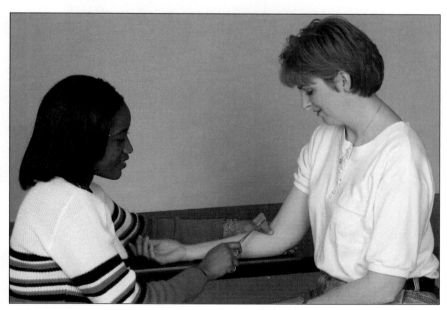

**(c)** Biceps reflex

**FIGURE 29.2** Demonstrate each of the following reflexes: (*a*) patellar reflex; (*b*) calcaneal reflex; (*c*) biceps reflex; (*d*) triceps reflex; and (*e*) plantar reflex. ((*a–e*) ©J & J Photography)

**(d)** Triceps reflex

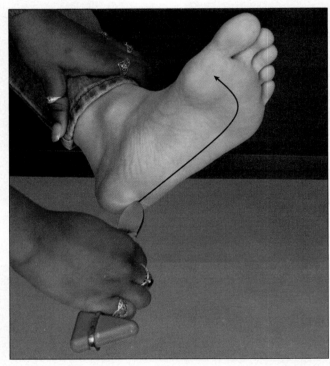

**(e)** Plantar reflex

**FIGURE 29.2**  *Continued.*

Name _____

Date _____

Section _____

The A corresponds to the indicated Learning Outcome(s) found at the beginning of the laboratory exercise.

# Reflex Arc and Reflexes

## PART A ASSESSMENTS

Complete the following statements:

1. Neural _____ are routes followed by nerve impulses as they pass through the nervous system. A

2. Interneurons in a withdrawal reflex are located in the _____ . A

3. _____ are automatic subconscious responses to external or internal stimuli. A

4. Effectors of a reflex arc are glands and _____ . A

5. A patellar reflex employs only _____ and motor neurons. A

6. The effector muscle of the patellar reflex is the _____ . A

7. The sensory stretch receptors (muscle spindles) of the patellar reflex are located in the _____ muscle. A

8. The patellar reflex helps the body to maintain upright _____ . A

9. The sensory receptors of a withdrawal reflex are located in the _____ . A

10. _____ muscles in the limbs are the effectors of a withdrawal reflex. A

11. The normal plantar reflex results in _____ of toes. A

12. Stroking the sole of the foot in infants results in dorsiflexion and toes that spread apart, called the _____ reflex. A

1. Complete the following table: **A**

| Reflex Tested | Response Observed | Degree of Response (hypoactive, normal, or hyperactive) | Effector Muscle Involved |
|---|---|---|---|
| Patellar (knee-jerk) | | | |
| Calcaneal (ankle-jerk) | | | |
| Biceps | | | |
| Triceps | | | |
| Plantar | | | Gastrocnemius, soleus, and flexor digitorum longus |

2. List the major events that occur in the patellar (knee-jerk) reflex, from the striking of the patellar ligament to the resulting response. **3** _____

_____

_____

_____

_____

## ASSESS

**CRITICAL THINKING**

What characteristics do the reflexes you demonstrated have in common? **A**

_____

_____

_____

# 30

# Receptors and General Senses

## MATERIALS NEEDED

Textbook
Marking pen (washable)
Millimeter ruler
Bristle or sharp pencil
Forceps (fine points) or measuring calipers
Blunt metal probes
Three beakers (250 mL)
Warm tap water or 45°C (113°F) water bath
Cold water (ice water)
Thermometer
Prepared microscope slides of tactile (Meissner's)
    and lamellated (Pacinian) corpuscles
Compound light microscope

## PURPOSE OF THE EXERCISE

To review the characteristics of sensory receptors and general senses and to investigate some of the general senses associated with the skin.

## LEARNING OUTCOMES

After completing this exercise, you should be able to

(1) Associate types of sensory receptors with general senses throughout the body.

(2) Determine and record the distribution of touch, warm, and cold receptors in various regions of the skin.

(3) Measure the two-point threshold of various regions of the skin.

Sensory receptors are sensitive to changes that occur within the body and its surroundings. Each type of receptor is particularly sensitive to a distinct kind of environmental change and is much less sensitive to other forms of stimulation. When receptors are stimulated, they initiate nerve impulses that travel into the central nervous system. The raw form in which these receptors send information to the brain is called *sensation*. The way our brains interpret this information is called *perception*.

The sensory receptors found widely distributed throughout skin, muscles, joints, and visceral organs are associated with **general senses.** These senses include touch, pressure, temperature, pain, and the senses of muscle movement and body position.

Sensory receptors that are more specialized and confined to the head are associated with **special senses.** Laboratory Exercises 31 through 35 describe the special senses.

 **PRACTICE**

### PROCEDURE A—Receptors

1. Review the sections entitled "Receptors, Sensation, and Perception" and "General Senses" in chapter 12 of the textbook.
2. Complete Part A of Laboratory Report 30.

 **LEARN: ACTIVITY**

Observe the tactile (Meissner's) corpuscle with the microscope set up by the laboratory instructor. This type of receptor is abundant in the superficial dermis in outer regions of the body, such as in the fingertips, soles, lips, and external genital organs. It is responsible for the sensation of light touch (see fig. 30.1 and chapter 6 of the textbook).

Observe the lamellated (Pacinian) corpuscle in the second demonstration microscope. This corpuscle is composed of many layers of connective tissue cells and has a nerve fiber in its central core. Lamellated corpuscles are numerous in the hands, feet, joints, and external genital organs. They are responsible for the sense of deep pressure (fig. 30.2). How are tactile and lamellated corpuscles similar?

_____

_____

How are they different?

_____

_____

FIGURE 30.1 Tactile (Meissner's) corpuscles, such as this one, are responsible for the sensation of light touch (250×). (©Ed Reschke)

FIGURE 30.2 Lamellated (Pacinian) corpuscles, such as this one, are responsible for the sensation of deep pressure (100×). (©Ed Reschke)

 **PRACTICE**

### PROCEDURE B—Sense of Touch

1. Investigate the distribution of touch receptors in your laboratory partner's skin. To do this, follow these steps:
   a. Use a marking pen and a millimeter ruler to prepare a square with 2.5 cm on each side on the skin on your partner's inner wrist, near the palm.
   b. Divide the square into smaller squares with 0.5 cm on a side, producing a small grid.
   c. Ask your partner to rest the marked wrist on the tabletop and to keep his/her eyes closed throughout the remainder of the experiment.
   d. Press the end of a bristle on the skin in some part of the grid, using just enough pressure to cause the bristle to bend. A sharp pencil can be used as an alternate device.
   e. Ask your partner to report whenever the touch of the bristle is felt. Record the results in Part B of the laboratory report.
   f. Continue this procedure until you have tested twenty-five different locations on the grid. Move randomly through the grid to help prevent anticipation of the next stimulation site.
2. Test two other areas of exposed skin in the same manner, and record the results in Part B of the laboratory report.
3. Answer the questions in Part B of the laboratory report.

 **PRACTICE**

### PROCEDURE C—Two-Point Threshold

1. Test your partner's ability to recognize the difference between one and two points of skin being stimulated simultaneously. To do this, follow these steps:
   a. Have your partner place a hand with the palm up on the table and close his/her eyes.
   b. Hold the tips of a forceps tightly together and gently touch the skin on your partner's index finger.
   c. Ask your partner to report if it feels like one or two points are touching the finger.
   d. Allow the tips of the forceps or the points of the measuring calipers to spread so they are 1 mm apart, press both points against the skin simultaneously, and ask your partner to report as before.
   e. Repeat this procedure, allowing the tips of the forceps to spread more each time until your partner can feel both tips being pressed against the skin. The minimum distance between the tips of the forceps when both can be felt is called the *two-point threshold*. As soon as you are able to distinguish two points, two separate receptors are being stimulated instead of only one receptor. Areas of skin with a small threshold are more sensitive than areas with large two-point thresholds.
   f. Record the two-point threshold for the skin of a fingertip in Part C of the laboratory report.

2. Repeat this procedure to determine the two-point threshold of the palm, the back of the hand, the back of the neck, the forearm, and the leg. Record the results in Part C of the laboratory report.

3. Answer the questions in Part C of the laboratory report.

 **PRACTICE**

**PROCEDURE D—Sense of Temperature**

1. Investigate the distribution of *warm receptors* in your partner's skin. To do this, follow these steps:

   **a.** Mark a square with 2.5 cm sides on your partner's palm.

   **b.** Prepare a grid by dividing the square into smaller squares, 0.5 cm on a side.

   **c.** Have your partner rest the marked palm on the table and close his/her eyes.

   **d.** Heat a blunt metal probe by placing it in a beaker of warm (hot) water (about 40–45°C/104–113°F) for a minute or so. *(Be sure the probe does not get so hot that it burns the skin.)* Use a thermometer to monitor the appropriate warm water from the tap or the water bath.

   **e.** Wipe the probe dry and touch it to the skin on some part of the grid.

   **f.** Ask your partner to report if the probe feels warm. Then record the results in Part D of the laboratory report.

   **g.** Keep the probe warm, and repeat the procedure until you have randomly tested several different locations on the grid.

2. Investigate the distribution of *cold receptors* by repeating the procedure. Use a blunt metal probe that has been cooled by placing it in ice water for a minute or so. Record the results in Part D of the laboratory report.

3. Answer the questions in Part D of the laboratory report.

 **LEARN: LAB IN MOTION**

Prepare three beakers of water of different temperatures. One beaker should contain warm water (about 40°C/104°F), one should be room temperature (about 22°C/72°F), and one should contain cold water (about 10°C/50°F). Place the index finger of one hand in the warm water and, at the same time, place the index finger of the other hand in the cold water for about 2 minutes. Then, simultaneously move both index fingers into the water at room temperature. What temperature do you sense with each finger? How do you explain the resulting perceptions?

_____

_____

_____

_____

# Notes

The Ⓐ corresponds to the indicated Learning Outcome(s) found at the beginning of the laboratory exercise.

# Receptors and General Senses

## 🔄 PART A ASSESSMENTS

Complete the following statements:

1. _____ are receptors that are sensitive to changes in the concentrations of chemicals. Ⓐ

2. Whenever tissues are damaged, _____ receptors are likely to be stimulated. Ⓐ

3. Receptors that are sensitive to temperature changes are called as _____ . Ⓐ

4. _____ are sensitive to changes in the intensity of light energy. Ⓐ

5. A sensation may seem to fade away when receptors are continuously stimulated as a result of _____ . Ⓐ

6. Tactile (Meissner's) corpuscles are responsible for the sense of light _____ . Ⓐ

7. Lamellated (Pacinian) corpuscles are responsible for the sense of deep _____ . Ⓐ

8. Warm receptors are most sensitive to temperatures between _____ . Ⓐ

9. Cold receptors are most sensitive to temperatures between _____ . Ⓐ

10. Widely distributed sensory receptors throughout the body are associated with _____ senses, in contrast to special senses. Ⓐ

## 🔄 PART B ASSESSMENTS

1. Record a "+" to indicate where the bristle was felt and a "0" to indicate where it was not felt. ②

Skin of wrist

2. Show the distribution of touch receptors in two other regions of skin. ②

Region tested _____

Region tested _____

**3.** Answer the following questions:

    **a.** How do you describe the pattern of distribution for touch receptors in the regions of the skin you tested? $\boxed{2}$

_____

_____

    **b.** How does the concentration of touch receptors seem to vary from region to region? $\boxed{2}$ _____

_____

## PART C ASSESSMENTS

**1.** Record the two-point threshold, in millimeters, for skin in each of the following regions: $\boxed{3}$

Fingertip _____

Palm _____

Back of hand _____

Back of neck _____

Forearm _____

Leg _____

**2.** Answer the following questions:

    **a.** What region of the skin tested has the greatest ability to discriminate two points? $\boxed{3}$ _____

    **b.** What region of the skin has the least sensitivity to this test? $\boxed{3}$ _____

    **c.** What is the significance of the observations in questions _a_ and _b_? $\boxed{3}$ _____

_____

_____

## PART D ASSESSMENTS

**1.** Record a "+" to indicate where warm was felt and a "0" to indicate where it was not felt. $\boxed{2}$

Skin of palm

**2.** Record a "+" to indicate where cold was felt and a "0" to indicate where it was not felt. $\boxed{2}$

Skin of palm

**3.** Answer the following questions:

    **a.** How do temperature receptors appear to be distributed in the skin of the palm? $\boxed{2}$ _____

_____

    **b.** Compare the distribution and concentration of warm and cold receptors in the skin of the palm. $\boxed{2}$ _____

_____

_____

# 31

# Smell and Taste

## MATERIALS NEEDED

*For Procedure A—Sense of Smell (Olfaction)*:
Textbook
Set of substances in stoppered bottles: cinnamon,
    sage, vanilla, garlic powder, oil of clove, oil of
    wintergreen, and perfume

*For Procedure B—Sense of Taste (Gustation)*:
Textbook
Paper cups (small)
Cotton swabs (sterile; disposable)
5% sucrose solution (sweet)
5% NaCl solution (salt)
1% acetic acid or unsweetened lemon juice (sour)
0.5% quinine sulfate solution or 0.1% Epsom salt
    solution (bitter)
1% monosodium glutamate (MSG) solution (umami)
Compound light microscope
Prepared microscope slides of olfactory epithelium
    and of taste buds
Pieces of apple, potato, carrot, and onion or packages
    of mixed flavors of LifeSavers

## ⚠ SAFETY

- Be aware of possible food allergies when selecting
  test solutions and foods to taste.
- Prepare fresh solutions for use in Procedure B.
- Wash your hands before starting the taste
  experiment.
- Wear disposable gloves when performing taste
  tests on your laboratory partner.
- Use a clean cotton swab for each test. Do not dip a
  used swab into a test solution.
- Dispose of used cotton swabs and paper towels
  as directed.
- Wash your hands before leaving the laboratory.

## PURPOSE OF THE EXERCISE

To review the structures of the organs of smell and taste and
to investigate the abilities of smell and taste receptors to dis-
criminate various chemical substances.

##  LEARNING OUTCOMES AP|R

After completing this exercise, you should be able to

1. Compare the characteristics of the smell (olfactory)
   receptors and taste (gustatory) receptors.
2. Explain how the senses of smell and taste function.
3. Record recognized odors and the time needed for
   olfactory sensory adaptation to occur.
4. Locate the distribution of taste receptors on the surface
   of the tongue and mouth cavity.

The senses of smell (olfaction) and taste (gustation) are
dependent upon chemoreceptors that are stimulated by
various chemicals dissolved in liquids. When these receptors
are stimulated by appropriate chemicals, action potentials
travel to the brain for interpretation. The receptors of smell are
found in the olfactory organs, which are located in the superior
parts of the nasal cavity and in a portion of the nasal septum.
The receptors of taste occur in the taste buds, which are sensory organs primarily found on the surface of the tongue, but
receptor cells are also distributed in other areas of the mouth
cavity and pharynx. Taste sensations are grouped into five recognized categories: sweet, sour, salty, bitter, and umami. The
sweet sensation is produced from sugars, the sour sensation
from acids, the salty sensation from inorganic salts, the bitter
sensation from alkaloids and spoiled foods, and the umami
sensation from aspartic and glutamic acids or a derivative such
as monosodium glutamate.

The senses of smell and taste function closely together,
because substances that are tasted often are smelled at the
same moment, and they play important roles in the selection
of foods. The taste and aroma of foods are also influenced
by appearance, texture, temperature, and the person's mood.
Chemicals are considered odorless and tasteless if receptor
sites for them are absent.

**PRACTICE**

### PROCEDURE A—Sense of Smell (Olfaction)

1. Review the section entitled "Sense of Smell" in chapter 12 of the textbook.
2. As a review activity, label figure 31.1.
3. Complete Part A of Laboratory Report 31.
4. Test your laboratory partner's ability to recognize the odors of the bottled substances available in the laboratory. To do this, follow these steps:
   a. Have your partner keep his/her eyes closed.
   b. Remove the stopper from one of the bottles and hold it about 4 cm under your partner's nostrils for about 2 seconds.
   c. Ask your partner to identify the odor, and then replace the stopper.
   d. Record your partner's response in Part B of the laboratory report.
   e. Repeat steps *b–d* for each of the bottled substances.
5. Repeat the preceding procedure, using the same set of bottled substances, but present them to your partner in a different sequence. Record the results in Part B of the laboratory report.

6. Wait 10 minutes and then determine the time it takes for your partner to experience olfactory sensory adaptation. To do this, follow these steps:
   a. Ask your partner to breathe in through the nostrils and exhale through the mouth.
   b. Remove the stopper from one of the bottles, and hold it about 4 cm under your partner's nostrils.
   c. Keep track of the time that passes until your partner is no longer able to detect the odor of the substance.
   d. Record the result in Part B of the laboratory report.
   e. Wait 5 minutes and repeat this procedure, using a different bottled substance.
   f. Test a third substance in the same manner.
   g. Record the results as before.
7. Complete Part B of the laboratory report.

 **LEARN: ACTIVITY**

Observe the olfactory epithelium in the demonstration microscope. The olfactory receptor cells are spindle-shaped, bipolar neurons with spherical nuclei. They also have six to eight cilia (olfactory hairs) at their apical ends. The supporting cells are pseudostratified columnar epithelial cells. However, in this region the tissue lacks goblet cells (see fig. 31.2 and chapter 5 of the textbook).

____ Cilia

____ Columnar epithelial cells

____ Cribriform plate of ethmoid bone

____ Nasal cavity

____ Nerve fibers within olfactory bulb

____ Olfactory nerves

____ Olfactory receptor cells

**FIGURE 31.1** Label this diagram of the olfactory organ by placing the correct numbers in the spaces provided. **A AP|R**

**FIGURE 31.2** Micrograph of olfactory epithelium (250×). Olfactory receptors have cilia (olfactory hairs) at their apical ends. **AP|R** (©Biophoto Associates/Science Source)

**PROCEDURE B—Sense of Taste (Gustation)**

1. Review the section entitled "Sense of Taste" in chapter 12 of the textbook.
2. As a review activity, label figure 31.3.
3. Complete Part C of the laboratory report.
4. Map the distribution of the receptors for the primary taste sensations on your partner's tongue. To do this, follow these steps:
   a. Ask your partner to rinse his/her mouth with water and then partially dry the surface of the tongue with a paper towel.
   b. Moisten a clean cotton swab with 5% sucrose solution, and touch several regions of your partner's tongue with the swab.
   c. Each time you touch the tongue, ask your partner to report what sensation is experienced.
   d. Test the tip, sides, and back of the tongue in this manner.

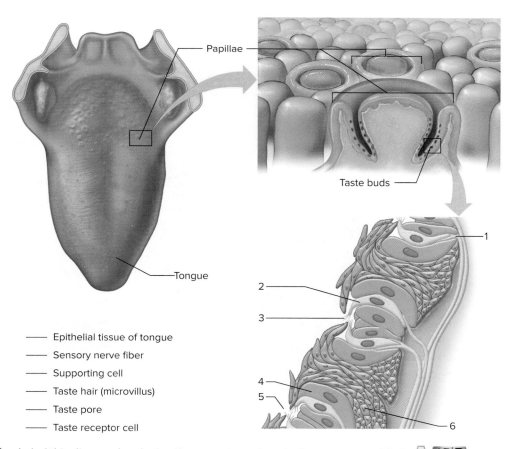

— Epithelial tissue of tongue
— Sensory nerve fiber
— Supporting cell
— Taste hair (microvillus)
— Taste pore
— Taste receptor cell

**FIGURE 31.3** Label this diagram by placing the correct numbers in the spaces provided. **A** **AP|R**

**e.** Test some other representative areas inside the mouth cavity such as the cheek, gums, and roof of the mouth (hard and soft palate) for a sweet sensation.

**f.** Record your partner's responses in Part D of the laboratory report.

**g.** Have your partner rinse his/her mouth and dry the tongue again, and repeat the preceding procedure, using each of the other four test solutions—NaCl, acetic acid, quinine or Epsom salt, and MSG solution. Be sure to use a fresh swab for each test substance and dispose of used swabs and paper towels as directed.

**5.** Complete Part D of the laboratory report.

 **LEARN: ACTIVITY**

Observe the oval-shaped taste bud in the demonstration microscope. Note the surrounding epithelial cells. The taste pore, an opening into the taste bud, may be filled with taste hairs (microvilli). Within the taste bud, there are supporting cells and thinner taste-receptor cells, which often have lightly stained nuclei (fig. 31.4).

 **LEARN: LAB IN MOTION**

Test your laboratory partner's ability to recognize the tastes of apple, potato, carrot, and onion. (A package of mixed flavors of LifeSavers is a good alternative.) To do this, follow these steps:

1. Have your partner close his/her eyes and hold the nostrils shut.
2. Place a small piece of one of the test substances on your partner's tongue.
3. Ask your partner to identify the substance without chewing or swallowing it.
4. Repeat the procedure for each of the other substances.

How do you explain the results of this experiment?

_____

_____

_____

_____

_____

_____

**FIGURE 31.4**    Micrograph of a taste bud (400×). Taste receptors are found in taste buds.  **AP|R** (©Ed Reschke)

Name _____

Date _____

Section _____

The  corresponds to the indicated Learning Outcome(s) found at the beginning of the laboratory exercise.

LABORATORY
REPORT

31

# Smell and Taste

## PART A ASSESSMENTS

Complete the following statements:

1. Olfactory, or smell, receptors are _____ neurons surrounded by columnar epithelial cells. **1**

2. The distal ends of the olfactory neurons are covered with hairlike _____. **1**

3. Before gaseous substances can stimulate the olfactory receptors, they must be dissolved in
_____ that surrounds the cilia. **1**

4. The axons of olfactory receptors pass through small openings in the _____ of the ethmoid bone. **2**

5. Olfactory bulbs lie on either side of the _____ of the ethmoid bone. **2**

6. The sensory impulses pass from the olfactory bulbs through the _____ to the interpreting centers of the brain. **2**

7. The olfactory interpreting centers are located deep within the temporal lobes and at the base of the
_____ lobes of the cerebrum. **2**

8. Olfactory sensations usually fade rapidly as a result of _____. **1**

9. Olfactory receptor neurons are the only parts of the nervous system that come in contact with the
_____ directly. **1**

10. A chemical would be considered _____ if a person lacked a particular receptor site on the cilia of the olfactory neurons. **1**

1. Record the results (as "+" if recognized; as "0" if unrecognized) from the tests of odor recognition in the following table: ◣3

| Substance Tested | Odor Reported | |
| --- | --- | --- |
| | First Trial | Second Trial |
| | | |
| | | |
| | | |
| | | |
| | | |
| | | |
| | | |
| | | |

2. Record the results of the olfactory sensory adaptation time in the following table: ◣3

| Substance Tested | Adaptation Time in Seconds |
| --- | --- |
| | |
| | |
| | |
| | |

3. Complete the following:

   **a.** How do you describe your partner's ability to recognize the odors of the substances you tested? /2\ _____

_____

_____

   **b.** Compare your experimental results with those of others in the class. Did you find any evidence to indicate that individuals may vary in their ability to recognize odors? Explain your answer. /2\ _____

_____

_____

## ASSESS

### CRITICAL THINKING

Does the time it takes for sensory adaptation to occur seem to vary with the substances tested? Explain your answer.

_____

_____

_____

## PART C ASSESSMENTS

Complete the following statements:

1. Taste receptor cells are modified _____ cells. /1\

2. The opening to a taste bud is called a(an) _____ . /1\

3. The _____ of a taste cell are its sensitive part. /1\

4. Before the taste of a substance can be detected, the substance must be dissolved in _____ . /1\

5. Substances that stimulate taste cells bind with _____ sites on the surfaces of taste hairs. /1\

6. Sour receptors are mainly stimulated by _____ . /1\

7. Salt receptors are mainly stimulated by ionized inorganic _____ . /1\

8. Alkaloids usually have a(an) _____ taste. /1\

1. *Taste receptor distribution.* Record a "+" to indicate where a taste sensation seemed to originate and a "0" if no sensation occurred when the spot was stimulated.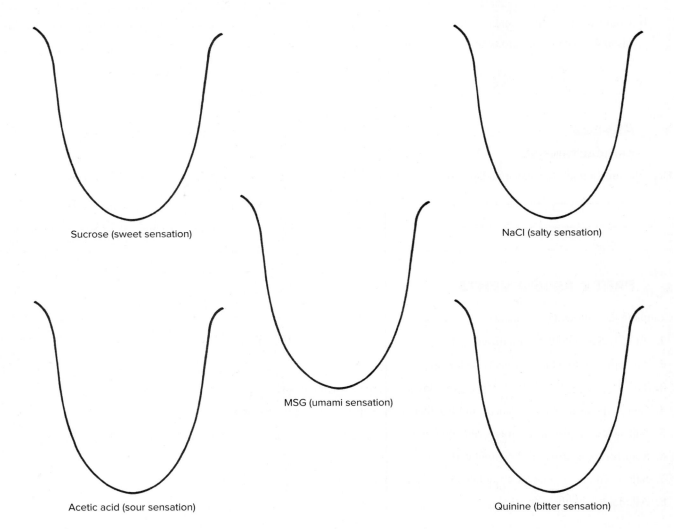

Sucrose (sweet sensation)

NaCl (salty sensation)

MSG (umami sensation)

Acetic acid (sour sensation)

Quinine (bitter sensation)

2. Complete the following:

   a. Describe how each type of taste receptor is distributed on the surface of your partner's tongue. _____

   _____

   _____

   b. Describe other locations inside the mouth where any sensations of sweet, salty, sour, bitter, or umami were located.

   _____

   _____

   c. How does your taste distribution map on the tongue compare to those of other students in the class? _____

   _____

   _____

# 32

# Ear and Hearing

## PURPOSE OF THE EXERCISE

To review the structural and functional characteristics of the ear and to conduct some ordinary hearing tests.

## LEARNING OUTCOMES AP|R

After completing this exercise, you should be able to

1. Locate the major structures of the ear.
2. Describe the functions of the structures of the ear.
3. Trace the pathway of sound vibrations from the tympanic membrane to the hearing receptors.
4. Conduct four ordinary hearing tests and summarize the results.

The ear is composed of outer (external), middle, and inner (internal) parts. The outer structures gather sound waves and direct them inward to the tympanic membrane. The parts of the middle ear, in turn, transmit vibrations from the tympanic membrane (eardrum) to the inner ear, where the hearing receptors are located. As they are stimulated, these receptors initiate impulses that are conducted over the vestibulocochlear nerve into the auditory cortex of the brain, where the impulses are interpreted and the sensations of hearing are created.

Although the ear is considered here as a hearing organ, it also has important functions for equilibrium (balance). The inner ear semicircular canals, utricle, and saccule have equilibrium functions. Equilibrium will be studied in Laboratory Exercise 33.

##  PRACTICE

### PROCEDURE A—Structure and Function of the Ear AP|R

1. Review the section entitled "Sense of Hearing" in chapter 12 of the textbook.
2. As a review activity, label figures 32.1 and 32.2, and study figure 32.3.
3. Examine the dissectible model of the ear and locate the following features:

   **outer (external) ear**
   auricle (pinna)
   external acoustic meatus
   tympanic membrane (eardrum)
   **middle ear**
   tympanic cavity
   auditory ossicles (fig. 32.4)
   malleus
   incus
   stapes
   oval window
   tensor tympani
   stapedius
   **auditory tube (pharyngotympanic tube; eustachian tube)**
   **inner (internal) ear**
   bony (osseous) labyrinth
   membranous labyrinth
   cochlea
   semicircular canals and ducts
   vestibule
   utricle
   saccule
   **vestibulocochlear nerve**
   vestibular nerve (balance branch)
   cochlear nerve (hearing branch)
4. Complete Parts A and B of Laboratory Report 32.

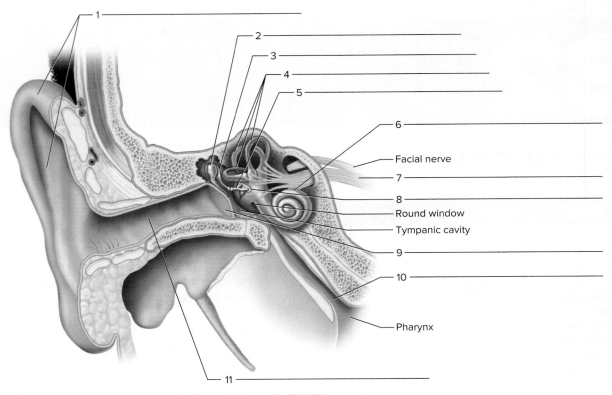

**FIGURE 32.1** Label the major structures of the ear.

Labels in figure:
1
2
3
4
5
6
Facial nerve
7
8
Round window
Tympanic cavity
9
10
Pharynx
11

## LEARN: ACTIVITY

Observe the section of the cochlea in the demonstration microscope. Locate one of the turns of the cochlea, and using figures 32.3 and 32.5 as a guide, identify the *scala vestibuli, cochlear duct (scala media), scala tympani, vestibular membrane, basilar membrane,* and *spiral organ.*

## PRACTICE

### PROCEDURE B—Hearing Tests

Perform the following tests in a quiet room, using your laboratory partner as the test subject.

1. *Auditory acuity test.* To conduct this test, follow these steps:
   a. Have the test subject sit with eyes closed.
   b. Pack one of the subject's ears with cotton.
   c. Hold a ticking watch close to the open ear, and slowly move it straight out and away from the ear.
   d. Have the subject indicate when the sound of the ticking can no longer be heard.
   e. Use a meterstick to measure the distance in centimeters from the ear to the position of the watch.
   f. Repeat this procedure to test the acuity of the other ear.
   g. Record the test results in Part C of the laboratory report.

2. *Sound localization test.* To conduct this test, follow these steps:
   a. Have the subject sit with eyes closed.
   b. Hold the ticking watch somewhere within the audible range of the subject's ears, and ask the subject to point to the watch.
   c. Move the watch to another position and repeat the request. In this manner, determine how accurately the subject can locate the watch when it is in each of the following positions: in front of the head, behind the head, above the head, on the right side of the head, and on the left side of the head.
   d. Record the test results in Part C of the laboratory report.

3. *Rinne test.* This test is done to assess possible conduction deafness by comparing bone and air conduction. To conduct this test, follow these steps:
   a. Obtain a tuning fork and strike it with a rubber hammer or on the heel of your hand, causing it to vibrate.
   b. Place the end of the fork's handle against the subject's mastoid process behind one ear. Have the prongs of the fork pointed downward and away from the ear, and be sure nothing is touching them (fig. 32.6a). The sound sensation is that of bone conduction. If no sound is experienced, nerve deafness exists.
   c. Ask the subject to indicate when the sound is no longer heard.

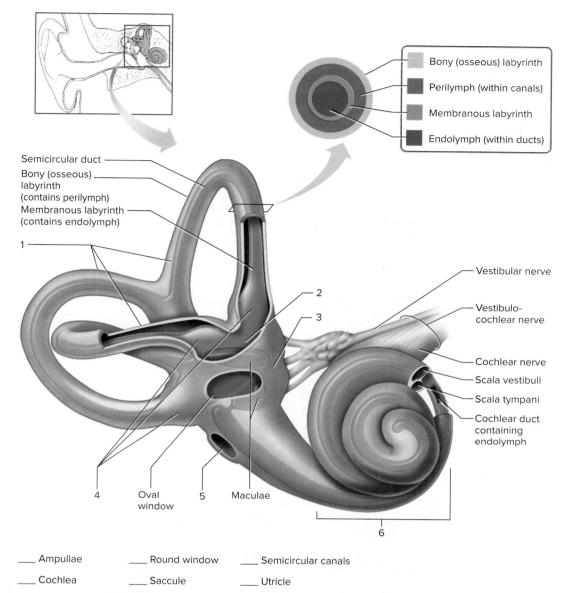

**Bony (osseous) labyrinth**

**Perilymph (within canals)**

**Membranous labyrinth**

**Endolymph (within ducts)**

Semicircular duct

Bony (osseous) labyrinth (contains perilymph)

Membranous labyrinth (contains endolymph)

1

Vestibular nerve

Vestibulo-cochlear nerve

2

3

Cochlear nerve

Scala vestibuli

Scala tympani

Cochlear duct containing endolymph

4          Oval window          5          Maculae

6

___ Ampullae          ___ Round window          ___ Semicircular canals

___ Cochlea          ___ Saccule          ___ Utricle

**FIGURE 32.2**    Label the structures of the inner ear by placing the correct numbers in the spaces provided.

**d.** Then quickly remove the fork from the mastoid process and position it in the air close to the opening of the nearby external acoustic meatus (fig. 32.6*b*).

If hearing is normal, the sound (from air conduction) will be heard again; if there is conductive impairment, the sound will not be heard. Conductive impairment involves outer or middle ear defects. Hearing aids can improve hearing for conductive deafness because bone conduction transmits the sound into the inner ear. Surgery could possibly correct this type of defect.

**e.** Record the test results in Part C of the laboratory report.

**4.** *Weber test.* This test is used to distinguish possible conduction or sensory deafness. To conduct this test, follow these steps:

**a.** Strike the tuning fork with the rubber hammer.

**b.** Place the handle of the fork against the subject's forehead in the midline (fig. 32.7).

**c.** Ask the subject to indicate if the sound is louder in one ear than in the other or if it is equally loud in both ears.

If hearing is normal, the sound will be equally loud in both ears. If there is conductive impairment, the sound will appear louder in the affected ear. If some degree of sensory (nerve) deafness exists, the sound will be louder in the normal ear. The impairment involves the spiral organ or the cochlear nerve. Hearing aids will not improve sensory deafness.

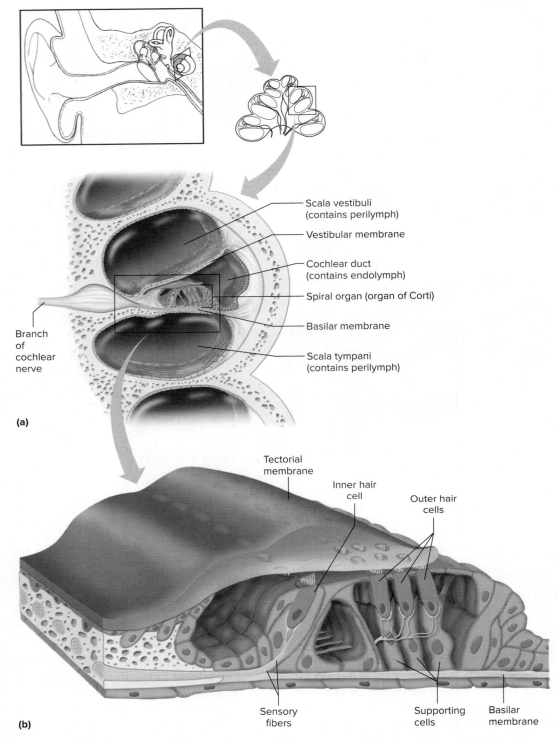

**(a)**

Scala vestibuli
(contains perilymph)

Vestibular membrane

Cochlear duct
(contains endolymph)

Spiral organ (organ of Corti)

Basilar membrane

Scala tympani
(contains perilymph)

Branch
of
cochlear
nerve

**(b)**

Tectorial
membrane

Inner hair
cell

Outer hair
cells

Sensory
fibers

Supporting
cells

Basilar
membrane

**FIGURE 32.3** Cochlea. (*a*) Cross section of the cochlea. (*b*) Spiral organ and the tectorial membrane.

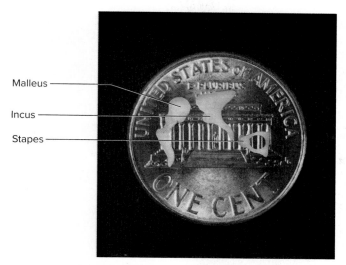

Malleus

Incus

Stapes

**FIGURE 32.4**   Middle ear bones (auditory ossicles) superimposed on a penny to show their relative size. The arrangement of the malleus, incus, and stapes in the enlarged photograph resembles the articulations in the middle ear.  (©J & J Photography)

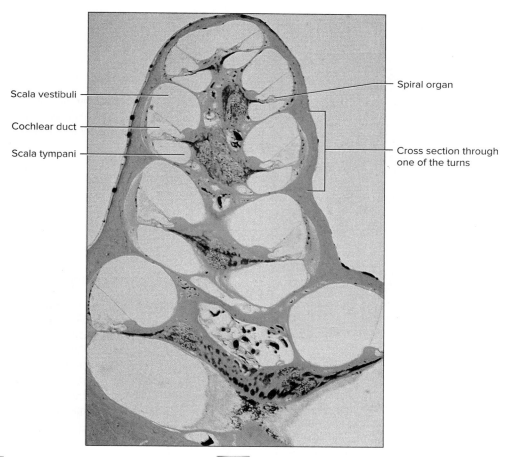

Scala vestibuli

Cochlear duct

Scala tympani

Spiral organ

Cross section through one of the turns

**FIGURE 32.5**   A section through the cochlea (22×).  **AP|R**  (©Biophoto Associates/Science Source)

(a)

(b)

**FIGURE 32.6** Rinne test: (*a*) first placement of vibrating tuning fork until sound is no longer heard; (*b*) second placement of tuning fork to assess air conduction. ((*a,b*) ©J & J Photography)

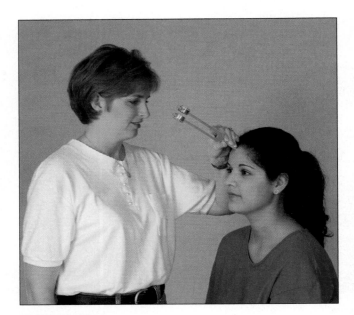

**FIGURE 32.7** Weber test. (©J & J Photography)

d. Have the subject experience the effects of conductive impairment by packing one ear with cotton and repeating the Weber test. Usually, the sound appears louder in the plugged (impaired) ear because extraneous sounds from the room are blocked out.

e. Record the test results in Part C of the laboratory report.

5. Complete Part C of the laboratory report.

 **ASSESS**

**CRITICAL THINKING**

Ear structures from the outer ear into the inner ear are progressively smaller. Using results obtained from the hearing tests, explain the advantage of this arrangement. 3

_____

_____

_____

_____

 **LEARN: ACTIVITY**

Ask the laboratory instructor to demonstrate the use of the audiometer. This instrument produces sound vibrations of known frequencies transmitted to one or both ears of a test subject through earphones. The audiometer can be used to determine the threshold of hearing for different sound frequencies, and, in the case of hearing impairment, it can be used to determine the percentage of hearing loss for each frequency.

Name _____

Date _____

Section _____

The  corresponds to the indicated Learning Outcome(s) found at the beginning of the laboratory exercise.

# Ear and Hearing

 **PART A ASSESSMENTS**

Match the terms in column A with the descriptions in column B. Place the letter of your choice in the space provided. **A₁ A₂**

### Column A

a. Auditory tube
b. Bony (osseous) labyrinth
c. Ceruminous gland
d. External acoustic meatus
e. Malleus
f. Membranous labyrinth
g. Scala tympani
h. Scala vestibuli
i. Stapes
j. Tectorial membrane
k. Tympanic cavity
l. Tympanic membrane (eardrum)

### Column B

_____ 1. Auditory ossicle attached to tympanic membrane

_____ 2. Air-filled space containing auditory ossicles within middle ear

_____ 3. Contacts hairs of hearing receptors

_____ 4. Leads from oval window to apex of cochlea and contains perilymph

_____ 5. S-shaped tube leading to tympanic membrane

_____ 6. Wax-secreting structure

_____ 7. Cone-shaped, semitransparent membrane attached to malleus

_____ 8. Auditory ossicle attached to oval window

_____ 9. Tube containing endolymph within bony (osseous) labyrinth

_____ 10. Bony canal of inner ear in temporal bone

_____ 11. Connects middle ear and pharynx

_____ 12. Extends from apex of cochlea to round window and contains perilymph

**PART B ASSESSMENTS**

Label the structures indicated in the micrograph of the spiral organ in figure 32.8. **A₁**

**Terms:**
Basilar membrane
Cochlear duct
Hair cells
Scala tympani
Tectorial membrane

**FIGURE 32.8** Label the structures associated with this spiral organ region of a cochlea, using the terms provided (400×). **AP|R** (©Biophoto Associates/Science Source)

1. Results of auditory acuity test: ⒜

| Ear Tested | Audible Distance (cm) |
|---|---|
| Right | |
| Left | |

2. Results of sound localization test: ⒜

| Actual Location | Reported Location |
|---|---|
| Front of the head | |
| Behind the head | |
| Above the head | |
| Right side of the head | |
| Left side of the head | |

3. Results of experiments using tuning forks: ⒜

| Test | Left Ear (normal or impaired) | Right Ear (normal or impaired) |
|---|---|---|
| Rinne | | |
| Weber | | |

4. Summarize the results of the hearing tests you conducted on your laboratory partner. ⒜

_____

_____

_____

_____

# Ear and Equilibrium

## MATERIALS NEEDED

Textbook
Swivel chair
Bright light
Compound light microscope
Prepared microscope slide of semicircular duct
(cross section through ampulla)

## ⚠ SAFETY

- Do not pick subjects who have frequent motion sickness.
- Have four people surround the subject in the swivel chair, in case the person falls from vertigo or loss of balance.
- Stop your experiment if the subject becomes nauseated.

## PURPOSE OF THE EXERCISE

To review the structure and function of the organs of equilibrium and to conduct some tests of equilibrium.

##  LEARNING OUTCOMES

After completing this exercise, you should be able to

(1) Locate the organs of static and dynamic equilibrium and describe their functions.

(2) Explain how vision contributes in the maintenance of equilibrium.

(3) Conduct and record the results of the Romberg and Bárány tests of equilibrium.

he sense of equilibrium involves two sets of sensory organs. One set helps to maintain the stability of the head and body when they are motionless or during linear acceleration and produces a sense of static (gravitational) equilibrium. The other set is concerned with balancing the head and body

when they are moved suddenly or during angular acceleration and produces a sense of dynamic (rotational) equilibrium.

The organs associated with the sense of static equilibrium are located within the utricle and saccule of the vestibules of the inner ears. Those associated with the sense of dynamic equilibrium are found within the ampullae of the three semicircular ducts of the inner ear.

##  PRACTICE

### PROCEDURE A—Organs of Equilibrium

1. Review the section entitled "Sense of Equilibrium" in chapter 12 of the textbook.
2. Complete Part A of Laboratory Report 33.

##  LEARN: ACTIVITY

Observe the cross section of the semicircular duct through the ampulla in the demonstration microscope. Note the crista ampullaris (figs. 33.1 and 33.2) projecting into the lumen of the membranous labyrinth, which in a living person is filled with endolymph. The space between the membranous and osseous (bony) labyrinths is filled with perilymph.

##  PRACTICE

### PROCEDURE B—Tests of Equilibrium

Perform the following tests, using a person as a test subject who is not easily disturbed by dizziness or rotational movement. Also have some other students standing close by to help prevent the test subject from falling during the tests. *The tests should be stopped immediately if the test subject begins to feel uncomfortable or nauseated.*

1. *Vision and equilibrium test.* To demonstrate the importance of vision in the maintenance of equilibrium, follow these steps:
   a. Have the test subject stand erect on one foot for 1 minute with eyes open.

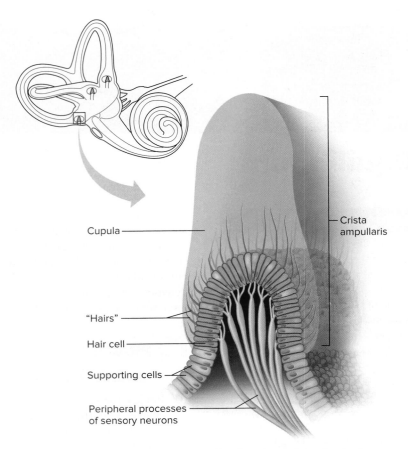

Cupula

Crista ampullaris

"Hairs"

Hair cell

Supporting cells

Peripheral processes of sensory neurons

**FIGURE 33.1**   The crista ampullaris is located within the ampulla of each semicircular duct.

Ampulla

Cupula

Hair

Hair cell

Region of sensory nerve fibers

**FIGURE 33.2**   A micrograph of a crista ampullaris (400×). (©Biophoto Associates/Science Source)

   **b.** Observe the subject's degree of unsteadiness.
   **c.** Repeat the procedure with the subject's eyes closed. *Be prepared to prevent the subject from falling.*
   **d.** Answer the questions related to the vision and equilibrium test in Part B of the laboratory report.
**2.** *Romberg test.* The purpose of this test is to evaluate how the organs of static equilibrium in the vestibule enable one to maintain balance. To conduct this test, follow these steps:
   **a.** Position the test subject close to a chalkboard with the back toward the board.
   **b.** Place a bright light in front of the subject so that a shadow of the body is cast on the board.
   **c.** Have the subject stand erect with feet close together and eyes staring straight ahead for a period of 3 minutes.
   **d.** During the test, make marks on the chalkboard along the edge of the shadow of the subject's shoulders to indicate the range of side-to-side swaying.
   **e.** Measure the maximum sway in centimeters and record the results in Part B of the laboratory report.
   **f.** Repeat the procedure with the subject's eyes closed.
   **g.** Position the subject so one side is toward the chalkboard.
   **h.** Repeat the procedure with the eyes open.
   **i.** Repeat the procedure with the eyes closed.

The Romberg test is used to evaluate a person's ability to integrate sensory information from proprioceptors and receptors within the organs of static equilibrium and to relay appropriate motor impulses to postural muscles. A person who shows little unsteadiness when standing with feet together and eyes open, but who becomes unsteady when the eyes are closed, has a positive Romberg test.

3. *Bárány test.* The purpose of this test is to evaluate the effects of rotational acceleration on the semicircular ducts and dynamic equilibrium. To conduct this test, follow these steps:

   **a.** Have the test subject sit on a swivel chair with eyes open and focused on a distant object, the head tilted forward about 30° and the hands gripped firmly to the seat. Position four people around the chair for safety. *Be prepared to prevent the subject and the chair from tipping over.*

   **b.** Rotate the chair ten rotations within 20 seconds.

   **c.** Abruptly stop the movement of the chair. The subject will still have the sensation of continuous movement and might experience some dizziness (vertigo).

   **d.** Have the subject look forward and immediately note the nature of the eye movements and their direction. (Such reflex eye twitching movements are called *nystagmus.*) Also note the time it takes for the nystagmus to cease. Nystagmus will continue until the cupula is returned to an original position.

   **e.** Record your observations in Part B of the laboratory report.

   **f.** Allow the subject several minutes of rest; then repeat the procedure with the subject's head tilted nearly 90° onto one shoulder.

   **g.** After another rest period, repeat the procedure with the subject's head bent forward so that the chin is resting on the chest.

   In this test, when the head is tilted about 30°, the lateral semicircular ducts receive maximal stimulation, and the nystagmus is normally from side to side. When the head is tilted at 90°, the superior ducts are stimulated, and the nystagmus is up and down. When the head is bent forward with the chin on the chest, the posterior ducts are stimulated, and the nystagmus is rotary. A membranous semicircular duct occupies the space within the osseous (bony) semicircular canal within a temporal bone. Observe a semicircular canal located in a temporal bone (fig. 33.3).

4. Complete Part B of the laboratory report.

Border of segment removed from temporal bone

Temporal bone (sectioned)

Semicircular canal within bony labyrinth of inner ear

**FIGURE 33.3**   Semicircular canal superimposed on a penny to show its relative size. A semicircular duct of the same shape would occupy the semicircular canal within the temporal bone. (©J & J Photography)

# Notes

The Ⓐ corresponds to the indicated Learning Outcome(s) found at the beginning of the laboratory exercise.

# Ear and Equilibrium

## ♻ PART A ASSESSMENTS

Complete the following statements:

1. The organs of static equilibrium are located within two expanded chambers within the vestibule called the _____ and the saccule. Ⓐ

2. All of the balance organs are located within the _____ bone of the skull. Ⓐ

3. The receptor cells of these organs are found in the wall of the membranous labyrinth in a structure called the _____ . Ⓐ

4. Otoliths are small grains composed of _____ . Ⓐ

5. Sensory impulses travel from the organs of equilibrium to the brain on the _____ nerve. Ⓐ

6. The sensory organ of a semicircular duct lies within a swelling of the duct called the _____ . Ⓐ

7. The sensory organ within the ampulla of a semicircular duct is called a(an) _____ . Ⓐ

8. The _____ of this sensory organ consists of a dome-shaped, gelatinous mass. Ⓐ

9. When the head is moved, the fluid inside the membranous portion of a semicircular duct tends to remain stationary because of the _____ of the fluid. Ⓐ

10. Parts of the _____ of the brain interpret impulses from the equilibrium receptors. Ⓐ

## ♻ PART B ASSESSMENTS

1. Vision and equilibrium test results:
   a. When the eyes are open, what sensory organs provide information needed to maintain equilibrium? ②

   _____

   _____

   b. When the eyes are closed, what sensory organs provide such information? ② _____

   _____

   _____

**2.** Romberg test results:

**a.** Record the test results in the following table: $\boxed{3}$

| Conditions | Maximal Movement (cm) |
|---|---|
| Back toward board, eyes open | |
| Back toward board, eyes closed | |
| Side toward board, eyes open | |
| Side toward board, eyes closed | |

**b.** Did the test subject's unsteadiness increase when the eyes were closed? _____ What is the significance of this observation? $\boxed{2}$ _____

_____

_____

**c.** Why would you expect a person with impairment of the organs of equilibrium to become more unsteady when the eyes are closed? $\boxed{2}$ _____

_____

_____

**3.** Bárány test results:

**a.** Record the test results in the following table: $\boxed{3}$

| Position of Head | Description of Eye Movements | Time for Movement to Cease |
|---|---|---|
| Tilted 30 degrees forward | | |
| Tilted 90 degrees onto shoulder | | |
| Tilted forward, chin on chest | | |

**b.** Summarize the results of this test. $\boxed{3}$ _____

_____

_____

## ASSESS

### CRITICAL THINKING

What additional sensory information would you expect persons with impairment of organs of equilibrium to use to supplement their relative lack of some sensory information?

_____

_____

_____

# 34

# Eye Structure

## MATERIALS NEEDED

Textbook
Dissectible eye model
Compound light microscope
Prepared microscope slide of a mammalian eye
(sagittal section)
Sheep or beef eye (fresh or preserved)
Dissecting tray
Dissecting instruments—forceps, sharp scissors, and
dissecting needle
Ophthalmoscope

## ⚠ SAFETY

- Do not allow light to reach the macula lutea region during the eye exam for longer than 1 second at a time.
- Wear disposable gloves and safety goggles when working on the eye dissection.
- Dispose of tissue remnants and gloves as instructed.
- Wash the dissecting tray and instruments as instructed.
- Wash your laboratory table.
- Wash your hands before leaving the laboratory.

## PURPOSE OF THE EXERCISE

To review the structure and function of the eye and to dissect a mammalian eye.

##  LEARNING OUTCOMES AP|R

After completing this exercise, you should be able to

1. Locate the major structures of an eye.
2. Describe the functions of the structures of an eye.
3. Trace the structures through which light passes as it travels from the cornea to the retina.
4. Dissect a mammalian eye and locate its major features.
5. Recognize the anatomical terms pertaining to the eye structure.

The eye contains photoreceptors, modified neurons located on its inner wall. Other parts of the eye provide protective functions or make it possible to move the eyeball. Still other structures focus light entering the eye so that a sharp image is projected onto the receptor cells. The impulses generated when the receptors are stimulated are conducted along the optic nerves to the brain, which interprets the impulses and creates the sensation of sight.

##  PRACTICE

**PROCEDURE A—Structure and Function of the Eye** AP|R

1. Review the sections entitled "Visual Accessory Organs," "Structure of the Eye," and "Photoreceptors" in chapter 12 of the textbook.
2. As a review activity, label figures 34.1–34.3.
3. Complete Part A of Laboratory Report 34.
4. Examine the dissectible model of the eye and locate the following features:

**eyelid**
**conjunctiva**
**orbicularis oculi**
**levator palpebrae superioris**
**lacrimal apparatus**
    lacrimal gland
    canaliculi
    lacrimal sac
    nasolacrimal duct
**extrinsic muscles**
    superior rectus
    inferior rectus
    medial rectus
    lateral rectus
    superior oblique
    inferior oblique
**trochlea (pulley)**
**outer (fibrous) tunic (layer)**
    sclera
    cornea

Nasal cavity

Inferior nasal concha

**FIGURE 34.1** Label the structures of the lacrimal apparatus. 🅰

Levator palpebrae superioris

Pulley (trochlea)

**FIGURE 34.2** Label the extrinsic muscles of the right eye (lateral view). 🅰 **AP** **R**

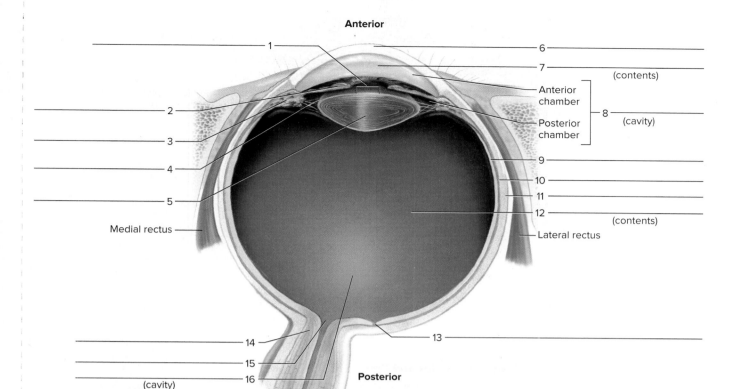

**Anterior**

1
2
3
4
5
Medial rectus

6
7
(contents)
Anterior chamber
Posterior chamber
8 (cavity)
9
10
11
12
(contents)
Lateral rectus
13

14
15
16
(cavity)

**Posterior**

**FIGURE 34.3**  Label the structures indicated in this transverse section of the right eye (superior view).  Ⓐ **AP|R**

**middle (vascular) tunic (layer)**
    choroid coat
    ciliary body
        ciliary processes
        ciliary muscle
        suspensory ligaments
    iris
    pupil (opening in center of iris)
**inner (nervous) tunic (layer)**
    retina
        macula lutea
        fovea centralis
        optic disc
    optic nerve
**lens**
**anterior cavity**
    anterior chamber
    posterior chamber
    aqueous humor
**posterior cavity**
    vitreous humor

5. Reexamine the labeled figures 34.1–34.3 and study the functions and nerve innervations of the muscles of the eyelids and eyes in table 34.1.
6. Obtain a microscope slide of a mammalian eye section, and locate as many of the features in step 4 as possible.
7. Using low-power magnification, observe the posterior portion of the eye wall, and locate the choroid coat and retina (fig 34.4).
8. Using high-power magnification, examine the retina portion of the eye wall in more detail, and note its layered structure (fig. 34.4). Locate the following:

**layer of nerve fiber** (innermost layer of the retina)
**layer of ganglion cells**
**layer of bipolar neurons**
**layer of photoreceptors (rods and cones)**
**pigmented epithelium** (outermost layer of the retina)

Choroid coat

Retinal pigment epithelium

Photoreceptors (rods and cones)

Bipolar neurons

Ganglion cell layer

Nerve fiber layer

Light

**FIGURE 34.4** The cells of the retina are arranged in distinct layers (75x). **AP|R** (©Ed Reschke)

**TABLE 34.1 Muscles Associated with the Eyelids and Eyes**

| Skeletal Muscles | Innervation | Function |
|---|---|---|
| Muscles of the eyelids | | |
| Orbicularis oculi | Facial nerve (VII) | Closes eye |
| Levator palpebrae superioris | Oculomotor nerve (III) | Opens eye |
| Extrinsic muscles of the eyes | | |
| Superior rectus | Oculomotor nerve (III) | Rotates eye upward and toward midline |
| Inferior rectus | Oculomotor nerve (III) | Rotates eye downward and toward midline |
| Medial rectus | Oculomotor nerve (III) | Rotates eye toward midline |
| Lateral rectus | Abducens nerve (VI) | Rotates eye away from midline |
| Superior oblique | Trochlear nerve (IV) | Rotates eye downward and away from midline |
| Inferior oblique | Oculomotor nerve (III) | Rotates eye upward and away from midline |
| **Smooth Muscles** | **Innervation** | **Function** |
| Ciliary muscle | Oculomotor nerve (III) parasympathetic fibers | Relax suspensory ligaments |
| Iris, circular muscle set | Oculomotor nerve (III) parasympathetic fibers | Constrict pupil |
| Iris, radial muscle set | Sympathetic fibers | Dilate pupil |

## LEARN: LAB IN MOTION

Use an ophthalmoscope to examine the interior of your laboratory partner's eye. This instrument consists of a set of lenses held in a rotating disc, a light source, and some mirrors that reflect the light into the test subject's eye (fig. 34.5).

The examination should be conducted in a dimly lit room. Have your partner seated and staring straight ahead at eye level. Move the rotating disc of the ophthalmoscope so that the O appears in the lens selection window. Hold the instrument in your right hand with the end of your index finger on the rotating disc. Direct the light at a slight angle from a distance of about 15 cm into the subject's right eye (fig. 34.6). The light beam should pass along the inner edge of the pupil. Look through the instrument, and you should see a reddish, circular area—the interior of the eye. Rotate the disc of lenses to higher values until sharp focus is achieved.

Move the ophthalmoscope to within about 5 cm of the eye being examined, *being very careful that the instrument does not touch the eye,* and again rotate the lenses to sharpen the focus (fig. 34.6). Locate the optic disc and the blood vessels that pass through it. Also locate the macula lutea by having your partner stare directly into the light of the instrument (fig. 34.7). *Do not allow light to reach the macula lutea region for longer than 1 second at a time.*

Examine the subject's iris by viewing it from the side and by using a lens with a +15 or +20 value.

Eyepiece

Rotating disc

Lens selection window

Rheostat control for light intensity

Handle of ophthalmoscope

**FIGURE 34.5** The ophthalmoscope is used to examine the interior of the eye. (©J & J Photography)

**FIGURE 34.6** Rotate the disc of lenses until sharp focus is achieved and then move the ophthalmoscope to within 5 cm of the eye to examine the optic disc of the right eye.

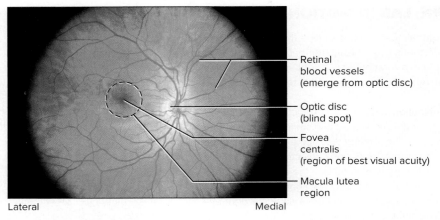

Retinal
blood vessels
(emerge from optic disc)

Optic disc
(blind spot)

Fovea
centralis
(region of best visual acuity)

Macula lutea
region

Lateral                                    Medial

**FIGURE 34.7**   Retina of the left eye as viewed through the pupil using an ophthalmoscope. Only small portions of the retina are able to be viewed at a time.  AP|R (©Science Source)

 **PRACTICE**

**PROCEDURE B—Eye Dissection**

1. Obtain a mammalian eye, place it in a dissecting tray, and dissect it as follows:
   a. Trim away the fat and other connective tissues but leave the stubs of the *extrinsic muscles* and the *optic nerve*. This nerve projects outward from the posterior region of the eyeball.
   b. The *conjunctiva* lines the inside of the eyelid and is reflected over the anterior surface of the eye, except for the cornea. Lift some of this thin membrane away from the eye with forceps and examine it.

   c. Locate and observe the *cornea, sclera,* and *iris.* Also note the *pupil* and its shape. The cornea from a fresh eye is transparent; when preserved, it becomes opaque.
   d. Use sharp scissors to make a frontal section of the eye. To do this, cut through the wall about 1 cm from the margin of the cornea and continue all the way around the eyeball. Try not to damage the internal structures of the eye (fig. 34.8).
   e. Gently separate the eyeball into anterior and posterior portions. Usually, the jellylike vitreous humor will remain in the posterior portion, and the lens may adhere to it. Place the parts in the dissecting tray with their contents facing upward.

Margin of cornea

**FIGURE 34.8**   Prepare a frontal section of the eye.

**f.** Examine the anterior portion of the eye, and locate the *ciliary body,* which appears as a dark, circular structure. Also note the *iris* and the *lens* if it remained in the anterior portion. The lens is normally attached to the ciliary body by many *suspensory ligaments,* which appear as delicate, transparent threads that detach as a lens is removed.

**g.** Use a dissecting needle to gently remove the lens, and examine it. If the lens is still transparent, hold it up and look through it at something in the distance and note that the lens inverts the image. The lens of a preserved eye is usually too opaque for this experience. If the lens of the human eye becomes opaque, the defect is called a cataract (fig. 34.9).

**h.** Examine the posterior portion of the eye. Note the *vitreous humor.* This jellylike mass helps to hold the lens in place anteriorly and to hold the *retina* against the choroid coat.

**i.** Carefully remove the vitreous humor and examine the retina. This layer will appear as a thin, nearly colorless to cream-colored membrane that detaches easily from the choroid coat. Compare the structures identified to figure 34.10.

**j.** Locate the *optic disc*—the point where the retina is attached to the posterior wall of the eyeball and where the optic nerve originates. There are no receptor cells in the optic disc, so this region is also called the "blind spot."

**k.** Note the iridescent area of the choroid coat beneath the retina. This colored surface in ungulates (mammals having hoofs) is called the *tapetum fibrosum.* It reflects light back through the retina, an action thought to aid the night vision of some animals. The tapetum fibrosum is lacking in the human eye.

**2.** Discard the tissues of the eye as directed by the laboratory instructor.

**3.** Complete Parts B and C of the laboratory report.

 **ASSESS**

### CRITICAL THINKING

A strong blow to the head might cause the retina to detach. From observations made during the eye dissection, explain why this can happen.

_____

_____

_____

_____

_____

**FIGURE 34.9** Human lens showing a cataract and one type of lens implant (intraocular lens replacement) on tips of fingers to show their relative size. A normal lens is transparent. (©J & J Photography)

Choroid coat (reflected)

Sclera

Retina (partly detached)

Tapetum fibrosum (iridescent)

Cornea
(posterior surface)

Pupil (opening in
center of iris)

Iris

Ciliary body

Posterior
portion

Anterior
portion

Vitreous humor (transparent)

Lens

**FIGURE 34.10** Internal structures of the preserved beef eye after dissection. (©J & J Photography)

The Ⓐ corresponds to the indicated Learning Outcome(s) found at the beginning of the laboratory exercise.

# Eye Structure

## 🔄 PART A ASSESSMENTS

Match the terms in column A with the descriptions in column B. Place the letter of your choice in the space provided. Ⓐ1 Ⓐ2

| Column A | Column B |
|---|---|
| **a.** Aqueous humor | _____ **1.** Posterior five-sixths of middle (vascular) tunic |
| **b.** Choroid coat | _____ **2.** White part of outer (fibrous) tunic |
| **c.** Ciliary muscle | _____ **3.** Transparent anterior portion of outer tunic |
| **d.** Conjunctiva | _____ **4.** Inner lining of eyelid |
| **e.** Cornea | _____ **5.** Secretes tears |
| **f.** Iris | _____ **6.** Fills posterior cavity of eye |
| **g.** Lacrimal gland | _____ **7.** Area where optic nerve exits the eye |
| **h.** Optic disc | _____ **8.** Smooth muscle that controls the pupil size and light entering the eye |
| **i.** Retina | _____ **9.** Fills anterior and posterior chambers of the anterior cavity of the eye |
| **j.** Sclera | |
| **k.** Suspensory ligament | _____ **10.** Contains photoreceptor cells called rods and cones |
| **l.** Vitreous humor | _____ **11.** Connects lens to ciliary body |
| | _____ **12.** Cause lens to change shape |

Complete the following:

**13.** List the structures and fluids through which light passes as it travels from the cornea to the retina. Ⓐ3 _____

_____

_____

_____

**14.** List three ways in which rods and cones differ in structure or function. Ⓐ2 _____

_____

_____

_____

Complete the following:

1. Which tunic/layer of the eye was the most difficult to cut?  _____

2. What kind of tissue do you think is responsible for this quality of toughness?  _____

3. How do you compare the shape of the pupil in the dissected eye with your pupil?  _____

_____

4. Where was the aqueous humor in the dissected eye?  _____

_____

5. What is the function of the dark pigment in the choroid coat? **2** _____

_____

6. Describe the lens of the dissected eye.  _____

_____

7. Describe the vitreous humor of the dissected eye.  _____

_____

**PART C ASSESSMENTS**

Locate the 23 anatomical terms pertaining to the "Eye Structure." **5**

| W | C | O | N | E | S | E | O | F | P | R | K |
|---|---|---|---|---|---|---|---|---|---|---|---|
| R | O | M | U | H | S | G | L | U | O | V | M |
| I | N | N | E | R | N | X | P | D | T | E | Z |
| L | J | F | Q | U | E | I | S | P | D | E | K |
| A | U | D | G | W | L | B | U | I | P | I | R |
| T | N | I | F | O | V | E | A | R | A | N | M |
| E | C | O | R | W | V | L | K | I | N | F | S |
| R | T | R | E | V | R | E | N | S | I | E | G |
| A | I | O | B | L | I | Q | U | E | T | R | D |
| L | V | H | W | G | L | A | N | D | E | I | G |
| M | A | C | U | L | A | A | E | N | R | O | C |
| D | I | L | E | Y | E | S | C | L | E | R | A |

Find and circle words pertaining to Laboratory Exercise 34. The words may be horizontal, vertical, diagonal, or backward.

8 terms (conjunctiva, eyelid, fovea, gland, humor, macula, nerve, oblique)

3 main eye layers; 3 rectus eye muscles; 9 eyeball structures.

# 35

# Visual Tests and Demonstrations

## MATERIALS NEEDED

Snellen eye chart
3" × 5" card (plain)
3" × 5" card with word typed in center
Astigmatism chart
Meterstick
Metric ruler
Pen flashlight
Ichikawa's or Ishihara's color plates for colorblindness test

## PURPOSE OF THE EXERCISE

To conduct and interpret the tests for visual acuity, astigmatism, accommodation, color vision, the blind spot, and reflexes of the eye.

##  LEARNING OUTCOMES

After completing this exercise, you should be able to

1. Conduct and interpret the tests used to evaluate visual acuity, astigmatism, the ability to accommodate for close vision, and color vision.

2. Describe four conditions that can lead to defective vision.

3. Demonstrate and describe the blind spot, photopupillary reflex, accommodation pupillary reflex, and convergence reflex.

Normal vision (emmetropia) results when light rays from objects in the external environment are refracted by the cornea and lens of the eye and focused onto the photoreceptors of the retina. If a person has *nearsightedness (myopia),* the image focuses in front of the retina because the eyeball is too long; nearby objects are clear, but distant objects are blurred. If a person has *farsightedness (hyperopia),* the image focuses behind the retina because the eyeball is too short; distant objects are clear, but nearby objects are blurred. A concave lens is required to correct vision for nearsightedness; a convex lens is required to correct vision for farsightedness.

If a defect occurs from unequal curvatures of the cornea or lens, a condition called *astigmatism* results. Focusing in different planes cannot occur simultaneously. Corrective cylindrical lenses allow proper refraction of light onto the retina to compensate for the unequal curvatures.

As part of a natural aging process, the lens's elasticity decreases, which degrades our ability to focus on nearby objects. The near point of accommodation test is used to determine the ability to accommodate. Often a person will use "reading glasses" to compensate for this condition. Hereditary defects in color vision result from a lack of certain cones needed to absorb certain wavelengths of light. Special color test plates are used to diagnose colorblindness.

As you conduct this laboratory exercise, it does not matter in what order the visual tests or visual demonstrations are performed. The tests and demonstrations do not depend upon the results of any of the others. If you wear glasses, perform the tests with and without the corrective lenses. If you wear contact lenses, it is not necessary to run the tests under both conditions; however, indicate in the laboratory report that all tests were performed with contact lenses in place.

##  PRACTICE

### PROCEDURE A—Visual Tests

Perform the following visual tests using your laboratory partner as a test subject. If your partner usually wears glasses, test each eye with and without the glasses.

1. *Visual acuity test.* Visual acuity (sharpness of vision) can be measured by using a Snellen eye chart (fig. 35.1). This chart consists of several sets of letters in different sizes printed on a white card. The letters near the top of the chart are relatively large, and those in each lower set become smaller. At one end of each set of letters is an acuity value in the form of a fraction. One of the sets near the bottom of the chart, for example, is marked 20/20. The normal eye can clearly see these letters from the standard distance of 20 feet and thus is said to have 20/20 vision. The letter at the top of the chart is marked 20/200. The normal eye can read letters of this size from a distance of 200 feet. Thus, an eye able to read only the top letter of the chart from a distance

| | | |
|---|---|---|
| $\frac{20}{200}$ | **E** | 1 |
| $\frac{20}{100}$ | **F P** | 2 |
| $\frac{20}{70}$ | **T O Z** | 3 |
| $\frac{20}{50}$ | **L P E D** | 4 |
| $\frac{20}{40}$ | **P E C F D** | 5 |
| $\frac{20}{30}$ | **E D F C Z P** | 6 |
| $\frac{20}{25}$ | F E L O P Z D | 7 |
| $\frac{20}{20}$ | D E F P O T E C | 8 |
| $\frac{20}{15}$ | L E F O D P C T | 9 |
| $\frac{20}{13}$ | F D P L T C E O | 10 |
| $\frac{20}{10}$ | P E Z O L C F T D | 11 |

**FIGURE 35.1** The Snellen eye chart looks similar to this but is somewhat larger.

of 20 feet is said to have 20/200 vision. This person has less than normal vision. A line of letters near the bottom of the chart is marked 20/15. The normal eye can read letters of this size from a distance of 15 feet, but a person might be able to read it from 20 feet. This person has better than normal vision.

To conduct the visual acuity test, follow these steps:

a. Hang the Snellen eye chart on a well-illuminated wall at eye level.

b. Have your partner stand 20 feet in front of the chart, gently cover your partner's left eye with a 3" × 5" card, and ask your partner to read the smallest set of letters possible using the right eye.

c. Record the visual acuity value for that set of letters in Part A of Laboratory Report 35.

d. Repeat the procedure, using the left eye for the test.

2. *Astigmatism test.* Astigmatism is a condition that results from a defect in the curvature of the cornea or lens. As a consequence, some portions of the image projected on the retina are sharply focused, but other portions are blurred. Astigmatism can be evaluated by using an

**FIGURE 35.2** Astigmatism is evaluated using a chart such as this one.

astigmatism chart (fig. 35.2). This chart consists of sets of black lines radiating from a central spot like the spokes of a wheel. To a normal eye, these lines appear sharply focused and equally dark; however, if the eye has an astigmatism, some sets of lines appear sharply focused and dark, whereas others are blurred and less dark.

To conduct the astigmatism test, follow these steps:

a. Hang the astigmatism chart on a well-illuminated wall at eye level.

b. Have your partner stand 20 feet in front of the chart, and gently cover your partner's left eye with a 3" × 5" card. Ask your partner to focus on the spot in the center of the radiating lines using the right eye and report which lines, if any, appear more sharply focused and darker.

c. Repeat the procedure, using the left eye for the test.

d. Record the results in Part A of the laboratory report.

3. *Accommodation test.* Accommodation is the changing of the shape of the lens that occurs when the normal eye is focused for near (close) vision. It involves a reflex in which muscles of the ciliary body are stimulated to contract, releasing tension on the suspensory ligaments that are fastened to the lens capsule. This allows the capsule to rebound elastically, causing the surface of the lens to become more spherical (convex). The ability to accommodate is likely to decrease with age because the tissues involved tend to lose their elasticity. This common aging condition, called *presbyopia,* can be corrected with reading glasses or bifocal lenses.

To evaluate the ability to accommodate, follow these steps:

a. Hold the end of a meterstick against your partner's chin so that the stick extends outward at a right angle to the plane of the face (fig. 35.3).

**FIGURE 35.3** To determine the near point of accommodation, slide the 3" × 5" card along the meterstick toward your partner's open eye until it reaches the closest location where your partner can still see the word sharply focused. (©J & J Photography)

b. Have your partner close the left eye.

c. Hold a 3" × 5" card with a word typed in the center at the distal end of the meterstick.

d. Slide the card along the stick toward your partner's open right eye, and locate the *point closest to the eye* where your partner can still see the letters of the word sharply focused. This distance is called the *near point of accommodation,* and it tends to increase with age (table 35.1).

e. Repeat the procedure with the right eye closed.

f. Record the results in Part A of the laboratory report.

4. *Color vision test.* Some people exhibit defective color vision because they lack certain cones, usually those sensitive to the reds or greens. This trait is an X-linked (sex-linked) inheritance, so the condition is more prevalent in males (7%) than in females (0.4%). People who lack or possess decreased sensitivity to the red-sensitive cones possess protanopia colorblindness; those who lack or are less sensitive to green-sensitive cones possess deuteranopia colorblindness. The colorblindness condition is often more of a deficiency or a weakness than a condition of blindness. Laboratory Exercise 53 describes the genetics for colorblindness.

To conduct the color vision test, follow these steps:

a. Examine the color test plates in Ichikawa's or Ishihara's book to test for any red-green color vision deficiency. Also examine figure 35.4.

b. Hold the test plates approximately 30 inches from the subject in bright light. All responses should occur within 3 seconds.

c. Compare your responses with the correct answers in Ichikawa's or Ishihara's book. Determine the percentage of males and females in your class who exhibit any color-deficient vision. If an individual exhibits

**TABLE 35.1 Near Point of Accommodation**

| Age (years) | Average Near Point (cm) |
|:---:|:---:|
| 10 | 7 |
| 20 | 10 |
| 30 | 13 |
| 40 | 20 |
| 50 | 45 |
| 60 | 90 |

color-deficient vision, determine if the condition is protanopia or deuteranopia.

d. Record the class results in Part A of the laboratory report.

5. Complete Part A of the laboratory report.

 **PRACTICE**

**PROCEDURE B—Visual Demonstrations**

Perform the following demonstrations with the help of your laboratory partner.

1. *Blind-spot demonstration.* There are no photoreceptors in the *optic disc,* located where the nerve fibers of the retina leave the eye and enter the optic nerve. Consequently, this region of the retina is commonly called the *blind spot.*

To demonstrate the blind spot, follow these steps:

a. Close your left eye, hold figure 35.5 about 35 cm away from your face, and stare at the "+" sign in the figure with your right eye.

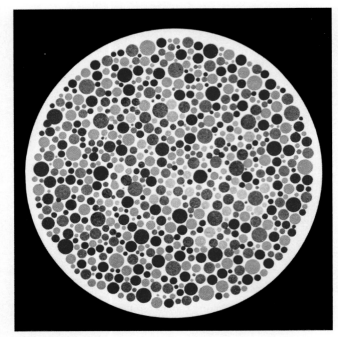

**FIGURE 35.4** Sample color test plate to diagnose possible colorblindness. A person with normal color vision perceives the embedded pattern of the number 74; some colorblind individuals only see dots. Tests for colorblindness cannot be conducted with this sample test plate reproduction. For accurate testing, the original plates should be used. (©Steve Allen/Getty Images RF)

b. Move the figure closer to your face as you continue to stare at the "+" until the dot on the figure suddenly disappears. This happens when the image of the dot is focused on the optic disc. Measure the right eye distance using a metric ruler or a meterstick.

c. Repeat the procedures with your right eye closed. This time, stare at the dot using your left eye, and the "+" will disappear when the image falls on the optic disc. Measure the distance.

d. Record the results in Part B of the laboratory report.

 **ASSESS**

**CRITICAL THINKING**

Explain why small objects are not lost from our vision under normal visual circumstances.

_____

_____

_____

_____

**FIGURE 35.5** Blind-spot demonstration.

2. *Photopupillary reflex.* The smooth muscles of the iris control the size of the pupil. For example, when the intensity of light entering the eye increases, a photopupillary reflex is triggered, and the circular muscles of the iris are stimulated to contract. As a result, the size of the pupil decreases, and less light enters the eye. This reflex is an autonomic response and protects the photoreceptor cells of the retina from damage.

To demonstrate this photopupillary reflex, follow these steps:

a. Ask your partner to sit with his/her hands thoroughly covering the eyes for 2 minutes.

b. Position a pen flashlight close to one eye with the light shining on the hand that covers the eye.

c. Ask your partner to remove the hand quickly.

d. Observe the pupil and note any change in its size.

e. Have your partner remove the other hand, but keep that uncovered eye shielded from extra light.

f. Observe both pupils and note any difference in their sizes.

3. *Accommodation pupillary reflex.* The pupil constricts as a normal accommodation reflex to focusing on close objects. This reduces peripheral light rays and allows the refraction of light to occur closer to the center portion of the lens, resulting in a clearer image.

To demonstrate the accommodation pupillary reflex, follow these steps:

a. Have your partner stare for several seconds at a dimly illuminated object in the room that is more than 20 feet away.

b. Observe the size of the pupil of one eye. Then hold a pencil about 25 cm in front of your partner's face, and have your partner stare at it.

c. Note any change in the size of the pupil.

4. *Convergence reflex.* The eyes converge as a normal convergence reflex to focusing on close objects. The convergence reflex orients the eyeballs toward the object so the image falls on the fovea centralis of both eyes.

To demonstrate the convergence reflex, follow these steps:

a. Repeat the procedure outlined for the accommodation pupillary reflex.

b. Note any change in the position of the eyeballs as your partner changes focus from the distant object to the pencil.

5. Complete Part B of the laboratory report.

Name _____

Date _____

Section _____

The A corresponds to the indicated Learning Outcome(s) found at the beginning of the laboratory exercise.

LABORATORY
REPORT

35

# Visual Tests and Demonstrations

## PART A ASSESSMENTS

1. Visual acuity test results: A 1

| Eye Tested | Acuity Values |
|---|---|
| Right eye | |
| Right eye with glasses (if applicable) | |
| Left eye | |
| Left eye with glasses (if applicable) | |

2. Astigmatism test results: A 1

| Eye Tested | Darker Lines |
|---|---|
| Right eye | |
| Right eye with glasses (if applicable) | |
| Left eye | |
| Left eye with glasses (if applicable) | |

3. Accommodation test results: A 1

| Eye Tested | Near Point (cm) |
|---|---|
| Right eye | |
| Right eye with glasses (if applicable) | |
| Left eye | |
| Left eye with glasses (if applicable) | |

**4.** Color vision test results: **A**

| Condition | Males | | | Females | | |
| --- | --- | --- | --- | --- | --- | --- |
| | Class Number | Class Percentage | Expected Percentage | Class Number | Class Percentage | Expected Percentage |
| Normal color vision | | | 93 | | | 99.6 |
| Deficient red-green color vision | | | 7 | | | 0.4 |
| Protanopia (lack red-sensitive cones) | | | Less frequent type | | | Less frequent type |
| Deuteranopia (lack green-sensitive cones) | | | More frequent type | | | More frequent type |

**5.** Complete the following: **2**

**a.** What is meant by 20/70 vision? _____

_____

**b.** What is meant by 20/10 vision? _____

_____

**c.** What visual problem is created by astigmatism? _____

_____

**d.** Why does the near point of accommodation often increase with age? _____

_____

**e.** Describe the eye defect that causes color-deficient vision. _____

_____

## PART B ASSESSMENTS

**1.** Blind-spot results: **3**

   **a.** Right eye distance _____

   **b.** Left eye distance _____

**2.** Complete the following:

   **a.** Explain why an eye has a blind spot. **3** _____

   _____

   **b.** Describe the photopupillary reflex. **3** _____

   _____

   **c.** What difference did you note in the size of the pupils when one eye was exposed to bright light and the other eye was shielded from the light? **3** _____

   _____

   **d.** Describe the accommodation pupillary reflex. **3** _____

   _____

   **e.** Describe the convergence reflex. **3** _____

   _____

# 36

# Endocrine Histology and Diabetic Physiology

## MATERIALS NEEDED

*For Procedure A—Endocrine Histology:*
Textbook
Human torso model
Compound light microscope
Prepared microscope slides of the following:
    Pituitary gland
    Thyroid gland
    Parathyroid gland
    Adrenal gland
    Pancreas

*For Procedure B—Insulin Shock:*
500 mL beaker
Live small fish (1" to 1.5" goldfish, guppy, rosy red feeder, or other)
Small fish net
Insulin (regular U-100) (HumulinR in 10 mL vials has 100 units/mL; store in refrigerator—do not freeze)
Syringes for U-100 insulin
10% glucose solution
Clock with second hand or timer
Compound light microscope
Prepared microscope slides of the following:
    Normal human pancreas (stained for alpha and beta cells)
    Human pancreas of diabetic

## ⚠ SAFETY

- Review all Laboratory Safety Guidelines in Appendix 1 of your laboratory manual.
- Wear disposable gloves when handling the fish and syringes.
- Dispose of the used syringe and needle in the puncture-resistant container.
- Wash your hands before leaving the laboratory.

## PURPOSE OF THE EXERCISE

To examine microscopically the tissues of endocrine glands and to observe behavioral changes that occur during insulin shock.

##  LEARNING OUTCOMES AP|R

After completing this exercise, you should be able to

1. Name and locate the major endocrine glands.
2. Associate the principal hormones and functions of the major endocrine glands.
3. Sketch and label tissue sections from the pituitary gland, thyroid gland, parathyroid glands, adrenal glands, and pancreas.
4. Compare the causes, symptoms, and treatments for type 1 and type 2 diabetes mellitus.
5. Observe and record the behavioral changes caused by insulin shock.
6. Observe and record the recovery from insulin shock when sugar is provided to cells.

The endocrine system consists of ductless glands that act together with parts of the nervous system to help control body activities. The endocrine glands secrete hormones that are transported in body fluids and affect cells that have appropriate receptor molecules. In this way, hormones influence the rate of metabolic reactions, the transport of substances through cell membranes, the regulation of water and electrolyte balances, and many other functions. By controlling cellular activities, endocrine glands play important roles in the maintenance of homeostasis.

The primary "fuel" for cells is sugar (glucose). Insulin produced by the beta cells of the pancreatic islets has a primary regulation role in the transport of blood sugar across the cell membranes of most cells of the body. Some individuals do not produce enough insulin;

others do not have adequate transport of glucose into the body cells when these target cells lose insulin receptors. This rather common disease is called *diabetes mellitus.* Diabetes mellitus is a functional disease that can have its onset either during childhood or later in life.

The protein hormone insulin has several functions: it stimulates the liver and skeletal muscles to synthesize glycogen from glucose; it inhibits the conversion of noncarbohydrates into glucose; it promotes the facilitated diffusion of glucose across cell membranes of cells possessing insulin receptors (adipose tissue, liver, and skeletal muscle); it decreases blood sugar (glucose); it increases protein synthesis; and it promotes fat storage in adipose cells. In a normal person, as blood sugar increases during nutrient absorption after a meal, the rising glucose levels directly stimulate beta cells of the pancreas to secrete insulin. This increase in insulin prevents blood sugar from sudden surges (hyperglycemia) by promoting glycogen production in the liver and an increase in the entry of glucose into muscle and adipose cells.

Between meals and during sleep, blood glucose levels decrease and insulin decreases. When insulin concentration decreases, more glucose is available to enter cells that lack insulin receptors. Neurons, liver cells, kidney cells, and red blood cells either lack insulin receptors or express limited insulin receptors, and they absorb and utilize glucose without the need for insulin. These cells depend upon blood glucose concentrations to enable cellular respiration and ATP production. When insulin is given as a medication for diabetes mellitus, blood glucose levels may drop drastically (hypoglycemia) if the individual does not eat. This can have significant negative effects on the nervous system, resulting in insulin shock.

*Type 1* diabetes mellitus usually has a rapid onset relatively early in life, but it can occur in adults. It occurs when insulin is no longer produced; thus, sometimes it is referred to as *insulin-dependent diabetes mellitus (IDDM).* This autoimmune disorder occurs when the immune system destroys beta cells, resulting in a decrease in insulin production. This disorder affects about 10% to 15% of diabetics. The treatment for type 1 diabetes mellitus is to administer insulin, usually by injection.

*Type 2* diabetes mellitus has a gradual onset, usually diagnosed in people over age forty. It is sometimes called *non-insulin-dependent diabetes mellitus (NIDDM).* This disease results as body cells become less sensitive to insulin or lose insulin receptors and thus cannot respond to insulin even if insulin levels remain normal. This situation is called *insulin resistance.* Risk factors for the onset of type 2 diabetes mellitus, which afflicts about 85% to 90% of diabetics, include heredity, obesity, and lack of exercise. Obesity is a major indicator for predicting type 2 diabetes; a weight loss of a mere 10 pounds will increase the body's sensitivity to insulin as a means of prevention. The treatments for type 2 diabetes mellitus include a diet that avoids foods that stimulate insulin production, weight control, exercise, and

medications. (The causes and treatments for type 2 diabetes closely resemble those for coronary artery disease.)

 **PRACTICE**

### PROCEDURE A—Endocrine Histology

1. Review the sections entitled "Pituitary Gland," "Thyroid Gland," "Parathyroid Glands," "Adrenal Glands," "Pancreas," and "Other Endocrine Glands" in chapter 13 of the textbook.
2. As a review activity, label figure 36.1.
3. Complete Part A of Laboratory Report 36.
4. Examine the human torso model and locate the following:

   **hypothalamus**
   **pituitary stalk (infundibulum)**
   **pituitary gland (hypophysis)**
   > anterior lobe (adenohypophysis)
   > posterior lobe (neurohypophysis)

   **thyroid gland**
   **parathyroid glands**
   **adrenal glands (suprarenal glands)**
   > adrenal medulla
   > adrenal cortex

   **pancreas**
   **pineal gland**
   **thymus**
   **ovaries**
   **testes**

5. Examine the microscopic tissue sections of the following glands, and identify the features described:

   **Pituitary gland.** The pituitary gland is located in the sella turcica of the sphenoid bone. To examine the pituitary tissue, follow these steps:
   a. Observe the tissues using low-power magnification (fig. 36.2).
   b. Locate the *pituitary stalk (infundibulum),* the *anterior lobe* (the largest part of the gland), and the *posterior lobe.* The anterior lobe makes hormones and appears glandular; the posterior lobe stores and releases hormones made by the hypothalamus and is nervous in appearance.
   c. Observe an area of the anterior lobe with high-power magnification. Locate a cluster of relatively large cells and identify some *acidophil cells,* which contain pink-stained granules, and some *basophil cells,* which contain blue-stained granules. These acidophil and basophil cells are the hormone-secreting cells. The hormones secreted include TSH, ACTH, prolactin, growth hormone, FSH, and LH.

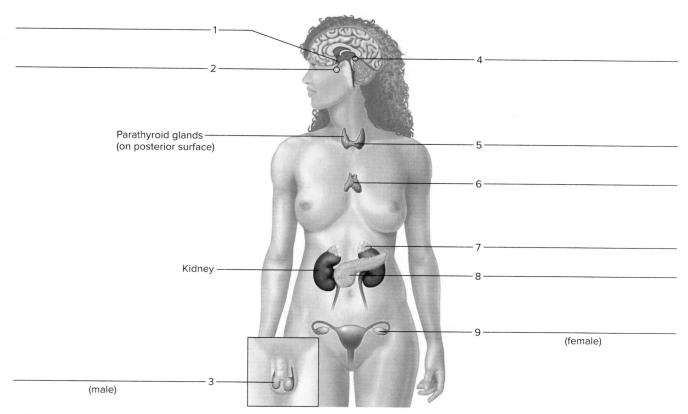

Parathyroid glands
(on posterior surface)

Kidney

1

2

4

5

6

7

8

9

3

(male)

(female)

**FIGURE 36.1** Label the major endocrine glands. Ⓐ

Hypothalamus

Pituitary stalk

Posterior lobe

Anterior lobe

Sphenoid bone

**FIGURE 36.2** Micrograph of the pituitary gland (6×). AP|R (©Biophoto Associates/Science Source)

**d.** Observe an area of the posterior lobe with high-power magnification. Note the numerous unmyelinated nerve fibers present in this lobe. Also locate some *pituicytes,* a type of neuroglia, scattered among the nerve fibers. The posterior lobe stores and releases ADH and oxytocin, which are synthesized in the hypothalamus.

**e.** Prepare labeled sketches of representative portions of the anterior and posterior lobes of the pituitary gland in Part B of the laboratory report.

**Thyroid gland.** To examine the thyroid tissue, follow these steps:

**a.** Use low-power magnification to observe the tissue (fig. 36.3). Note the numerous *follicles,* each of which consists of a layer of cells surrounding a colloid-filled cavity.

**b.** Observe the tissue using high-power magnification. The cells forming the wall of a follicle are simple cuboidal epithelial cells. These cells secrete thyroid hormone ($T_3$ and $T_4$). The extrafollicular (parafollicular) cells secrete calcitonin.

**c.** Prepare a labeled sketch of a representative portion of the thyroid gland in Part B of the laboratory report.

**Parathyroid gland.** Typically, four parathyroid glands are attached to the posterior surface of the thyroid gland. To examine the parathyroid tissue, follow these steps:

**a.** Use low-power magnification to observe the tissue (fig. 36.4). The gland consists of numerous tightly packed secretory cells.

**b.** Switch to high-power magnification and locate two types of cells—a smaller form (chief cells) arranged in cordlike patterns and a larger form (oxyphil cells) that have distinct cell boundaries and are present in clusters. *Chief cells* secrete parathyroid hormone, whereas the function of *oxyphil cells* is not clearly understood.

**c.** Prepare a labeled sketch of a representative portion of the parathyroid gland in Part B of the laboratory report.

**Adrenal gland.** To examine the adrenal tissue, follow these steps:

**a.** Use low-power magnification to observe the tissue (fig. 36.5). Note the thin capsule of connective tissue that covers the gland. Just beneath the capsule, there is a relatively thick *adrenal cortex*. The central portion of the gland is the *adrenal medulla*. The cells of the cortex are in three poorly defined layers. Those of the outer layer (zona glomerulosa) are arranged irregularly; those of the middle layer (zona fasciculata) are in long cords; and those of the inner layer (zona reticularis) are arranged in an interconnected network of cords. The most noted hormones secreted from the cortex include aldosterone from the zona glomerulosa, cortisol from the zona fasciculata, and estrogens and androgens from the zona reticularis. The cells of the medulla are relatively large and irregularly shaped, and they often occur in clusters. The medulla cells secrete epinephrine (adrenaline) and norepinephrine (noradrenaline).

**b.** Using high-power magnification, observe each of the layers of the cortex and the cells of the medulla.

**c.** Prepare labeled sketches of representative portions of the adrenal cortex and medulla in Part B of the laboratory report.

**Pancreas.** To examine the pancreas tissue, follow these steps:

**a.** Use low-power magnification to observe the tissue (fig. 36.6). The gland largely consists of deeply stained exocrine cells arranged in clusters around secretory ducts. These exocrine cells (acinar cells) secrete pancreatic juice rich in digestive enzymes.

Follicular cells secrete thyroid hormone ($T_3$ and $T_4$)

Colloid of thyroglobulin stores thyroid hormone

Parafollicular cells or C cells secrete calcitonin

Follicle

**FIGURE 36.3** Micrograph of the thyroid gland (300×). **AP|R** (©Biophoto Associates/Science Source)

There are circular masses of differently stained cells scattered throughout the gland. These clumps of cells constitute the *pancreatic islets (islets of Langerhans),* and they represent the endocrine portion of the pancreas. The beta cells secrete insulin and the alpha cells secrete glucagon.

b. Examine an islet, using high-power magnification. Unless special stains are used, it is not possible to distinguish alpha and beta cells.

c. Prepare a labeled sketch of a representative portion of the pancreas in Part B of the laboratory report.

6. Complete Part C of the laboratory report.

**FIGURE 36.4** Micrograph of the parathyroid gland (65×). **AP|R** (©Biophoto Associates/Science Source)

**FIGURE 36.5** Micrograph of the adrenal cortex and the adrenal medulla (75×). **AP|R** (©Ed Reschke)

**FIGURE 36.6** Micrograph of the pancreas (200×). **AP|R** (©Dr.Kent M. Van De Graaff)

### PROCEDURE B—Insulin Shock

If too much insulin is present relative to the amount of glucose available to the cells, insulin shock can occur. In this procedure, insulin shock will be created in a fish by introducing insulin into the water with a fish. Insulin will be absorbed into the blood of the gills of the fish. When insulin in the blood reaches levels above normal, rapid hypoglycemia occurs. As hypoglycemia happens from the increased insulin absorption, the brain cells obtain less of the much-needed glucose, and epinephrine is secreted. Consequently, a rapid heart rate, anxiety, sweating, mental disorientation, impaired vision, dizziness, convulsions, and possible unconsciousness are complications during insulin shock. Although these complications take place in humans, some of them are visually noticeable in a fish that is in insulin shock. The most observable components of insulin shock in a fish are related to behavioral changes, including rapid and irregular movements of the entire fish and increased gill cover and mouth movements.

1. Review the sections entitled "Pancreas" and "Diabetes Mellitus—Clinical Application" in chapter 13 of the textbook, as well as the introduction to this laboratory exercise.
2. Complete the table in Part D of the laboratory report.
3. A fish can be used to observe normal behavior, induced insulin shock, and recovery from insulin shock. The pancreatic islets are located in the pyloric ceca or other scattered regions in fish. The behavioral changes of fish when they experience insulin shock and recovery mimic those of humans. Observe a fish in 200 mL of aquarium water in a 500 mL beaker. Make your observations of swimming, gill cover (operculum), and mouth movements for 5 minutes. Record your observations in Part E of the laboratory report.
4. Add 200 units of room-temperature insulin slowly, using the syringe, into the water with the fish to induce insulin shock. Record the time or start the timer when the insulin is added. Watch for changes of the swimming, gill cover, and mouth movements as insulin diffuses into the blood at the gills. Signs of insulin shock might include rapid and irregular movements. Record your observations of any behavioral changes and the time of the onset in Part E of the laboratory report.
5. After definite behavioral changes are observed, use the small fish net to move the fish into another 500 mL beaker containing 200 mL of a 10% room-temperature glucose solution. Record the time when the fish is transferred into this container. Make observations of any recovery of normal behaviors, including the amount of time involved. Record your observations in Part E of the laboratory report.
6. When the fish appears to be fully recovered, return it to its normal container. Dispose of the syringe according to directions from your laboratory instructor.
7. Complete Part E of the laboratory report.

 **LEARN: ACTIVITY**

Examine a normal stained pancreas under high-power magnification (fig. 36.6). Locate the clumps of cells that constitute the *pancreatic islets (islets of Langerhans)* that represent the endocrine portions of the organ. A pancreatic islet contains four distinct cells that secrete different hormones: *alpha cells,* which secrete glucagon; *beta cells,* which secrete insulin; *delta cells,* which secrete somatostatin; and *F cells,* which secrete pancreatic polypeptide. Some of the specially stained pancreatic islets allow for distinguishing the different types of cells (figs. 36.7 and 36.8). The normal pancreatic islet has about 80% beta cells concentrated in the central region of the islet, about 15% alpha cells near the periphery of the islet, about 5% delta cells, and a small number of F cells. There are complex interactions among the four hormones affecting blood sugar, but in general, glucagon will increase blood sugar, whereas insulin will lower blood sugar; thus, these two hormones are considered antagonistic in their function. Somatostatin acts as a paracrine secretion to inhibit alpha and beta cell secretions, and pancreatic polypeptide inhibits secretions from delta cells and digestive enzyme secretions from the pancreatic acini.

Examine the pancreas of a diabetic under high-power magnification (fig. 36.9). Locate a pancreatic islet and note the number of beta cells as compared to a normal number of beta cells in the normal pancreatic islet. Prepare a labeled sketch of a pancreatic islet showing indications of diabetes mellitus in Part B of the laboratory report.

Describe the differences that you observed between a normal pancreatic islet and a pancreatic islet of a person with diabetes mellitus.

_____

_____

_____

_____

_____

Secretory granules of
exocrine (acinar) cells

Capillary

Stained beta cells
within pancreatic islet

**FIGURE 36.7** Micrograph of a specially stained pancreatic islet surrounded by exocrine (acinar) cells (500×). The insulin-
containing beta cells are stained dark purple. **AP|R** (©J & J Photography)

Stained beta cells

Stained alpha cells

**FIGURE 36.8** Micrograph of a pancreatic islet using immunofluorescence stains to visualize alpha and beta cells. Only the
areas of interest are visible in the fluorescence microscope. **AP|R** (©McGraw-Hill Education/Al Telser, photographer)

Exocrine
(acinar) cells

Pancreatic
islets

**FIGURE 36.9** Micrograph of pancreas showing indications of changes from diabetes mellitus (100×). The changes from diabetes mellitus include areas of fibrosis and amyloid replacement of islet cells. (©Michael Abbey/Science Source)

Name _____

Date _____

Section _____

The 🅐 corresponds to the indicated Learning Outcome(s) found at the beginning of the laboratory exercise.

# Endocrine Histology and Diabetic Physiology

### 🔄 PART A ASSESSMENTS

Complete the following:

1. Name six hormones secreted by the anterior lobe of the pituitary gland. 🄰 _____

_____

2. Name two hormones secreted by the posterior lobe of the pituitary gland. 🄰 _____

3. Name the pituitary hormone responsible for the following actions: 🄰

    **a.** Causes kidneys to conserve water. _____

    **b.** Stimulates cells to increase in size and divide more rapidly. _____

    **c.** Stimulates secretion from thyroid gland. _____

    **d.** Causes contraction of uterine wall muscles. _____

    **e.** Stimulates secretion from adrenal cortex. _____

    **f.** Stimulates milk production. _____

4. Name two thyroid hormones that affect metabolic rate. 🄰 _____

5. Name a hormone secreted by the thyroid gland that acts to lower blood calcium. 🄰 _____

6. Name a hormone that acts to raise blood calcium. 🄰 _____

7. Name three target organs of parathyroid hormone. 🄰 _____

8. Name two hormones secreted by the adrenal medulla. 🄰 _____

9. List five different effects produced by these medullary hormones. 🄰 _____

_____

10. Name the most important mineralocorticoid secreted by the adrenal cortex. 🄰 _____

11. List two actions of this mineralocorticoid. 🄰 _____

_____

12. Name the most important glucocorticoid secreted by the adrenal cortex. 🄰 _____

13. List three actions of this glucocorticoid. 🄰 _____

_____

14. Distinguish the two hormones secreted by the alpha and beta cells of the pancreatic islets. 🄰 _____

_____

**CRITICAL THINKING**

Briefly explain how the actions of pancreatic hormones complement one another. **2**

_____

_____

_____

**PART B ASSESSMENTS**

Sketch and label representative portions of the following endocrine glands: **3**

| | |
|---|---|
| Pituitary gland (____×)<br>(anterior and posterior lobes) | Thyroid gland (____×) |
| Parathyroid gland (____×) | Adrenal gland (____×)<br>(cortex and medulla) |
| Pancreatic islet (____×)<br>(normal) | Pancreatic islet (____×)<br>(showing changes from diabetes mellitus) |

Match the endocrine gland in column A with a characteristic of the gland in column B. Place the letter of your choice in the space provided. **A** **2**

**Column A**

a. Adrenal cortex
b. Adrenal medulla
c. Hypothalamus
d. Pancreatic islets
e. Parathyroid gland
f. Pituitary gland
g. Thymus
h. Thyroid gland

**Column B**

_____ 1. Located in sella turcica of sphenoid bone

_____ 2. Contains alpha and beta cells

_____ 3. Contains colloid-filled cavities

_____ 4. Attached to posterior surface of thyroid gland

_____ 5. Secretes cortisol and aldosterone

_____ 6. Attached to pituitary gland by a stalk

_____ 7. Gland inside another gland near kidneys

_____ 8. Located in mediastinum

## PART D ASSESSMENTS

Complete the missing parts of the table on diabetes mellitus: **4**

| Characteristic | Type 1 Diabetes | Type 2 Diabetes |
|---|---|---|
| Onset age | Early age or adult | |
| Onset of symptoms | | Slow |
| Percentage of diabetics | | 85%–90% |
| Natural insulin levels | Below normal | |
| Beta cells of pancreatic islets | | Not destroyed |
| Pancreatic islet cell antibodies | Present | |
| Risk factors of having the disease | Heredity | |
| Typical treatments | Insulin administration | |
| Untreated blood sugar levels | | Hyperglycemia |

## PART E ASSESSMENTS

1. Record your observations of normal behavior of the fish.

   a. Swimming movements: _____

   _____

   b. Gill cover movements: _____

   _____

   c. Mouth movements: _____

   _____

**2.** Record your observations of the fish after insulin was added to the water. **5**

   **a.** Recorded time that insulin was added: _____

   **b.** Swimming movements:

   _____

   **c.** Gill cover movements:

   _____

   **d.** Mouth movements:

   _____

   **e.** Elapsed time until the insulin shock symptoms occurred: _____

**3.** Record your observations after the fish was transferred into a glucose solution. **6**

   **a.** Recorded time that fish was transferred into the glucose solution: _____

   **b.** Describe any changes in the behavior of the fish that indicate a recovery from insulin shock. _____

   _____

   _____

   _____

   _____

   **c.** Elapsed time until indications of a recovery from insulin shock occurred: _____

## ASSESS

### CRITICAL THINKING

Justify the importance for a type 1 diabetic to regulate insulin administration, meals, and exercise. **4**

_____

_____

_____

_____

_____

_____

_____

# 37

# Blood Cells and Blood Typing

## PURPOSE OF THE EXERCISE

To review the characteristics of blood cells, to examine them microscopically, to perform a differential white blood cell count, to determine the ABO blood type of a blood sample, and to observe an Rh blood-typing test.

##  LEARNING OUTCOMES AP|R

After completing this exercise, you should be able to

1. Describe the structure and function of red blood cells, white blood cells, and platelets.
2. Identify and sketch red blood cells, five types of white blood cells, and platelets.
3. Perform and interpret the results of a differential white blood cell count.
4. Analyze the basis of ABO blood typing.
5. Match the ABO type of a blood sample.
6. Explain the basis of Rh blood typing.
7. Interpret how the Rh type of a blood sample is determined.

**B**lood is a type of connective tissue whose cells are suspended in a liquid extracellular matrix. These cells are formed mainly in red bone marrow, and they include *red blood cells, white blood cells,* and some cellular fragments called *platelets* (fig. 37.1).

Red blood cells contain hemoglobin and transport gases between the body cells and the lungs, white blood cells defend the body against infections, and platelets play an important role in stoppage of bleeding (hemostasis).

Blood typing involves identifying protein substances called *antigens* present on the outer surface of red blood cell membranes. Although many different antigens are associated with human red blood cells, only a few of them are of clinical importance. These include the antigens of the ABO group and those of the Rh group.

To determine which antigens are present, a blood sample is mixed with blood-typing sera that contain known types of antibodies. If a particular antibody contacts a corresponding antigen, a reaction occurs, and the red blood cells clump together (agglutination). Thus, if blood cells are mixed with serum containing antibodies that react with antigen A and the cells clump together, antigen A must be present in those cells.

**Neutrophils (three of many variations)**
- Fine light-purple granules
- Nucleus single to five lobes (highly variable)
- Immature neutrophils, called bands, have a single C-shaped nucleus
- Mature neutrophils, called segs, have a lobed nucleus
- Often called polymorphonuclear leukocytes when older

**Eosinophils (three of many variations)**
- Coarse reddish granules
- Nucleus usually bilobed

**Basophils (three of many variations)**
- Coarse deep blue to almost black granules
- Nucleus often almost hidden by granules

**Lymphocytes (three of many variations)**
- Slightly larger than RBCs
- Thin rim of nearly clear cytoplasm
- Nearly round nucleus appears to fill most of cell in smaller lymphocytes
- Larger lymphocytes hard to distinguish from monocytes

**Monocytes (three of many variations)**
- Largest WBC; 2–3× larger than RBCs
- Cytoplasm nearly clear
- Nucleus round, kidney-shaped, oval, or lobed

**Platelets (several variations)**
- Cell fragments
- Single to small clusters

**Erythrocytes (several variations)**
- Lack nucleus (mature cell)
- Biconcave discs
- Thin centers appear almost hollow

**FIGURE 37.1** Micrographs of blood cells illustrating some of the numerous variations of each type. Appearance characteristics for each cell pertain to a thin blood film using Wright's stain (1,000×). **AP|R** (©McGraw-Hill Education/Al Telser, photographer)

**PRACTICE**

### PROCEDURE A—Types of Blood Cells

1. Review the section entitled "Blood Cells" in chapter 14 of the textbook.
2. Complete Part A of Laboratory Report 37.
3. Refer to figure 37.1 as an aid in identifying the various types of blood cells. Study the functions of the blood cells listed in table 37.1. Use the prepared slide of blood and locate each of the following:

   **red blood cell (erythrocyte)**
   **white blood cell (leukocyte)**
   > granulocytes
   >> neutrophil
   >> eosinophil
   >> basophil
   > agranulocytes
   >> lymphocyte
   >> monocyte

   **platelet (thrombocyte)**

**TABLE 37.1 Cellular Components of Blood**

| Component | Function |
|---|---|
| Red blood cell (erythrocyte) | Transports oxygen and carbon dioxide |
| White blood cell (leukocyte) | Destroys pathogenic micro-organisms and parasites, removes worn cells, provides immunity |
| *Granulocytes—Have Granular Cytoplasm* | |
| Neutrophil | Phagocytizes small particles |
| Eosinophil | Destroys parasites, helps control inflammation and allergic reactions |
| Basophil | Releases heparin and histamine |
| *Agranulocytes—Lack Granular Cytoplasm* | |
| Lymphocyte | Provides immunity |
| Monocyte | Phagocytizes large particles |
| Platelet (thrombocyte) | Helps control blood loss from injured blood vessels |

4. In Part B of the laboratory report, prepare sketches of single blood cells to illustrate each type. Pay particular attention to the relative size, nuclear shape, and color of granules in the cytoplasm (if present). The sketches should be accomplished using either the high-power objective or the oil immersion objective of the compound light microscope.

 **PRACTICE**

### PROCEDURE B—Differential White Blood Cell Count

A differential white blood cell count is performed to determine the percentage of each of the various types of white blood cells present in a blood sample. The test is useful because the relative proportions of white blood cells may change in particular diseases, as indicated in table 37.2. Neutrophils, for example, usually increase during bacterial infections, whereas eosinophils may increase during certain parasitic infections and allergic reactions.

1. To make a differential white blood cell count, follow these steps:
   a. Using high-power magnification or an oil immersion objective, focus on the cells at one end of a prepared blood slide where the cells are well distributed.
   b. Slowly move the blood slide back and forth, following a path that avoids passing over the same cells twice (fig. 37.2).
   c. Each time you encounter a white blood cell, identify its type and record it in Part C of the laboratory report.
   d. Continue searching for and identifying white blood cells until you have recorded 100 cells in the data table. *Percent* means "parts of 100" for each type of white blood cell, so the total number observed is equal to its percentage in the blood sample.
2. Complete Part C of the laboratory report.

**TABLE 37.2 Differential White Blood Cell Count**

| Cell Type | Normal Value (percent) | Elevated Levels May Indicate |
|---|---|---|
| Neutrophil | 54–62 | Bacterial infections, stress |
| Lymphocyte | 25–33 | Mononucleosis, whooping cough, viral infections |
| Monocyte | 3–9 | Malaria, tuberculosis, fungal infections |
| Eosinophil | 1–3 | Allergic reactions, autoimmune diseases, parasitic worms |
| Basophil | <1 | Cancers, chicken pox, hypothyroidism |

**FIGURE 37.2** Move the blood slide back and forth to avoid passing the same cells twice.

 **LEARN: ACTIVITY**

Obtain a prepared slide of pathological blood that has been stained with Wright's stain. Perform a differential white blood cell count using this slide, and compare the results with the values for normal blood listed in table 37.2. What differences do you note?

_____

_____

_____

_____

_____

_____

**FIGURE 37.3** Slide prepared for ABO blood typing.

**PRACTICE** _____

#### PROCEDURE C—ABO Blood Typing

1. Review the section entitled "ABO Blood Group" in chapter 14 of the textbook.
2. Complete Part D of the laboratory report.
3. Perform the ABO blood type test using the blood-typing kit. To do this, follow these steps:
   a. Obtain a clean microscope slide and mark across its center with a wax pencil to divide it into right and left halves. Also write "Anti-A" near the edge of the left half and "Anti-B" near the edge of the right half (fig. 37.3).
   b. Place a small drop of blood on each half of the microscope slide. Work quickly so that the blood will not have time to clot.
   c. Add a drop of anti-A serum to the blood on the left half and a drop of anti-B serum to the blood on the right half. Note the color coding of the anti-A and anti-B typing sera. To avoid contaminating the serum, avoid touching the blood with the serum while it is in the dropper; instead, allow the serum to fall from the dropper onto the blood.
   d. Use separate toothpicks to stir each sample of serum and blood together, and spread each over an area about as large as a quarter. Dispose of toothpicks in an appropriate container.
   e. Examine the samples for clumping of red blood cells (agglutination) after 2 minutes.
   f. See table 37.3 and figure 37.4 for aid in interpreting the test results.
   g. Discard contaminated materials as directed by the laboratory instructor.
4. Complete Part E of the laboratory report.

**TABLE 37.3   Possible Reactions of ABO Blood-Typing Sera and Donor Compatibilities**

| REACTIONS | | | | Permissible | |
| Anti-A Serum | Anti-B Serum | Blood Type | Preferred Donor Type | Donor in Limited Amounts | Incompatible Donor |
|---|---|---|---|---|---|
| Clumping | No clumping | Type A | A | O | B, AB |
| No clumping | Clumping | Type B | B | O | A, AB |
| Clumping | Clumping | Type AB (universal recipient) | AB | A, B, O | None |
| No clumping | No clumping | Type O (universal donor) | O | No alternative types | A, B, AB |

Note: The inheritance of the ABO blood groups is described in Laboratory Exercise 53.

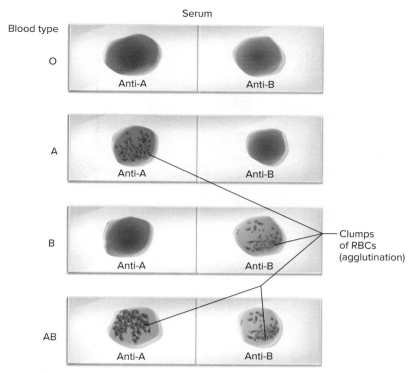

Serum

Blood type

O     Anti-A     Anti-B

A     Anti-A     Anti-B

B     Anti-A     Anti-B

AB     Anti-A     Anti-B

Clumps of RBCs (agglutination)

**FIGURE 37.4** Four possible results of the ABO test.

 **ASSESS**

### CRITICAL THINKING

Judging from the observations of the blood-typing results, suggest which components in the anti-A and anti-B sera caused clumping (agglutination).

_____

_____

_____

_____

 **PRACTICE**

#### DEMONSTRATION—Rh Blood Typing

1. Review the section entitled "Rh Blood Group" in chapter 14 of the textbook.
2. Complete Part F of the laboratory report.
3. To determine the Rh blood type of a blood sample, follow these steps:
   a. Thoroughly wash hands with soap and water, and dry them with paper towels.
   b. Cleanse the end of the middle finger with an alcohol swab, and let the finger dry in the air.
   c. Lance the tip of a finger using a sterile disposable blood lancet. Place a small drop of blood in the center of a clean microscope slide. Cover the lanced finger location with a bandage.
   d. Add a drop of anti-D serum to the blood and mix them together with a clean toothpick.
   e. Place the slide on the plate of a slide warming table (Rh blood-typing box) prewarmed to 45°C (113°F) (fig. 37.5).
   f. Slowly rock the table (box) back and forth to keep the mixture moving, and watch for clumping (agglutination) of the blood cells. When clumping occurs in anti-D serum, the clumps usually are smaller than those that appear in anti-A or anti-B sera, so they may be less obvious. However, if clumping occurs, the blood is called Rh positive; if no clumping occurs *within 2 minutes,* the blood is called Rh negative.
   g. Discard all contaminated materials in appropriate containers.
4. Complete Parts G and H of the laboratory report.

Plate heated to
45°C (113°F)

Mixture of blood
and anti-D serum

Rh TYPING BOX

**FIGURE 37.5**   Slide warming box (table) used for Rh blood typing.

Name _____

Date _____

Section _____

The Ⓐ corresponds to the indicated Learning Outcome(s) found at the beginning of the laboratory exercise.

# Blood Cells and Blood Typing

## PART A ASSESSMENTS

Complete the following:

1. Red blood cells (RBCs) are also called _____ . Ⓐ

2. The functions of red blood cells are _____ . Ⓐ

3. _____ is the oxygen-carrying substance in a red blood cell. Ⓐ

4. White blood cells are also called _____ . Ⓐ

5. White blood cells with granular cytoplasm are called _____ . Ⓐ

6. White blood cells lacking granular cytoplasm are called _____ . Ⓐ

7. Normally, the most numerous white blood cells are _____ . Ⓐ

8. A platelet, a fragment of a cell, lacks a(an) _____ . Ⓐ

## PART B ASSESSMENTS

Sketch a single blood cell of each type in the following spaces. Use colored pencils to represent the stained colors of the cells. Label any features that can be identified. ②

Red blood cell and platelet ( _____×)

Neutrophil ( _____×)

Lymphocyte ( _____×)

Monocyte ( _____×)

Eosinophil ( _____×)

Basophil ( _____×)

## PART C ASSESSMENTS

1. *Differential White Blood Cell Count Data Table.* As you identify white blood cells, record them in the table by using a tally system, such as 卌 ||. Place tally marks in the "Number Observed" column, and total each of the five WBCs when the differential count is completed. Obtain a total of all five WBCs counted to determine the percentage of each WBC type. 3

| Type of WBC | Number Observed | Total | Percent |
|---|---|---|---|
| Neutrophil | | | |
| Lymphocyte | | | |
| Monocyte | | | |
| Eosinophil | | | |
| Basophil | | | |
| | | **Total of Column** | |

2. How do the results of your differential white blood cell count compare with the normal values listed in table 37.2? 3

_____

_____

_____

## ASSESS

### CRITICAL THINKING

Explain the difference between a differential white blood cell count and a total white blood cell count. 3

_____

_____

_____

_____

## PART D ASSESSMENTS

Complete the following statements:

1. The antigens of the ABO blood group are located on the red blood cell _____ . 4

2. The blood of every person contains one of (how many possible?) _____ combinations of antigens. 4

3. Type A blood contains antigen _____ . 4

4. Type B blood contains antigen _____ . 4

5. Type A blood contains _____ antibody in the plasma. 4

6. Type B blood contains _____ antibody in the plasma. 4

7. Persons with ABO blood type _____ are sometimes called universal recipients. 4

8. Persons with ABO blood type _____ are sometimes called universal donors. 4

## PART E ASSESSMENTS

Complete the following:

1. What was the ABO type of the blood tested? **5** _____

2. What ABO antigens are present on the red blood cells of this type of blood? **5** _____

3. What ABO antibodies are present in the plasma of this type of blood? **5** _____

4. If a person with this blood type needed a blood transfusion, what ABO type(s) of blood could be received safely? **5**

   _____

5. If a person with this blood type were serving as a blood donor, what ABO blood type(s) could receive the blood

   safely? **5** _____

## PART F ASSESSMENTS

Complete the following statements:

1. The Rh blood group was named after the _____ . **6**

2. Of the antigens in the Rh group, the most important is _____ . **6**

3. If red blood cells lack Rh antigens, the blood is called _____ . **6**

4. If an Rh-negative person sensitive to Rh-positive blood receives a transfusion of Rh-positive blood, the donor's cells are

   likely to _____ . **6**

5. An Rh-negative woman who might be carrying a(an) _____ fetus is given an injection

   of RhoGAM to prevent hemolytic disease of the fetus and newborn (erythroblastosis fetalis). **6**

## PART G ASSESSMENTS

Complete the following:

1. What was the Rh type of blood tested in the demonstration? **7** _____

2. What Rh antigen is present on the red blood cells of this type of blood? **7** _____

3. If a person with this blood type needed a blood transfusion, what type of blood could be received safely? **7**

   _____

4. If a person with this blood type were serving as a blood donor, a person with what type of blood could receive the blood

   safely? **7** _____

Examine figure 37.6. The figure contains test results for both ABO type and Rh type performed in this laboratory exercise. Observe whether clumping (agglutination) occurred to answer the questions.

| Anti-A serum | Anti-B serum | Anti-D serum | Blood only (control) |

**FIGURE 37.6** Blood test results adding anti-A serum, anti-B serum, and anti-D serum to separate drops of blood. An unaltered drop of blood is shown as a control; it does not indicate any clumping (agglutination) reaction. (©Image Source/Getty Images RF)

1. This individual has ABO type _____ . **5**
2. This individual has Rh type _____ . **7**

# 38

# Heart Structure

## MATERIALS NEEDED

Textbook
Dissectible human heart model
Preserved sheep or other mammalian heart
Dissecting tray
Dissecting instruments
Disposable gloves
Colored pencils (red and blue)

## ⚠ SAFETY

- Wear disposable gloves when working on the heart dissection.
- Save or dispose of the dissected heart as instructed.
- Wash the dissecting tray and instruments as instructed.
- Wash your laboratory table.
- Wash your hands before leaving the laboratory.

## PURPOSE OF THE EXERCISE

To review the structural characteristics of the human heart and to examine the major features of a mammalian heart.

##  LEARNING OUTCOMES AP|R

After completing this exercise, you should be able to

1. Identify and label the major structural features of the human heart and closely associated blood vessels.

2. Match heart structures with appropriate locations and functions.

3. Compare the features of the human heart with those of another mammal.

The heart is a muscular pump located within the mediastinum and resting upon the diaphragm. It is enclosed by the lungs, thoracic vertebrae, and sternum, and attached at its superior end (the base) are several large blood vessels. Its inferior end extends downward to the left and terminates as a bluntly pointed apex.

The heart and the proximal ends of the attached blood vessels are enclosed by a double-layered pericardium. The innermost layer of this membrane (visceral pericardium) consists of a thin covering closely applied to the surface of the heart, whereas the outer layer (parietal pericardium with fibrous pericardium) forms a tough, protective sac surrounding the heart. Between the parietal and visceral layers of the pericardium is a space, the pericardial cavity, that contains a small volume of serous (pericardial) fluid.

##  PRACTICE

### PROCEDURE A—The Human Heart

1. Review the section entitled "The Heart" in chapter 15 of the textbook.
2. As a review activity, label figures 38.1–38.3.
3. Complete Part A of Laboratory Report 38.
4. Examine the human heart model, and locate the following features:

**heart**
  base (superior region where blood vessels emerge)
  apex (inferior, rounded end)
**pericardium (pericardial sac)**
  fibrous pericardium (outer layer)
  parietal pericardium (inner lining of fibrous pericardium)
  visceral pericardium (epicardium)
**pericardial cavity** (between parietal and visceral pericardial membranes)
**myocardium** (cardiac muscle)
**endocardium** (lines heart chambers)
**atria**
  right atrium
  left atrium
  auricles
**ventricles**
  right ventricle
  left ventricle

atrioventricular orifices
atrioventricular valves (AV valves)
  tricuspid valve (right atrioventricular valve)
  mitral valve (bicuspid valve; left atrioventricular valve)
semilunar valves
  pulmonary valve
  aortic valve
chordae tendineae
papillary muscles
atrioventricular sulcus
interventricular sulci
  anterior sulcus
  posterior sulcus
superior vena cava
inferior vena cava
pulmonary trunk

pulmonary arteries
pulmonary veins
aorta
left coronary artery
  circumflex artery
  anterior interventricular (descending) artery
right coronary artery
  posterior interventricular artery
  marginal artery
cardiac veins
  great cardiac vein
  middle cardiac vein
  small cardiac vein
coronary sinus (for return of blood from cardiac veins into right atrium)

5. Label the human heart model in figure 38.4.

Brachiocephalic veins

Right pulmonary artery

Right pulmonary veins

(chamber)
Auricle of right atrium

(chamber)

Marginal artery

Brachiocephalic artery
Left common carotid artery
Left subclavian artery

1

2

Ligamentum arteriosum

Left pulmonary artery

7

Left pulmonary veins

Auricle of left atrium

8

9

3

4

5

6

10

(chamber)

Anterior interventricular (descending) artery

Adipose tissue

Apex of the heart

**FIGURE 38.1**    Label this anterior view of the human heart.  1  AP|R

Right pulmonary artery

Right pulmonary veins

(chamber)

Left auricle

Great cardiac vein

(chamber)

(chamber)

Coronary sinus

Middle cardiac vein

Posterior interventricular artery

(chamber)

**FIGURE 38.2** Label this posterior view of the human heart. 🅐

Brachiocephalic artery

Left common carotid artery

Left subclavian artery

Brachiocephalic veins

Right pulmonary artery

(chamber)

(valve)

(valve)

Right pulmonary veins

(valve)

(chamber)

Opening of coronary sinus

(valve)

(chamber)

(chamber)

Epicardium

Myocardium

Endocardium

**FIGURE 38.3** Label this frontal section of the human heart. The arrows indicate the direction of blood flow. 🅐 **AP|R**

Labels on figure:

1
2
SA node
Openings of coronary arteries
3
Fossa ovalis in interatrial septum
Opening of coronary sinus
4
5
Inferior vena cava
6
7

Ligamentum arteriosum
Pulmonary trunk (cut)
Openings of right pulmonary veins
8
Openings of left pulmonary veins
9
Chordae tendineae
Papillary muscle
10
Muscular ridges
11

**Terms:**

| | |
|---|---|
| Aorta | Mitral valve |
| Aortic valve | Right atrium |
| Apex | Right ventricle |
| Interventricular septum | Superior vena cava |
| Left atrium | Tricuspid valve |
| Left ventricle | |

**FIGURE 38.4** Identify the features indicated on this anterior view of a frontal section of a human heart model, using the terms provided. (*Note:* The pulmonary valve is not shown on the portion of the model photographed.) 🅰 **AP|R** (©McGraw-Hill Education/Al Telser, photographer)

 **LEARN: ACTIVITY**

Use red and blue colored pencils to color the blood vessels in figure 38.3. Use red to illustrate a blood vessel with oxygen-rich (oxygenated) blood, and use blue to illustrate a blood vessel with oxygen-poor (deoxygenated) blood. You can check your work by referring to the corresponding figures in the textbook, presented in full color.

 **PRACTICE**

**PROCEDURE B—Dissection of a Sheep Heart**

1. Obtain a preserved sheep heart. Rinse it in water thoroughly to remove as much of the preservative as possible. Also run water into the large blood vessels to force any blood clots out of the heart chambers. The structures of a sheep heart are similar to those of a human heart and will therefore be used for further exploration. Although sheep are quadrupeds, human terminology will be used for the sheep heart.

2. Place the heart in a dissecting tray with its anterior surface up (fig. 38.5), and proceed as follows:

   a. Although the relatively thick *pericardial sac* probably is missing, look for traces of this membrane around the origins of the large blood vessels.

   b. Locate the *visceral pericardium,* which appears as a thin, transparent layer on the surface of the heart. Use a scalpel to remove a portion of this layer and expose the *myocardium* beneath. Also note the abundance of fat along the paths of various blood vessels. This adipose tissue occurs in the loose connective tissue that underlies the visceral pericardium.

   c. Identify the following:

   **right atrium**

   **right ventricle**

   **left atrium**

   **left ventricle**

   **atrioventricular (coronary) sulcus**

   **anterior interventricular sulcus**

   d. Carefully remove the fat from the anterior interventricular sulcus, and expose the blood vessels that pass along this groove. They include a branch of the *left coronary artery (anterior interventricular artery)* and a *cardiac vein.*

3. Examine the posterior surface of the heart (fig. 38.6), and proceed as follows:

   a. Identify the *atrioventricular (coronary) sulcus* and the *posterior interventricular sulcus.*

   b. Locate the stumps of two relatively thin-walled veins that enter the right atrium. Demonstrate this connection by passing a slender probe through them. The upper vessel is the *superior vena cava,* and the lower one is the *inferior vena cava.*

4. Open the right atrium. To do this, follow these steps:

   a. Insert a blade of the scissors into the superior vena cava and cut downward through the atrial wall (fig. 38.6).

   b. Open the chamber, locate the *right atrioventricular valve (tricuspid valve),* and examine its cusps.

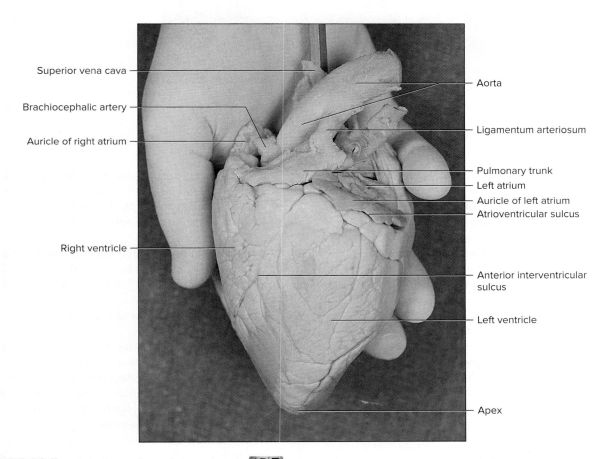

**FIGURE 38.5**   Anterior surface of sheep heart.   AP|R   (©McGraw-Hill Education/Al Telser, photographer)

**FIGURE 38.6** Posterior surface of sheep heart. To open the right atrium, insert a blade of the scissors into the superior (anterior in sheep) vena cava and cut downward. (©McGraw-Hill Education/Al Telser, photographer)

**c.** Also locate the opening to the *coronary sinus* between the valve and the inferior vena cava.

**d.** Run some water through the right atrioventricular valve to fill the chamber of the right ventricle.

**e.** Gently squeeze the ventricles, and watch the cusps of the valve as the water moves up against them.

**5.** Open the right ventricle as follows:

   **a.** Continue cutting downward through the right atrioventricular valve and the right ventricular wall until you reach the apex of the heart.

   **b.** Locate the *chordae tendineae* and the *papillary muscles*.

   **c.** Find the opening to the *pulmonary trunk,* and use the scissors to cut upward through the wall of the right ventricle. Follow the pulmonary trunk until you have exposed the *pulmonary valve.*

   **d.** Examine the valve and its cusps.

**6.** Open the left side of the heart. To do this, follow these steps:

   **a.** Insert the blade of the scissors through the wall of the left atrium and cut downward on the anterior portion to the apex of the heart.

**b.** Open the left atrium, and locate the four openings of the *pulmonary veins.* Pass a slender probe through each opening, and locate the stump of its vessel.

**c.** Examine the *left atrioventricular valve (bicuspid valve)* and its cusps.

**d.** Also examine the left ventricle, and compare the thickness of its wall with that of the right ventricle.

**7.** Locate the aorta, which leads away from the left ventricle, and proceed as follows:

   **a.** Compare the thickness of the aortic wall with that of a pulmonary artery.

   **b.** Use scissors to cut along the length of the aorta to expose the *aortic valve* at its base.

   **c.** Examine the cusps of the valve, and locate the openings of the *coronary arteries* just distal to them.

**8.** As a review, locate and identify the stumps of each of the major blood vessels associated with the heart.

**9.** Discard or save the specimen as directed by the laboratory instructor.

**10.** Complete Part B of the laboratory report.

Name _____

Date _____

Section _____

The 🅰 corresponds to the indicated Learning Outcome(s) found at the beginning of the laboratory exercise.

LABORATORY
REPORT

**38**

# Heart Structure

## ⟳ PART A ASSESSMENTS

Match the terms in column A with the descriptions in column B. Place the letter of your choice in the space provided. 🅰

| Column A | Column B |
|---|---|
| **a.** Aorta | _____ **1.** Structure from which chordae tendineae originate |
| **b.** Cardiac vein | _____ **2.** Prevents blood movement from right ventricle to right atrium |
| **c.** Coronary artery | _____ **3.** Membranes around heart |
| **d.** Coronary sinus | _____ **4.** Prevents blood movement from left ventricle to left atrium |
| **e.** Endocardium | _____ **5.** Gives rise to left and right pulmonary arteries |
| **f.** Mitral valve | _____ **6.** Drains blood from myocardium into right atrium |
| **g.** Myocardium | _____ **7.** Inner lining of heart chamber |
| **h.** Papillary muscle | _____ **8.** Layer largely composed of cardiac muscle tissue |
| **i.** Pericardial cavity | _____ **9.** Space containing serous fluid to reduce friction during heartbeats |
| **j.** Pericardial sac | _____ **10.** Drains blood from myocardial capillaries |
| **k.** Pulmonary trunk | _____ **11.** Supplies blood to heart muscle |
| **l.** Tricuspid valve | _____ **12.** Distributes blood to body organs (systemic circuit) except lungs |

## ⟳ PART B ASSESSMENTS

Complete the following:

**1.** Compare the structure of the right atrioventricular (tricuspid) valve with that of the pulmonary valve. 🅰

_____

_____

_____

_____

**2.** Describe the action of the right atrioventricular (tricuspid) valve when you squeezed the water-filled right ventricle. 🅰

_____

_____

_____

_____

**3.** Describe the function of the chordae tendineae and the papillary muscles. **3** _____

_____

_____

_____

**4.** What is the significance of the difference in thickness between the wall of the aorta and the wall of the pulmonary trunk? **3** _____

_____

_____

_____

**5.** List the correct pathway through which blood must flow in relation to the region of the heart. Assume blood is currently in a vena cava and will eventually enter the aorta. Include the major blood vessels (arteries and veins) attached directly to a heart chamber, the four heart chambers, and the four heart valves in your list. A completed figure 38.3 would be a helpful reference to use as you compose the answer. **1** **2** **3** _____

_____

_____

_____

_____

## ASSESS

### CRITICAL THINKING

What is the significance of the difference in thickness of the ventricular walls? **3**

_____

_____

_____

_____

# 39

# Cardiac Cycle

## MATERIALS NEEDED

Textbook

*For Procedure A—Heart Sounds:*
Stethoscope
Alcohol swabs

*For Procedure B—Electrocardiogram:*
Electrocardiograph (or other instrument for recording an ECG)
Cot or table
Alcohol swabs
Electrode paste (gel)
Electrodes and cables
Lead selector switch

## PURPOSE OF THE EXERCISE

To review the events of a cardiac cycle, to become acquainted with normal heart sounds, and to record an electrocardiogram.

## LEARNING OUTCOMES AP|R

After completing this exercise, you should be able to

① Interpret the major events of a cardiac cycle.

② Associate the sounds produced during a cardiac cycle with the valves closing.

③ Correlate the components of a normal ECG pattern with the phases of a cardiac cycle.

④ Record and interpret an electrocardiogram.

**A** set of atrial contractions while the ventricular walls relax, followed by ventricular contractions while the atrial walls relax, constitutes a *cardiac cycle.* Such a cycle is accompanied by blood pressure changes within the heart chambers, movement of blood into and out of the chambers, and opening and closing of heart valves. These closing valves produce vibrations in the tissues and thus create the sounds associated with the heartbeat. If backflow of blood occurs when a heart valve is closed, creating a turbulence noise, the condition is known as a murmur.

The regulation and coordination of the cardiac cycle involves the *cardiac conduction system.* The pathway of electrical signals originates from the *SA (sinoatrial) node* located in the right atrium near the entrance of the superior vena cava. The stimulation of the heartbeat and the heart rate originate from the SA node, so it is often called the *pacemaker.* As the signals pass through the atrial walls toward the *AV (atrioventricular) node,* contractions of the atria take place. When the AV node has been signaled, the rapid continuation of the electrical signals occurs through the *AV bundle,* passes through the *right and left bundle branches* within the interventricular septum, and terminates via the *Purkinje fibers* throughout the ventricular walls. Once the myocardium has been stimulated, ventricular contractions happen, completing one cardiac cycle. The sympathetic and parasympathetic subdivisions of the autonomic nervous system influence the activity of the pacemaker under various conditions. An increased rate results from sympathetic responses; a decreased rate results from parasympathetic responses.

A number of electrical changes also occur in the myocardium as it contracts and relaxes. These changes can be detected by using metal electrodes and an instrument called an *electrocardiograph.* The recording produced by the instrument is an *electrocardiogram,* or *ECG (EKG).*

## PRACTICE

### PROCEDURE A—Heart Sounds

1. Review the sections entitled "Cardiac Cycle" and "Heart Sounds" in chapter 15 of the textbook.
2. Complete Part A of Laboratory Report 39.
3. Listen to your heart sounds. To do this, follow these steps:
   a. Obtain a stethoscope, and clean its earpieces and the diaphragm by using alcohol swabs.
   b. Fit the earpieces into your ear canals so that the angles are positioned in the forward direction.
   c. Firmly place the diaphragm (bell) of the stethoscope on the chest over the fifth intercostal space near the apex of the heart (fig. 39.1) and listen to the sounds. This is a good location to hear the first sound (*lubb*) of a cardiac cycle when the AV valves are closing, which occurs during ventricular *systole* (contraction).

**d.** Move the diaphragm to the second intercostal space, just to the left of the sternum, and listen to the sounds from this region. You should be able to hear the second sound (*dupp*) of the cardiac cycle clearly when the semilunar valves are closing, which occurs during ventricular *diastole* (relaxation).

**e.** It is possible to hear sounds associated with the aortic and pulmonary valves by listening from the second intercostal space on either side of the sternum. The aortic valve sound comes from the right and the pulmonary valve sound from the left. The sound associated with the mitral valve can be heard from the fifth intercostal space at the nipple line on the left. The sound of the tricuspid valve can be heard at the fourth intercostal space just to the left of the sternum (fig. 39.1). The four points for listening (auscultating) the four heart valve sounds when closing are not directly superficial to the valve locations. This is because the sounds associated with the closing valves travel in a slanting direction to the surface of the chest wall where the stethoscope placements occur.

**4.** Inhale slowly and deeply, and exhale slowly while you listen to the heart sounds from each of the locations as before. Note any changes that have occurred in the sounds.

**5.** Exercise moderately outside the laboratory for a few minutes so that other students listening to heart sounds will not be disturbed. After the exercise period, listen to the heart sounds and note any changes that have occurred in them. *This exercise should be avoided by anyone with health risks.*

**6.** Complete Part B of the laboratory report.

 **PRACTICE**

### PROCEDURE B—Electrocardiogram

**1.** Review sections entitled "Cardiac Conduction System," "Electrocardiogram," and "Arrhythmias—Clinical Application" in chapter 15 of the textbook.

**2.** Complete Part C of the laboratory report.

**3.** The laboratory instructor will demonstrate the proper adjustment and use of the instrument available to record an electrocardiogram.

**4.** Record your laboratory partner's ECG. To do this, follow these steps:

**a.** Have your partner lie on a cot or table close to the electrocardiograph, remaining as relaxed and still as possible.

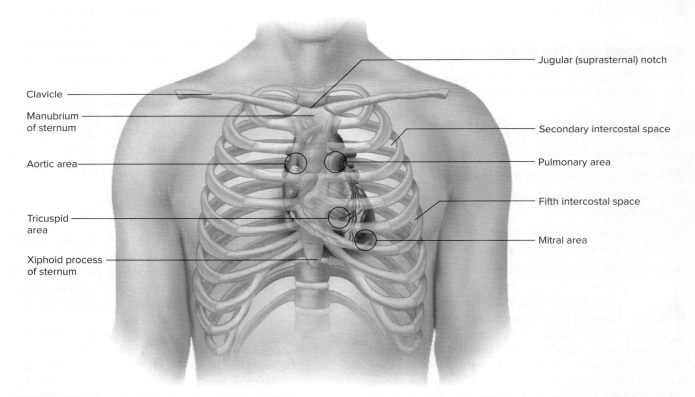

Clavicle

Manubrium of sternum

Aortic area

Tricuspid area

Xiphoid process of sternum

Jugular (suprasternal) notch

Secondary intercostal space

Pulmonary area

Fifth intercostal space

Mitral area

**FIGURE 39.1** The first sound (lubb) of a cardiac cycle can be heard by placing the diaphragm of a stethoscope over the fifth intercostal space near the apex of the heart. The second sound (dupp) can be heard over the second intercostal space, just left of the sternum. The thoracic regions circled indicate where the sounds of each heart valve are most easily heard. Note that where the sound is heard may differ slightly from the anatomical location of the structure. **AP|R**

**b.** Scrub the electrode placement locations with alcohol swabs (fig. 39.2). Apply a small quantity of electrode paste to the skin on the insides of the wrists and ankles. (Any jewelry on the wrists or ankles should be removed.)

**c.** Spread some electrode paste over the inner surfaces of four electrodes and attach one to each of the prepared skin areas (fig. 39.2). Make sure there is good contact between the skin and the metal of the electrodes. The electrode on the right ankle is the grounding system.

**d.** Attach the electrodes to the corresponding cables of a lead selector switch. When an ECG recording is made, only two electrodes are used at a time, and the selector switch allows various combinations of electrodes (leads) to be activated. Three standard limb leads placed on the two wrists and the left ankle are used for an ECG. This arrangement has become known as *Einthoven's triangle*,[1] which permits the recording of the potential difference between any two of the electrodes.

The standard leads I, II, and III are called bipolar leads because they are the potential difference

---

[1]Willem Einthoven (1860–1927), a Dutch physiologist, received the Nobel Prize for Physiology or Medicine for his work with electrocardiograms.

between two electrodes (a positive and a negative). Lead I measures the potential difference between the right wrist (negative) and the left wrist (positive). Lead II measures the potential difference between the right wrist and the left ankle, and lead III measures the potential difference between the left wrist and the left ankle. The right ankle is always the ground.

**e.** Turn on the recording instrument and adjust it as previously demonstrated by the laboratory instructor. The paper speed should be set at 2.5 cm/second. This is the standard speed for ECG recordings.

**f.** Set the lead selector switch to lead I (right wrist, left wrist electrodes), and record the ECG for 1 minute.

**g.** Set the lead selector switch to lead II (right wrist, left ankle electrodes), and record the ECG for 1 minute.

**h.** Set the lead selector switch to lead III (left wrist, left ankle electrodes), and record the ECG for 1 minute.

**i.** Remove the electrodes and clean the paste from the metal and skin.

**j.** Use figure 39.3 to label the ECG components of the results from leads I, II, and III. The PQ interval is often called the PR interval because the Q wave is often small or absent. The normal PQ interval is 0.12–0.20 seconds. The normal QRS complex duration is less than 0.10 second.

**5.** Complete Part D of the laboratory report.

**FIGURE 39.2** To record an ECG, attach electrodes to the wrists and ankles.

**FIGURE 39.3** Components of a normal ECG pattern with a time scale. **AP|R**

# Notes

302

Name _____

Date _____

Section _____

The  corresponds to the indicated Learning Outcome(s) found at the beginning of the laboratory exercise.

LABORATORY
REPORT

39

# Cardiac Cycle

## PART A ASSESSMENTS

Complete the following statements:

1. About _____ % of the blood from the atria passes into the ventricles before the atrial walls contract. **1**

2. The period during which a heart chamber is contracting is called _____ . **1**

3. The period during which a heart chamber is relaxing is called _____ . **1**

4. During ventricular contraction, the AV valves (tricuspid and mitral valves) are _____ . **2**

5. During ventricular relaxation, the AV valves are _____ . **2**

6. The pulmonary and aortic valves open when the pressure in the _____ exceeds the pressure in the pulmonary trunk and aorta. **1**

7. Heart sounds are due to _____ in heart tissues created by changes in blood flow. **2**

8. The first sound of a cardiac cycle occurs when the _____ are closing. **2**

9. The second sound of a cardiac cycle occurs when the _____ are closing. **2**

10. The sound created when blood leaks back through an incompletely closed valve is called a(an) _____ . **2**

## PART B ASSESSMENTS

Complete the following:

1. What changes did you note in the heart sounds when you inhaled deeply? **2**

_____

_____

2. What changes did you note in the heart sounds following the exercise period? **2**

_____

_____

**PART C ASSESSMENTS**

Complete the following statements:

1. The cells (fibers) of the cardiac conduction system are specialized _____ tissue. [3]

2. Normally, the _____ node serves as the pacemaker of the heart. [3]

3. The _____ node is located in the inferior portion of the interatrial septum. [3]

4. The large fibers on the distal side of the AV node make up the _____ . [3]

5. The fibers that carry cardiac impulses from the interventricular septum into the myocardium are called _____ fibers. [3]

6. A(An) _____ is a recording of electrical changes occurring in the myocardium during the cardiac cycle. [3]

7. Between cardiac cycles, cardiac muscle fibers remain _____ with no detectable electrical changes. [3]

8. The P wave corresponds to depolarization of the muscle fibers of the _____ . [3]

9. The QRS complex corresponds to depolarization of the muscle fibers of the _____ . [3]

10. The T wave corresponds to repolarization of the muscle fibers of the _____ . [3]

11. Why is atrial repolarization not observed in the ECG? [3] _____

_____

12. A rapid heartbeat is called _____ ; a slow heartbeat is called _____ . [4]

 **PART D ASSESSMENTS**

1. Attach a short segment of the ECG recording from each of the three leads you used, and label the waves of each. [4]

Lead I

Lead II




Lead III




**2.** What differences do you find in the ECG patterns of these leads? 🅰 _____

_____

_____

_____

_____

_____

**3.** How much time passed from the beginning of the P wave to the beginning of the QRS complex (PQ interval, or PR interval) in the ECG from lead I? [4] _____

**4.** What is the significance of this PQ (PR) interval? [4] _____

_____

_____

_____

_____

**5.** How can you determine heart rate from an electrocardiogram? [4] _____

_____

_____

_____

**6.** What was your heart rate as determined from the ECG? [4] _____

## ASSESS

### CRITICAL THINKING

If a person's heart rate were 72 beats per minute, determine the number of QRS complexes that would have appeared on an ECG during the first 30 seconds. [4]

_____

_____

_____

# 40

# Blood Vessel Structure, Arteries, and Veins

## MATERIALS NEEDED

Textbook
Dissectible human heart model
Human torso model
Anatomical charts of the cardiovascular system
Compound light microscope
Prepared microscope slides:
    Artery cross section
    Vein cross section
Live frog or goldfish
Frog Ringer's solution
Paper towel
Rubber bands
Frog board or heavy cardboard (with a 1-inch hole cut in one corner)
Dissecting pins
Thread
Masking tape
Ice
Hot plate
Thermometer

## ⚠ SAFETY

- Wear disposable gloves when handling the live frogs.
- Return the frogs to the location indicated after the experiment.
- Wash your hands before leaving the laboratory.

## PURPOSE OF THE EXERCISE

To review the structure and functions of blood vessels, to trace major circulatory pathways, and to locate the major arteries and veins.

##  LEARNING OUTCOMES AP|R

After completing this exercise, you should be able to

1. Distinguish the structures and functions of arteries, veins, and capillaries.
2. Identify the types of blood vessels in the web of a frog's foot.
3. Distinguish and trace the pulmonary and systemic circuits.
4. Locate the major arteries in these circuits on a diagram, chart, or model.
5. Locate the major veins in these circuits on a diagram, chart, or model.
6. Recognize the anatomical terms pertaining to the blood vessel structure, arteries, and veins.

The blood vessels form a closed system of tubes that carry blood to and from the heart, lungs, and body cells. These tubes include *arteries* and *arterioles* that transport blood away from the heart; *capillaries* in which exchanges of substances occur between the blood and surrounding tissues; and *venules* and *veins* that return blood to the heart.

The blood vessels of the cardiovascular system can be divided into two major pathways—the *pulmonary circuit* and the *systemic circuit*. Within each circuit, arteries transport blood away from the heart. After exchanges of gases, nutrients, and wastes have occurred between the blood and the surrounding tissues, veins return the blood to the heart.

Variations exist in anatomical structures among humans, especially in arteries and veins. The illustrations in this laboratory manual represent normal (meaning the most common variation) anatomy.

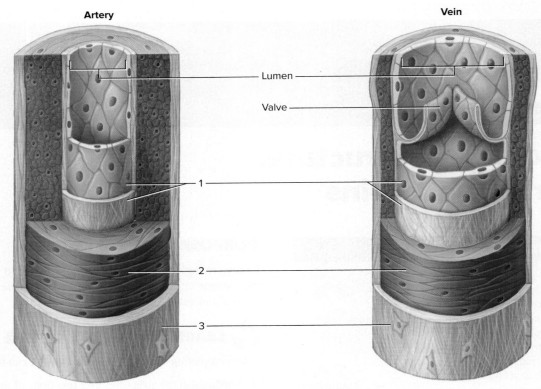

Artery                                                   Vein

Lumen

Valve

1

2

3

**FIGURE 40.1**     Label the tunics of the walls of this artery and vein. **1** **AP|R**

 **PRACTICE**

### PROCEDURE A—Blood Vessel Structure

1. Review the section entitled "Blood Vessels" in chapter 15 of the textbook.
2. As a review activity, label figure 40.1.
3. Complete Part A of Laboratory Report 40.
4. Obtain a microscope slide of an artery cross section, and examine it using low-power and high-power magnification. Identify the three distinct layers (tunics) of the arterial wall. The inner layer (**tunica interna**) is composed of an endothelium (simple squamous epithelium) and appears as a wavy line due to an abundance of elastic fibers that have recoiled just beneath it. The middle layer (**tunica media**) consists of numerous concentrically arranged smooth muscle cells with elastic fibers scattered among them. The outer layer (**tunica externa**) contains connective tissue rich in collagen fibers (fig. 40.2).
5. Prepare a labeled sketch of the arterial wall in Part B of the laboratory report.
6. Obtain a slide of a vein cross section and examine it as you did the artery cross section. Note the thinner wall and larger lumen relative to an artery of comparable

locations. Identify the three layers of the wall, and prepare a labeled sketch in Part B of the laboratory report.
7. Complete Part B of the laboratory report.
8. Observe the blood vessels in the webbing of a frog's foot. (As a substitute for a frog, a live goldfish could be used to observe circulation in the tail.) To do this, follow these steps:
   a. Obtain a live frog. Wrap its body in a moist paper towel, leaving one foot extending outward. Secure the towel with rubber bands, but be careful not to wrap the animal so tightly that it could be injured. Try to keep the nostrils exposed.
   b. Place the frog on a frog board or on a piece of heavy cardboard with the foot near the hole in one corner.
   c. Fasten the wrapped body to the board with masking tape.
   d. Carefully spread the web of the foot over the hole and secure it to the board with dissecting pins and thread (fig. 40.3). Keep the web moist with frog Ringer's solution.
   e. Secure the board on the stage of a microscope with heavy rubber bands, and position it so that the web is beneath the objective lens.

Endothelium
of inner layer
(tunica interna)

Middle layer
(tunica
media)

Outer layer
(tunica
externa)

Lumen of venule

Lumen of arteriole

**FIGURE 40.2**   Cross section of a small artery (arteriole) and a small vein (venule) (200×).   **AP|R**  (©McGraw-Hill Education/Al Telser, photographer)

 **LEARN: ACTIVITY**

Investigate the effect of temperature change on the blood vessels of the frog's foot by flooding the web with a small quantity of ice water. Observe the blood vessels with low-power magnification and note any changes in their diameters or the rate of blood flow. Remove the ice water and replace it with water heated to about 35°C (95°F). Repeat your observations. What do you conclude from this experiment?

_____

_____

_____

**f.** Focus on the web, using low-power magnification, and locate some blood vessels. Note the movement of the blood cells and the direction of the blood flow. You might notice that red blood cells of frogs are nucleated. Identify an arteriole, a capillary, and a venule.

**g.** Examine each of these vessels with high-power magnification.

**h.** When finished, return the frog to the location indicated by your instructor. The microscope lenses and stage will likely need cleaning after the experiment.

**9.** Complete Part C of the laboratory report.

 **PRACTICE**

### PROCEDURE B—Paths of Circulation

**1.** Review the sections entitled "Pulmonary Circuit" and "Systemic Circuit" in chapter 15 of the textbook.

**2.** As a review activity, label figure 40.4.

**3.** Locate the following blood vessels on the available anatomical charts, the dissectible human heart model, and the human torso model:

**pulmonary circuit**
    pulmonary trunk 1
    pulmonary arteries 2
    pulmonary veins 4

**systemic circuit**
    aorta 1
    superior vena cava 1
    inferior vena cava 1

 **ASSESS**

### CRITICAL THINKING

Why is the left ventricle wall thicker than the right ventricle wall?

_____

_____

_____

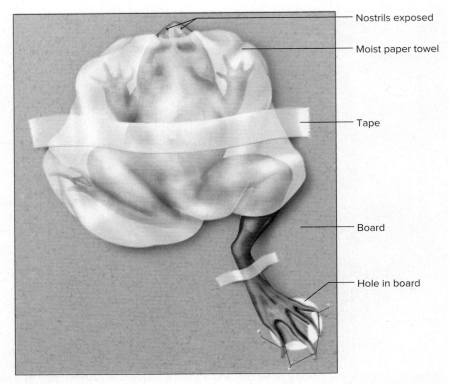

Nostrils exposed

Moist paper towel

Tape

Board

Hole in board

**FIGURE 40.3** Spread the web of the foot over the hole and secure it to the board with pins and thread.

 **PRACTICE**

### PROCEDURE C—Arterial System

1. Review the section entitled "Arterial System" in chapter 15 of the textbook.
2. As a review activity, label and study figures 40.5, 40.6, 40.7, 40.8, and 40.9.
3. Locate the following arteries of the systemic circuit on the charts and human torso model:

   **aorta**
   - ascending aorta
   - aortic arch (arch of aorta)
   - thoracic aorta
   - abdominal aorta

   **branches of the aorta**
   - coronary artery
   - brachiocephalic trunk (artery)
   - left common carotid artery
   - left subclavian artery
   - celiac trunk (artery)
   - superior mesenteric artery
   - renal artery
   - inferior mesenteric artery
   - common iliac artery

   **arteries to neck, head, and brain**
   - vertebral artery
   - common carotid artery
   - external carotid artery
   - internal carotid artery
   - facial artery

   **arteries of base of brain** (fig. 40.7)
   - vertebral artery
   - basilar artery
   - internal carotid artery
   - cerebral arterial circle (circle of Willis)

   **arteries to shoulder and upper limb**
   - subclavian artery
   - axillary artery
   - brachial artery
   - deep brachial artery
   - ulnar artery
   - radial artery

   **arteries to pelvis and lower limb**
   - common iliac artery
   - internal iliac artery
   - external iliac artery
   - femoral artery
   - popliteal artery
   - anterior tibial artery
   - dorsalis pedis artery (dorsal artery of foot)
   - posterior tibial artery

4. Complete Part D of the laboratory report.

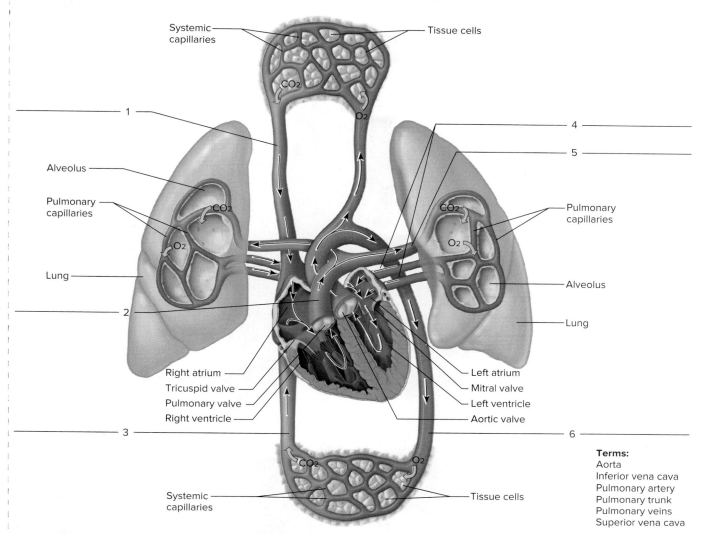

Systemic capillaries — Tissue cells

$CO_2$

$O_2$

1

Alveolus

Pulmonary capillaries

$CO_2$

$O_2$

Lung

2

Right atrium
Tricuspid valve
Pulmonary valve
Right ventricle

4

5

$CO_2$

$O_2$

Pulmonary capillaries

Alveolus

Lung

Left atrium
Mitral valve
Left ventricle
Aortic valve

3

6

$CO_2$

$O_2$

Systemic capillaries — Tissue cells

**Terms:**
Aorta
Inferior vena cava
Pulmonary artery
Pulmonary trunk
Pulmonary veins
Superior vena cava

**FIGURE 40.4** Label the major blood vessels associated with the pulmonary and systemic circuits, using the terms provided. The red colors indicate locations of oxygen-rich (oxygenated) blood; the blue colors indicate locations of oxygen-poor (deoxygenated) blood. (Structures are not drawn to scale.) 3 APR

1

2

3

4

5

6

7

8

9

10

11

Left gastric artery

Spleen

Splenic artery

Stomach

Superior mesenteric artery

Lumbar artery

Inferior mesenteric artery

Middle (median) sacral artery

Liver

Celiac trunk

Hepatic artery

Right gastric artery

Kidney

Gonadal artery

**FIGURE 40.5** Label the portions of the aorta and its principal branches. **4 AP|R**

**FIGURE 40.6**  Label the arteries supplying the right side of the neck and head. (The clavicle has been removed.) **A** **AP|R**

Occipital artery

Thyrocervical trunk

1

2

3

4

5

6

7

8

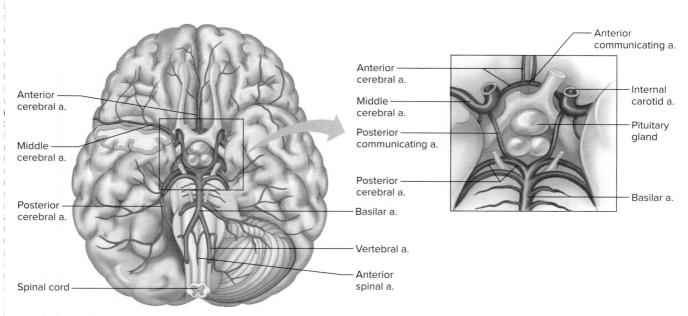

Anterior
cerebral a.

Middle
cerebral a.

Posterior
cerebral a.

Spinal cord

Basilar a.

Vertebral a.

Anterior
spinal a.

Anterior
cerebral a.

Middle
cerebral a.

Posterior
communicating a.

Posterior
cerebral a.

Anterior
communicating a.

Internal
carotid a.

Pituitary
gland

Basilar a.

**FIGURE 40.7**  View of inferior surface of the brain. The cerebral arterial circle (circle of Willis) is formed by the anterior and posterior cerebral arteries, which join the internal carotid arteries (a. stands for artery). **AP|R**

**FIGURE 40.8** Label the major arteries of the shoulder and upper limb. **4** **AP|R**

5 —————————————

1 —————

2 —————

6 —————————————

Inguinal ligament

3 —————

7 —————————————

(on posterior side)

4 —————

8 —————————————

**Terms:**
Abdominal aorta
Anterior tibial artery
Common iliac artery
Deep femoral artery
Dorsalis pedis artery
External iliac artery
Femoral artery
Internal iliac artery
Popliteal artery

9 —————————————

**Anterior view**

**FIGURE 40.9**   Label the major arteries supplying the pelvis and lower limb, using the terms provided.   A AP R

**PROCEDURE D—Venous System**

1. Review the section entitled "Venous System" in chapter 15 of the textbook.
2. As a review activity, label figures 40.10, 40.11, 40.12, and 40.13.
3. Locate the following veins of the systemic circuit on the charts and human torso model:

**veins from brain, head, and neck**
  external jugular vein
  internal jugular vein
  vertebral vein
  subclavian vein
  brachiocephalic vein
  superior vena cava

**veins from upper limb and shoulder**
  radial vein
  ulnar vein
  brachial vein
  basilic vein
  cephalic vein
  median cubital vein (antecubital vein)

axillary vein
subclavian vein

**veins of abdominal viscera**
  hepatic portal vein
  gastric vein
  superior mesenteric vein
  splenic vein
  inferior mesenteric vein
  hepatic vein
  renal vein

**veins from lower limb and pelvis**
  anterior tibial vein
  posterior tibial vein
  popliteal vein
  femoral vein
  great (long) saphenous vein
  small (short) saphenous vein
  external iliac vein
  internal iliac vein
  common iliac vein
  inferior vena cava

4. Complete Parts E and F of the laboratory report.

**FIGURE 40.10** Label the major veins associated with the head and neck. (The clavicle has been removed.)

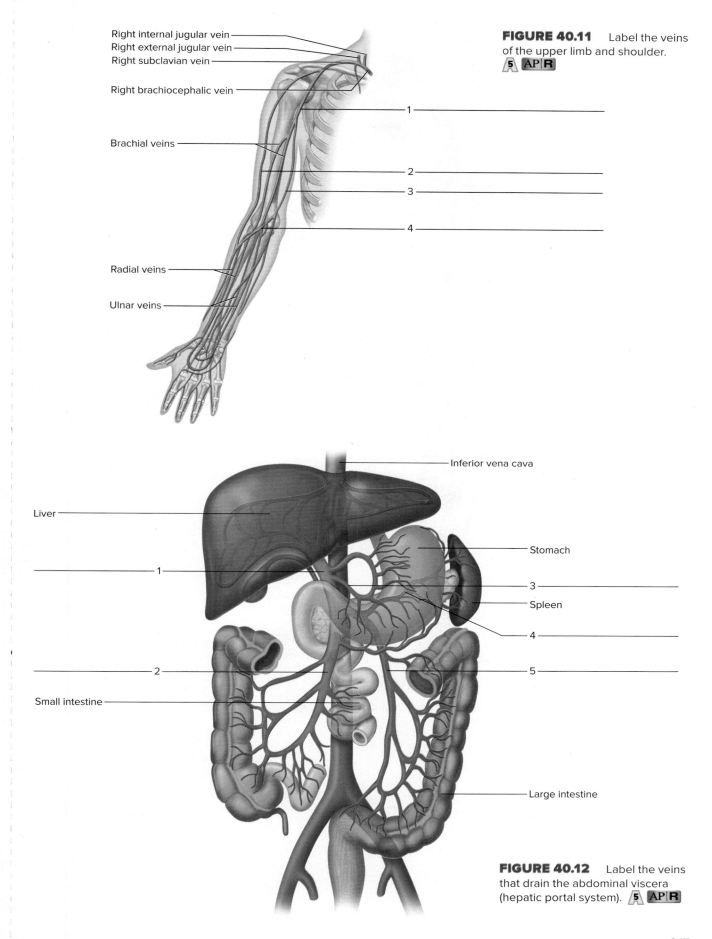

Right internal jugular vein
Right external jugular vein
Right subclavian vein
Right brachiocephalic vein

Brachial veins

Radial veins

Ulnar veins

1
2
3
4

**FIGURE 40.11**  Label the veins of the upper limb and shoulder. **A** AP|R

Inferior vena cava

Liver

Stomach

1

3

Spleen

4

2

5

Small intestine

Large intestine

**FIGURE 40.12**  Label the veins that drain the abdominal viscera (hepatic portal system). **A** AP|R

1 ————————
2 ————————
3 ————————
4 ————————
— Inguinal ligament
5 ————————
6 ————————
— Popliteal vein
(on posterior side)
— Anterior tibial veins
— Dorsalis pedis vein

**Anterior view**

**FIGURE 40.13** Label the veins of the pelvis and lower limb. Ⓐ **AP|R**

The Ⓐ corresponds to the indicated Learning Outcome(s) found at the beginning of the laboratory exercise.

# Blood Vessel Structure, Arteries, and Veins

## ♻ PART A ASSESSMENTS

Complete the following statements:

1. Simple squamous epithelial tissue called _____ forms the inner linings of the tunica interna of blood vessels. Ⓐ

2. The _____ of an arterial wall contains many smooth muscle cells. Ⓐ

3. The _____ of an arterial wall is largely composed of connective tissue. Ⓐ

4. When contraction of the smooth muscle in a blood vessel wall occurs, the vessel is referred to as being in a condition of _____ . Ⓐ

5. Relaxation of the smooth muscle in a blood vessel wall results in the vessel being in a condition of _____
_____ . Ⓐ

6. The smallest blood vessels are called _____ . Ⓐ

7. Filtration results when substances are forced through capillary walls by _____ pressure. Ⓐ

8. The presence of plasma proteins in the blood increases its _____ pressure as compared to tissue fluids. Ⓐ

9. _____ in certain veins close if the blood begins to back up in that vein. Ⓐ

## ♻ PART B ASSESSMENTS

1. Sketch and label a section of an artery wall and a vein wall. Ⓐ

| | |
|---|---|
| Artery wall (____×) | Vein wall (____×) |

**2.** Describe the differences you noted in the structures of the wall of an artery and a vein. Mention each of the three layers of the wall. 🔺1 _____

_____

_____

## ♻ ASSESS

**CRITICAL THINKING**

Explain the functional significance of the differences you noted in the structures of the arterial and venous walls. 🔺1

_____

_____

## ♻ PART C ASSESSMENTS

**1.** How did you distinguish between arterioles and venules when you observed the vessels in the web of the frog's foot? 🔺2

_____

_____

**2.** How did you recognize capillaries in the web? 🔺2 _____

**3.** What differences did you note in the rate of blood flow through the arterioles, capillaries, and venules? 🔺2

_____

_____

## ♻ PART D ASSESSMENTS

Match the arteries in column A with the descriptions in column B. Place the letter of your choice in the space provided. 🔺4

| Column A | Column B |
|---|---|
| **a.** Anterior tibial | _____ **1.** Jaw, teeth, and face |
| **b.** Brachial | _____ **2.** Kidney |
| **c.** Celiac | _____ **3.** Upper digestive tract, spleen, and liver |
| **d.** External carotid | _____ **4.** Foot and toes |
| **e.** Internal carotid | _____ **5.** Gluteal muscles |
| **f.** Internal iliac | _____ **6.** Biceps muscle |
| **g.** Phrenic | _____ **7.** Knee joint |
| **h.** Popliteal | _____ **8.** Adrenal gland |
| **i.** Renal | _____ **9.** Diaphragm |
| **j.** Suprarenal | _____ **10.** Brain |

Match each vein in column A with the vein it drains into from column B. Place the letter of your choice in the space provided.

| Column A | Column B |
|---|---|
| **a.** Anterior tibial | _____ **1.** Popliteal |
| **b.** Basilic | _____ **2.** Axillary |
| **c.** Brachiocephalic | _____ **3.** Inferior vena cava |
| **d.** Common iliac | _____ **4.** Subclavian |
| **e.** External jugular | _____ **5.** Brachial |
| **f.** Femoral | _____ **6.** Superior vena cava |
| **g.** Popliteal | _____ **7.** Femoral |
| **h.** Radial | _____ **8.** External iliac |

Locate the 20 anatomical terms pertaining to the "Blood Vessel Structure, Arteries, and Veins."

| | | | | | | | | | | |
|---|---|---|---|---|---|---|---|---|---|---|
| B | A | S | I | L | I | C | C | I | N | U | T |
| W | J | Y | K | S | F | A | C | I | A | L | H |
| V | U | R | A | U | M | G | L | O | A | N | O |
| F | G | A | R | O | K | A | R | E | Z | A | R |
| C | U | N | T | N | N | T | T | B | Z | R | A |
| E | L | O | E | E | A | I | L | I | A | C | C |
| P | A | M | R | H | L | A | T | R | O | P | I |
| H | R | L | I | P | V | C | A | I | L | E | C |
| A | W | U | O | A | X | I | L | L | A | R | Y |
| L | U | P | L | S | Y | S | T | E | M | I | C |
| I | H | V | E | R | T | E | B | R | A | L | G |
| C | A | P | I | L | L | A | R | Y | T | M | D |

Find and circle words pertaining to Laboratory Exercise 40. The words may be horizontal, vertical, diagonal, or backward.

7 terms (arteriole, capillary, portal, pulmonary, systemic, thoracic, tunic)

9 arteries; 4 veins.

Label the major arteries and veins indicated in figure 40.14. **3** **4** **5**

Internal jugular vein

6

7

Aortic arch

8

Pulmonary artery

9

Right lung

Liver

Right kidney

10

11

12

13

Urinary bladder

Deep femoral artery

14

**Terms:**

Abdominal aorta
Axillary vein
Brachiocephalic vein
Common carotid artery
Common iliac artery
Common iliac vein
External jugular vein

Femoral artery
Femoral vein
Great saphenous vein
Inferior vena cava
Pulmonary vein
Subclavian artery
Superior vena cava

**FIGURE 40.14** Label the major arteries and veins of the systemic and pulmonary circuits, using the terms provided.

# Pulse Rate and Blood Pressure

### MATERIALS NEEDED

Textbook
Clock with second hand
Sphygmomanometer
Stethoscope
Alcohol swabs

## PURPOSE OF THE EXERCISE

To examine the pulse, to determine the pulse rate, to measure blood pressure, and to investigate the effects of body position and exercise on pulse rate and blood pressure.

 **LEARNING OUTCOMES**

After completing this exercise, you should be able to

1. Determine pulse rate and correlate pulse characteristics.

2. Test the effects of various factors on pulse rate.

3. Describe and measure blood pressure using a sphygmomanometer.

4. Test the effects of various factors on blood pressure.

The surge of blood that enters the arteries each time the ventricles of the heart contract causes the elastic walls of these vessels to swell. Then, as the ventricles relax, the walls recoil. This alternate expanding and recoiling of an arterial wall can be felt as a pulse in vessels that run close to the surface of the body.

The force exerted by the blood pressing against the inner walls of arteries also creates blood pressure. This pressure reaches a maximum during ventricular contraction and then drops to its lowest level while the ventricles are relaxed.

 **PRACTICE**

### PROCEDURE A—Pulse Rate

1. Review the section entitled "Arterial Blood Pressure" in chapter 15 of the textbook.

2. Complete Part A of Laboratory Report 41.

3. Examine your laboratory partner's radial pulse. To do this, follow these steps:

   a. Have your partner sit quietly, remaining as relaxed as possible.

   b. Locate the pulse by placing your index and middle fingers over the radial artery on the anterior surface near the lateral side of the wrist. Do not use your thumb for sensing the pulse because you may feel a pulse coming from an artery in the thumb. The radial artery is most often used because it is easy and convenient to locate.

   c. To determine the pulse rate, count the number of pulses that occur in 1 minute. This can be accomplished by counting pulses in 30 seconds and multiplying that number by 2. (A pulse count for a full minute, although taking a little longer, would give a better opportunity to detect pulse characteristics and irregularities.)

   d. Note the characteristics of the pulse. That is, can it be described as regular or irregular, strong or weak, hard or soft? The pulse should be regular and the amplitude (magnitude) will decrease as the distance from the left ventricle increases. The strength of the pulse reveals some indications of blood pressure. Under high blood pressure, the pulse feels very hard and strong; under low blood pressure, the pulse feels weak and can be easily compressed.

   e. Record the pulse rate and pulse characteristics while sitting in Part B of the laboratory report.

4. Repeat the procedure and determine the pulse rate and pulse characteristics in each of the following conditions:

   a. Immediately after lying down

   b. 3–5 minutes after lying down

   c. Immediately after standing

   d. 3–5 minutes after standing quietly

   e. Immediately after 3 minutes of moderate exercise *(omit if the person has health problems)*

   f. 3–5 minutes after exercise has ended

   Record the pulse rates and pulse characteristics in Part B of the laboratory report.

5. Complete Part B of the laboratory report.

## LEARN: LAB IN MOTION

Determine the pulse rate and pulse characteristics in two additional locations on your laboratory partner. Locate and record the pulse rate from the common carotid artery and the dorsalis pedis artery. ⚠

Common carotid artery pulse rate _____

Dorsalis pedis artery pulse rate _____

Compare the amplitude of the pulse characteristics between these two pulse locations and interpret any of the variations noted. _____

_____

_____

_____

_____

## PRACTICE

### PROCEDURE B—Blood Pressure

1. Although blood pressure is present in all blood vessels, the standard location to record blood pressure is the brachial artery.
2. Measure your laboratory partner's arterial blood pressure. To do this, follow these steps:
   a. Obtain a sphygmomanometer and a stethoscope.
   b. Clean the earpieces and the diaphragm of the stethoscope with alcohol swabs.
   c. Have your partner sit quietly with upper limb resting on a table at heart level. Have the person remain as relaxed as possible.
   d. Locate the brachial artery at the antecubital space. Wrap the cuff of the sphygmomanometer around the arm so that its lower border is about 2.5 cm above the bend of the elbow. Center the bladder of the cuff in line with the *brachial pulse* (fig. 41.1).
   e. Palpate the *radial pulse*. Close the valve on the neck of the rubber bulb connected to the cuff, and pump air from the bulb into the cuff. Inflate the cuff while watching the sphygmomanometer, and note the pressure when the pulse disappears. (This is a rough estimate of the systolic pressure.) Immediately deflate the cuff. *Do not leave the cuff inflated for more than 1 minute.*
   f. Position the stethoscope over the brachial artery. Reinflate the cuff to a level 30 mm Hg higher than the point where the pulse disappeared during palpation.
   g. Slowly open the valve of the bulb until the pressure in the cuff drops at a rate of about 2 or 3 mm Hg per second.
   h. Listen for sounds (Korotkoff sounds) from the brachial artery. When the first loud tapping sound is heard, record the reading as the systolic pressure.

This indicates the pressure exerted against the arterial wall during systole.
   i. Continue to listen to the sounds as the pressure drops, and note the level when the last sound is heard. Record this reading as the diastolic pressure, which measures the constant arterial resistance.
   j. Release all of the pressure from the cuff.
   k. Repeat the procedure until you have two blood pressure measurements from each arm, allowing 2–3 minutes of rest between readings.
   l. Average your readings and enter them in the table in Part C of the laboratory report.
3. Measure and record your partner's blood pressure in each of the following conditions:
   a. 3–5 minutes after lying down
   b. 3–5 minutes after standing quietly
   c. Immediately after 3 minutes of moderate exercise *(omit if the person has health problems)*
   d. 3–5 minutes after exercise has ended
   Record the blood pressures in Part C of the laboratory report.
4. Complete Part C of the laboratory report.

**FIGURE 41.1** Blood pressure is commonly measured by using a sphygmomanometer (blood pressure cuff). The use of a column of mercury is the most accurate measurement, but due to environmental concerns, it has been replaced by alternative gauges and digital readouts.

Labels on figure:
- Aneroid pressure gauge
- Rubber inflatable cuff
- Antecubital space (location of brachial pulse and stethoscope placement)
- Air control valve
- Bulb
- Location of radial pulse

The 🅐 corresponds to the indicated Learning Outcome(s) found at the beginning of the laboratory exercise.

# Pulse Rate and Blood Pressure

## 🔄 PART A ASSESSMENTS

Complete the following statements:

1. The term *blood pressure* most commonly is used to refer to systemic _____ pressure. 🅐**3**

2. The maximum pressure achieved during ventricular contraction is called _____ pressure. 🅐**3**

3. The lowest pressure that remains in the arterial system during ventricular relaxation is called _____ pressure. 🅐**3**

4. The pulse rate is equal to the _____ rate. 🅐**1**

5. A pulse that feels full and is not easily compressed is produced by an elevated _____ . 🅐**1**

6. The instrument commonly used to measure systemic arterial blood pressure is called a(an) _____ . 🅐**3**

7. Blood pressure is expressed in units of _____ . 🅐**3**

8. The upper number of the fraction used to record blood pressure indicates the _____ pressure. 🅐**3**

9. The difference between the systolic pressure and diastolic pressure is called the _____ . 🅐**3**

10. The _____ artery in the arm is the standard systemic artery in which blood pressure is measured. 🅐**3**

## 🔄 PART B ASSESSMENTS

1. Enter your observations of pulse rates and pulse characteristics in the table. 🅐**1** 🅐**2**

| Test Subject | Pulse Rate (beats/min) | Pulse Characteristics |
|---|---|---|
| Sitting | | |
| Lying down | | |
| 3–5 minutes later | | |
| Standing | | |
| 3–5 minutes later | | |
| After exercise | | |
| 3–5 minutes later | | |

**2.** Summarize the effects of body position and exercise on the pulse rates and the pulse characteristics. **2** _____

_____

_____

_____

## PART C ASSESSMENTS

**1.** Enter the initial measurements of blood pressure from a relaxed sitting position in the table. **3**

| Reading | Blood Pressure in Right Arm | Blood Pressure in Left Arm |
|---------|-----------------------------|----------------------------|
| First   |                             |                            |
| Second  |                             |                            |
| Average |                             |                            |

**2.** Enter your test results in the table. **4**

| Test Subject | Blood Pressure |
|--------------|----------------|
| 3–5 minutes after lying down |  |
| 3–5 minutes after standing |  |
| After 3 minutes of moderate exercise |  |
| 3–5 minutes later |  |

**3.** Summarize the effects of body position and exercise on blood pressure. **4** _____

_____

_____

_____

**4.** Summarize any correlations between pulse rate and blood pressure from any of the experimental conditions. **2** **4**

_____

_____

_____

## ASSESS

### CRITICAL THINKING

When a pulse is palpated and counted, which pressure (systolic or diastolic) is characteristic at that moment? Explain your answer. **1**

_____

_____

_____

# 42

# Lymphatic System

## MATERIALS NEEDED

Textbook
Human torso model
Anatomical chart of the lymphatic system
Compound light microscope
Prepared microscope slides:
   Lymph node section
   Human thymus section
   Human spleen section

## PURPOSE OF THE EXERCISE

To review the structure of the lymphatic system and to observe the microscopic structure of a lymph node, the thymus, and the spleen.

 ## LEARNING OUTCOMES AP|R

After completing this exercise, you should be able to

(1) Trace the major lymphatic pathways in an anatomical chart or model.

(2) Locate and identify the major clusters of lymph nodes (lymph glands) in an anatomical chart or a model.

(3) Describe the structure and function of a lymphatic vessel, a lymph node, the thymus, and the spleen.

(4) Locate and sketch the major microscopic structures of a lymph node, the thymus, and the spleen.

The lymphatic system is a vast collection of cells and biochemicals that travel in lymphatic capillaries and vessels and the organs and glands that produce them. The system is closely associated with the cardiovascular system and includes a network of vessels that assist in the circulation of body fluids. These vessels provide pathways through which excess fluid can be transported away from interstitial spaces within most tissues and returned to the bloodstream. Without the lymphatic system, this fluid would accumulate in tissue spaces, producing edema.

    The organs of the lymphatic system also help defend the tissues against infections by filtering particles from lymph and by supporting the activities of lymphocytes that furnish immunity against specific disease-causing agents or pathogens.

 ## PRACTICE

### PROCEDURE A—Lymphatic Pathways AP|R

1. Review the section entitled "Lymphatic Pathways" in chapter 16 of the textbook.
2. As a review activity, label figure 42.1.
3. Complete Part A of Laboratory Report 42.
4. Observe the human torso model and the anatomical chart of the lymphatic system, and locate the following features:

**lymphatic vessels**

**lymph nodes**

**lymphatic trunks**

   lumbar trunk

   intestinal trunk

   intercostal trunk

   bronchomediastinal trunk

   subclavian trunk

   jugular trunk

**cisterna chyli**

**collecting ducts**

   thoracic (left lymphatic) duct

   right lymphatic duct

**internal jugular veins**

**subclavian veins**

 ## PRACTICE

### PROCEDURE B—Lymph Nodes

1. Review the section entitled "Lymph Nodes" in chapter 16 of the textbook.
2. As a review activity, label figure 42.2.
3. Complete Part B of the laboratory report.

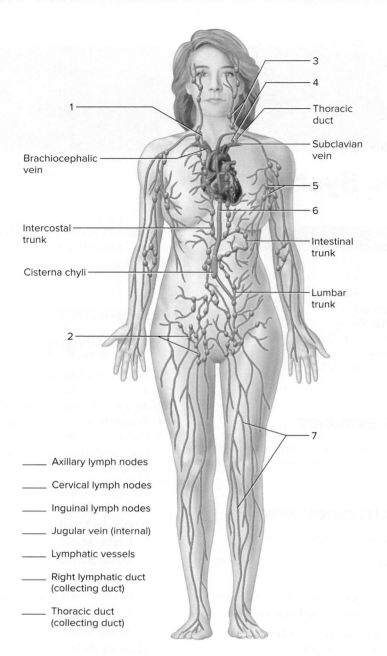

3

4

Thoracic
duct

1

Subclavian
vein

Brachiocephalic
vein

5

6

Intercostal
trunk

Intestinal
trunk

Cisterna chyli

Lumbar
trunk

2

7

_____ Axillary lymph nodes

_____ Cervical lymph nodes

_____ Inguinal lymph nodes

_____ Jugular vein (internal)

_____ Lymphatic vessels

_____ Right lymphatic duct
(collecting duct)

_____ Thoracic duct
(collecting duct)

**FIGURE 42.1** Label
the diagram by placing
the correct numbers in the
spaces provided. **A**

1

Lymph flow

2

3

5

6

4

Lymph flow

_____ Afferent lymphatic vessel

_____ Capsule

_____ Efferent lymphatic vessel

_____ Hilum

_____ Lymphatic nodule

_____ Lymphatic sinus

**FIGURE 42.2** Label this
diagram of a lymph node by placing
the correct numbers in the spaces
provided. **A** **AP|R**

4. Observe the anatomical chart of the lymphatic system and the human torso model, and locate the clusters of lymph nodes in the following regions:

**cervical region**
**axillary region**
**inguinal region**
**pelvic cavity**
**abdominal cavity**
**thoracic cavity**

5. Palpate the lymph nodes in your cervical region. They are located along the lower border of the mandible and between the ramus of the mandible and the sternocleidomastoid muscle. They feel like small, firm lumps.
6. Observe a human lymph node in figure 42.3a.
7. Obtain a prepared microscope slide of a lymph node and observe it, using low-power magnification (fig. 42.3b). Identify the *capsule* that surrounds the node and is mainly composed of collagen fibers, the *lymphatic nodules* that appear as dense masses near the surface of the node, and the *lymphatic sinuses* that appear as narrow spaces where lymph circulates between the nodules and the capsule.
8. Using high-power magnification, examine a nodule within the lymph node. The nodule contains densely packed *lymphocytes.*
9. Prepare a labeled sketch of a representative section of a lymph node in Part D of the laboratory report.

 **PRACTICE**

**PROCEDURE C—Thymus and Spleen**

1. Review the sections entitled "Thymus" and "Spleen" in chapter 16 of the textbook.
2. Locate the thymus and spleen in the anatomical chart of the lymphatic system and on the human torso model.
3. Complete Part C of the laboratory report.
4. Obtain a prepared microscope slide of human thymus and observe it, using low-power magnification (fig. 42.4). Note how the thymus is subdivided into *lobules* by *septa* of connective tissue that contain blood vessels. Identify the *capsule* of loose connective tissue that surrounds the thymus, the outer *cortex* of a lobule composed of densely packed cells and deeply stained, and the inner *medulla* of a lobule composed of loosely packed *lymphocytes* and epithelial cells and lightly stained.
5. Examine the cortex tissue of a lobule using high-power magnification. The cells of the cortex are composed of densely packed lymphocytes among some epithelial cells and *macrophages.* Some of these cortical cells may be undergoing mitosis, so their chromosomes may be visible.
6. Prepare a labeled sketch of a representative section of the thymus in Part D of the laboratory report.
7. Obtain a prepared slide of the human spleen and observe it, using low-power magnification (fig. 42.5).

(a)

(b)

**FIGURE 42.3** Lymph nodes: (*a*) photograph of a human lymph node on tip of finger; (*b*) micrograph of a lymph node (20×).
AP|R ((*a*) ©J & J Photography; (*b*) ©Alvin Telser/Science Photo)

Identify the *capsule* of dense connective tissue that surrounds the spleen. The tissues of the spleen include circular *nodules* of *white pulp* enclosed in a matrix of *red pulp*.

8. Using high-power magnification, examine a nodule of white pulp and red pulp. The cells of the white pulp are mainly *lymphocytes*. Also, there may be an arteriole centrally located in the nodule. The cells of the red pulp are mostly red blood cells with many lymphocytes and macrophages. The macrophages engulf and destroy cellular debris and bacteria.

9. Prepare a labeled sketch of a representative section of the spleen in Part D of the laboratory report.

**FIGURE 42.4**   Micrograph of a section of the thymus (15×). **AP|R** (©Biophoto Associates/Science Source)

**FIGURE 42.5**   Micrograph of a section of the spleen (40×). **AP|R** (©Biophoto Associates/Science Source)

Name _____

Date _____

Section _____

The  corresponds to the indicated Learning Outcome(s) found at the beginning of the laboratory exercise.

LABORATORY
REPORT

42

# Lymphatic System

## PART A ASSESSMENTS

Complete the following statements:

1. Lymphatic pathways begin as _____ that merge to form lymphatic vessels. **A**

2. The wall of a lymphatic capillary consists of a single layer of _____ cells. **3**

3. Once tissue (interstitial) fluid is inside a lymphatic capillary, the fluid is called _____ . **3**

4. Lymphatic vessels have walls similar to those of _____ but thinner. **3**

5. Lymphatic vessels contain _____ that help prevent the backflow of lymph. **A**

6. Lymphatic vessels usually lead to lymph _____ that filter the fluid being transported. **A**

7. The lymphatic trunk that drains the abdominal viscera is called the _____ trunk. **A**

8. The lymphatic trunk that drains the head and neck is called the _____ trunk. **A**

9. The _____ duct is the larger and longer of the two lymphatic collecting ducts. **A**

## PART B ASSESSMENTS

Complete the following statements:

1. Lymph nodes contain large numbers of white blood cells called _____ and macrophages that fight invading microorganisms. **3**

2. The indented region where blood vessels and nerves join a lymph node is called the _____ . **3**

3. Lymphatic _____ that contain germinal centers are the structural units of a lymph node. **3**

4. The spaces within a lymph node are called lymphatic _____ through which lymph circulates. **3**

5. Lymph enters a node through a(an) _____ lymphatic vessel. **3**

6. A connective tissue _____ encloses a lymph node. **3**

7. The aggregations of lymph nodules found within the mucosal lining of the small intestine are called _____ . **2**

8. The lymph nodes in the cervical region are associated with the lymphatic vessels that drain the _____ . **A**

9. The lymph nodes associated with the lymphatic vessels that drain the lower limbs are located in the _____ region. **2**

**PART C ASSESSMENTS**

Complete the following statements:

1. The thymus is located in the _____ , anterior to the aortic arch. 3

2. The thymus reaches its greatest size during infancy and early childhood and tends to decrease in size following _____ . 3

3. As a person ages, the thymus tissue is often replaced by _____ . 3

4. The lymphocytes of the thymus develop from precursor cells that originated in the _____ . 3

5. The hormones secreted by the thymus are called _____ . 3

6. The _____ is the largest of the lymphatic organs. 3

7. Blood vessels enter the spleen through the region called the _____ . 3

8. The sinuses within the spleen contain _____ . 3

9. The tiny islands (nodules) of tissue within the spleen that contain many lymphocytes constitute the _____ pulp. 3

10. The _____ pulp of the spleen contains large numbers of red blood cells, lymphocytes, and macrophages. 3

11. _____ within the spleen engulf and destroy foreign particles and cellular debris. 3

**PART D ASSESSMENTS**

Sketch and label structural features of the following lymphatic organs: 4

| | |
|---|---|
| **Lymph node (_____×)** | **Thymus (_____×)** |

**Spleen (_____×)**

# 43

# Digestive Organs

## MATERIALS NEEDED

Textbook
Human torso model
Head model, sagittal section
Skull with teeth
Teeth, sectioned
Tooth model, sectioned
Paper cup
Compound light microscope
Prepared microscope slides of the following:
    Salivary gland
    Esophagus
    Stomach (fundus)
    Pancreas (exocrine portion)
    Liver
    Small intestine (jejunum)
    Large intestine

## PURPOSE OF THE EXERCISE  AP|R

To review the structure and function of the digestive organs and to examine the tissues of these organs.

## LEARNING OUTCOMES  AP|R

After completing this exercise, you should be able to

1. Locate and label the major digestive organs and their major structures.
2. Describe the functions of these organs.
3. Examine and sketch the structures of a section of the small intestine.

The digestive system includes the organs associated with the alimentary canal and several accessory structures. The alimentary canal, a muscular tube, passes through the body from the opening of the mouth to the anus. It includes the mouth, pharynx, esophagus, stomach, small intestine, and large intestine. The function of the canal is to move substances throughout its length. It is specialized in various regions to store, digest, and absorb food materials and to eliminate the residues. The accessory organs, which include the salivary glands, liver, gallbladder, and pancreas, secrete products into the alimentary canal that aid digestive functions.

## PRACTICE

### PROCEDURE A—Mouth and Salivary Glands

1. Review the sections entitled "Mouth" and "Salivary Glands" in chapter 17 of the textbook.
2. As a review activity, label figures 43.1–43.3.
3. Examine the head model (sagittal section) and a skull. Locate the following structures:

**mouth**
**oral cavity**
**vestibule**
**tongue**
    lingual frenulum
    papillae
**palate**
    hard palate
    soft palate
    uvula
**palatine tonsils**
**gums (gingivae)**
**teeth**
    incisors
    canines (cuspids)
    premolars (bicuspids)
    molars

4. Examine a sectioned tooth and a tooth model. Locate the following features:

**crown**
    enamel
    dentin
**neck**
**root**
    pulp cavity
    cementum
    root canal

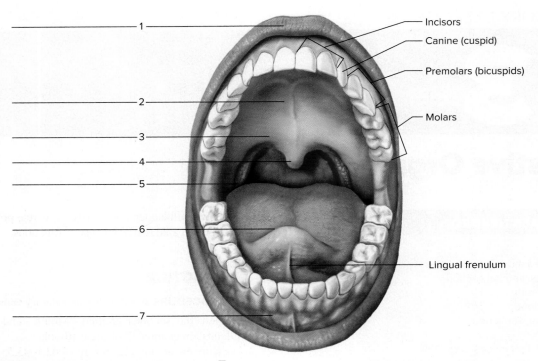

Incisors

Canine (cuspid)

Premolars (bicuspids)

Molars

Lingual frenulum

**FIGURE 43.1**   Label the features of the mouth. Ⓐ

Mandible (cut)

**FIGURE 43.2**   Label the features associated with the major salivary glands. Ⓐ APⓇR

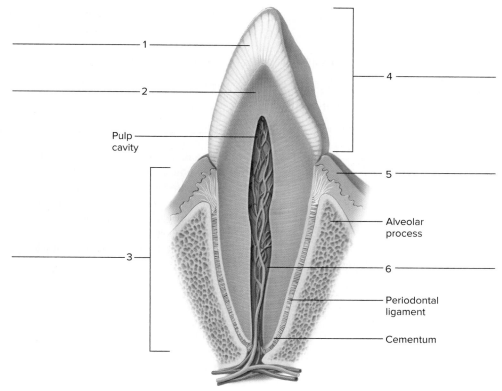

**FIGURE 43.3** Label the features of this canine (cuspid) tooth. 🅰

Labels in figure:
1
2
Pulp cavity
3
4
5
Alveolar process
6
Periodontal ligament
Cementum

5. Observe the head of the human torso model, and locate the following:

   **parotid salivary gland**
       parotid duct (Stensen's duct)
   **submandibular salivary gland**
       submandibular duct (Wharton's duct)
   **sublingual salivary gland**
       sublingual ducts (10–12)

6. Examine a microscopic section of a salivary gland, using low- and high-power magnification. Note that the mucous cells that produce mucus and the serous cells that produce enzymes are clustered around small ducts. Also note a larger secretory duct surrounded by lightly stained cuboidal epithelial cells (fig. 43.4).
7. Complete Part A of Laboratory Report 43.

Labels in figure: Mucous cell, Serous cell, Secretory duct

**FIGURE 43.4** Micrograph of the salivary gland (300×). (©Biophoto Associates/Science Source)

**PRACTICE**

**PROCEDURE B—Pharynx and Esophagus**

1. Review the section entitled "Pharynx and Esophagus" in chapter 17 of the textbook.
2. As a review activity, label figure 43.5.
3. Observe the human torso model, and locate the following features:

**pharynx**
    nasopharynx
    oropharynx
    laryngopharynx
    constrictor muscles
**epiglottis**
**esophagus**
**lower esophageal sphincter (cardiac sphincter)**

4. Have your laboratory partner take a swallow from a cup of water. Carefully watch the movements in the anterior region of the neck. What steps in the swallowing process did you observe?

_____

_____

_____

5. Examine a microscopic section of esophagus wall, using low-power magnification (fig. 43.6). The inner lining is composed of stratified squamous epithelium, and there are layers of muscle tissue in the wall. Locate some mucous glands in the submucosa. They appear as clusters of lightly stained cells.
6. Complete Part B of the laboratory report.

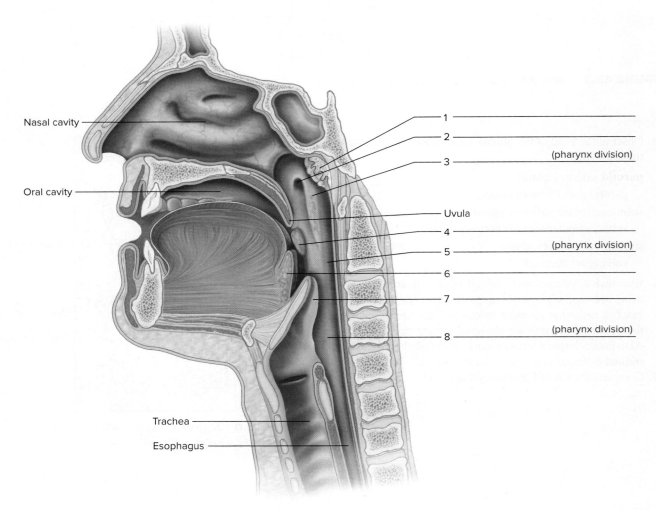

Nasal cavity

Oral cavity

1 _____

2 _____
(pharynx division)

3 _____

Uvula

4 _____
(pharynx division)

5 _____

6 _____

7 _____

8 _____
(pharynx division)

Trachea

Esophagus

**FIGURE 43.5** Label the features associated with the pharynx.  **AP|R**

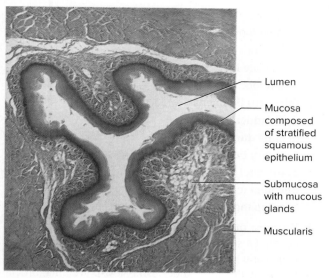

FIGURE 43.6 Micrograph of a cross section of the esophagus (10×). **AP|R** (©Ed Reschke)

Labels on figure 43.6:
- Lumen
- Mucosa composed of stratified squamous epithelium
- Submucosa with mucous glands
- Muscularis

 **PRACTICE**

**PROCEDURE C—Stomach**

1. Review the section entitled "Stomach" in chapter 17 of the textbook.
2. As a review activity, label figure 43.7.
3. Observe the human torso model, and locate the following features of the stomach:

**cardia**
**fundus**
**body**
**pylorus**
    pyloric antrum
    pyloric canal
**pyloric sphincter (valve)**
**lesser curvature**
**greater curvature**
**gastric folds (rugae)**

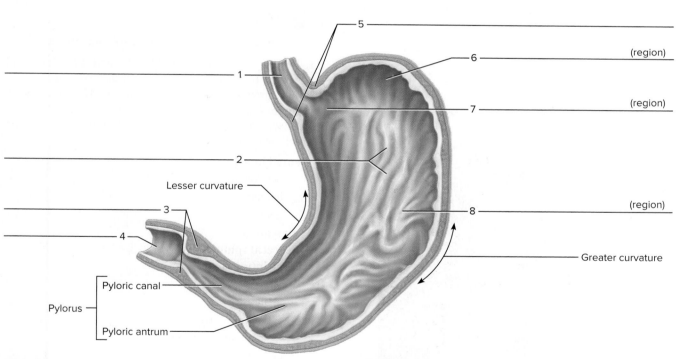

Labels on figure 43.7:
- 5
- 6 (region)
- 1
- 7 (region)
- 2
- Lesser curvature
- 3
- 4
- 8 (region)
- Pyloric canal
- Pylorus
- Pyloric antrum
- Greater curvature

FIGURE 43.7 Label the stomach and associated structures. **1 AP|R**

4. Examine a microscopic section of the stomach wall, using low-power magnification. Note how the inner lining of simple columnar epithelium dips inward to form gastric pits. The gastric glands are tubular structures that open into the gastric pits. Near the deep ends of these glands, you should be able to locate some intensely stained (bluish) chief cells and some lightly stained (pinkish) parietal cells (fig. 43.8). What are the functions of these cells?

_____

_____

_____

_____

5. Complete Part C of the laboratory report.

 **PRACTICE** _____

### PROCEDURE D—Pancreas and Liver

1. Review the sections entitled "Pancreas," "Liver and Gallbladder," and "Small Intestine" in chapter 17 of the textbook.
2. As a review activity, label figure 43.9.
3. Observe the human torso model, and locate the following structures:

**pancreas**
**pancreatic duct**

**FIGURE 43.8** Micrograph of the mucosa of the stomach wall (60×). **AP|R** (©Ed Reschke)

Labels on figure:
- Simple columnar epithelium
- Gastric pit
- Mucous cell
- Portion of a gastric gland
- Parietal cell
- Portion of a gastric gland
- Chief cell

**liver**
    right lobe
    quadrate lobe
    caudate lobe
    left lobe
**falciform ligament**
**gallbladder**
**hepatic ducts**
**common hepatic duct**
**cystic duct**
**bile duct (common bile duct)**
**hepatopancreatic sphincter**

4. Examine the pancreas slide, using low-power magnification. Observe the exocrine (acinar) cells that secrete pancreatic juice (fig. 43.10).
5. Examine the liver slide, using low-power magnification. Observe the hepatocytes within the hepatic lobules. Each lobule has a central vein and numerous hepatic sinusoids. A hepatic triad would possess a small bile duct and branches of a hepatic artery and a hepatic portal vein (fig. 43.11).

 **PRACTICE** _____

### PROCEDURE E—Small and Large Intestines

1. Review the sections entitled "Small Intestine" and "Large Intestine" in chapter 17 of the textbook.
2. As a review activity, label figure 43.12 and examine figure 43.13.
3. Observe the human torso model, and locate each of the following features:

**small intestine**
    duodenum
    jejunum
    ileum
**mesentery**
**ileocecal sphincter (valve)**
**large intestine**
    large intestinal wall
        haustra
        taeniae coli
        epiploic (fatty) appendages
    cecum
    appendix (vermiform appendix)
    ascending colon
    right colic (hepatic) flexure
    transverse colon
    left colic (splenic) flexure
    descending colon
    sigmoid colon
    rectum

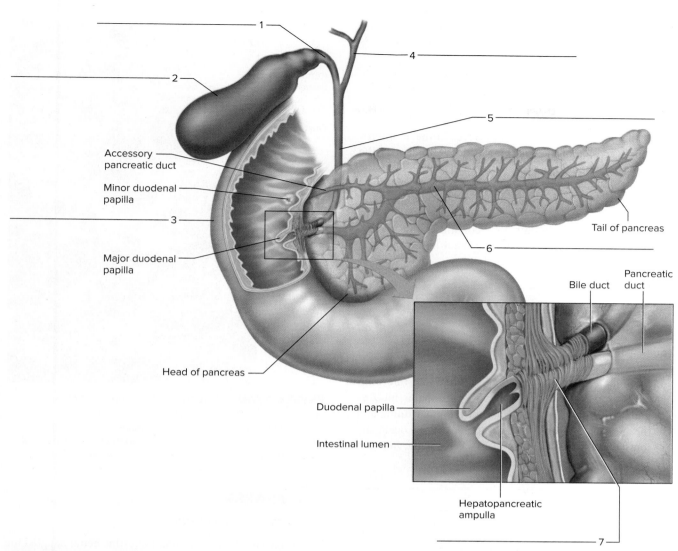

Accessory
pancreatic duct

Minor duodenal
papilla

Major duodenal
papilla

Head of pancreas

Tail of pancreas

Duodenal papilla

Intestinal lumen

Bile duct

Pancreatic
duct

Hepatopancreatic
ampulla

**FIGURE 43.9**   Label the features associated with the liver, gallbladder, and pancreas.  A  AP|R

anal canal

anal columns

**anal sphincter muscles**

external anal sphincter

internal anal sphincter

**anus**

4. Using low-power magnification, examine a microscopic section of the small intestine wall. Identify the mucosa, submucosa, muscularis, and serosa. Note the villi that extend into the lumen of the tube. Study a single villus, using high-power magnification. Note the core of connective tissue and the covering of simple columnar epithelium that contains some lightly stained goblet cells (fig. 43.14). What is the function of these villi?

_____

_____

_____

5. Prepare a labeled sketch of the wall of the small intestine in Part D of the laboratory report.

6. Examine a microscopic section of large intestine wall. Note the lack of villi. Also note the tubular mucous glands that open on the surface of the inner lining and the numerous lightly stained goblet cells. Locate the four layers of the wall (fig. 43.15). What is the function of the mucus secreted by these glands?

_____

_____

_____

7. Complete Part E of the laboratory report.

Connective tissue

Ducts

Exocrine acinar cells

Vein

**FIGURE 43.10**   Micrograph of exocrine portion of pancreas showing acinar cells and small pancreatic ducts (200×). The acinar cells secrete an alkaline pancreatic juice into small ducts that converge into the larger pancreatic duct. (©McGraw-Hill Education/Dennis Strete, photographer)

**ASSESS**

**CRITICAL THINKING**

How is the structure of the small intestine better adapted for absorption than that of the large intestine? **2**

_____

_____

_____

_____

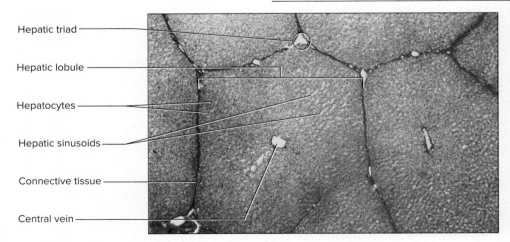

Hepatic triad

Hepatic lobule

Hepatocytes

Hepatic sinusoids

Connective tissue

Central vein

**FIGURE 43.11**   Micrograph of hepatic lobules of the liver (40×). (©Victor P.Eroschenko)

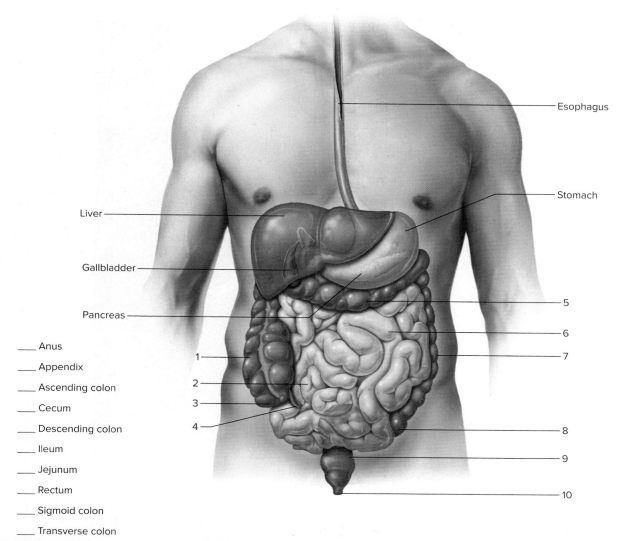

Liver ──

Gallbladder ──

Pancreas ──

_____ Anus

_____ Appendix

_____ Ascending colon

_____ Cecum

_____ Descending colon

_____ Ileum

_____ Jejunum

_____ Rectum

_____ Sigmoid colon

_____ Transverse colon

Esophagus

Stomach

5

6

7

1

2

3

4

8

9

10

**FIGURE 43.12** Label this diagram of the small and large intestines by placing the correct numbers in the spaces provided.
1 AP|R

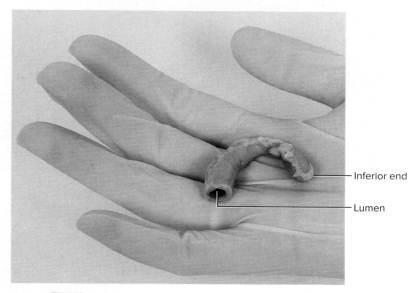

Inferior end

Lumen

**FIGURE 43.13** Normal appendix. AP|R (©J & J Photography)

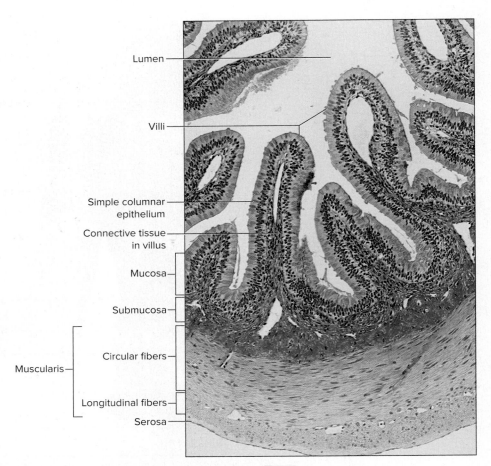

Lumen

Villi

Simple columnar epithelium

Connective tissue in villus

Mucosa

Submucosa

Muscularis

Circular fibers

Longitudinal fibers

Serosa

**FIGURE 43.14** Micrograph of the small intestine wall (40×). AP|R (©Ed Reschke)

Lumen

Mucosa with mucous glands

Submucosa

Muscularis (circular and longitudinal)

Serosa

**FIGURE 43.15** Micrograph of the large intestine wall (64×). AP|R (©Dennis Strete)

The △ corresponds to the indicated Learning Outcome(s) found at the beginning of the laboratory exercise.

# Digestive Organs

## PART A ASSESSMENTS

Match the terms in column A with the descriptions in column B. Place the letter of your choice in the space provided. △

**Column A**

a. Crown
b. Dentin
c. Incisor
d. Lingual frenulum
e. Molar
f. Oral cavity
g. Papillae
h. Periodontal ligament
i. Sublingual gland
j. Uvula
k. Vestibule

**Column B**

_____ 1. Bonelike substance beneath tooth enamel

_____ 2. Smallest of major salivary glands

_____ 3. Tooth specialized for grinding

_____ 4. Chamber between tongue and palate

_____ 5. Projections on tongue surface

_____ 6. Cone-shaped projection of soft palate

_____ 7. Attaches tooth to jaw

_____ 8. Chisel-shaped tooth

_____ 9. Space between the teeth, cheeks, and lips

_____ 10. Anchors tongue to floor of mouth

_____ 11. Portion of tooth projecting beyond gum

## PART B ASSESSMENTS

Complete the following:

1. The part of the pharynx superior to the soft palate is called the _____ . △

2. The middle part of the pharynx is called the _____ . △

3. The inferior portion of the pharynx is called the _____ . △

4. _____ is the main secretion of the esophagus. ②

5. Summarize the functions of the esophagus. ② _____

_____

_____

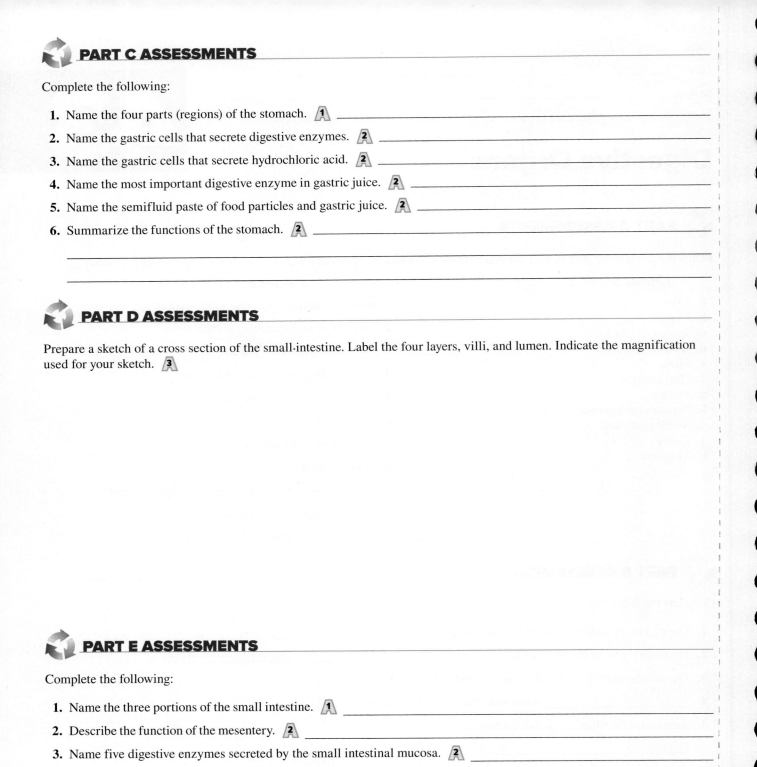

## PART C ASSESSMENTS

Complete the following:

1. Name the four parts (regions) of the stomach.  **1** _____

2. Name the gastric cells that secrete digestive enzymes.  **2** _____

3. Name the gastric cells that secrete hydrochloric acid.  **2** _____

4. Name the most important digestive enzyme in gastric juice.  **2** _____

5. Name the semifluid paste of food particles and gastric juice.  **2** _____

6. Summarize the functions of the stomach.  **2** _____

_____

_____

## PART D ASSESSMENTS

Prepare a sketch of a cross section of the small·intestine. Label the four layers, villi, and lumen. Indicate the magnification used for your sketch.  **3**

## PART E ASSESSMENTS

Complete the following:

1. Name the three portions of the small intestine.  **1** _____

2. Describe the function of the mesentery.  **2** _____

3. Name five digestive enzymes secreted by the small intestinal mucosa.  **2** _____

_____

4. Name the valve located between the small and the large intestine.  **1** _____

5. Summarize the functions of the small intestine.  **2** _____

_____

_____

6. Summarize the functions of the large intestine.  **2** _____

_____

# Action of a Digestive Enzyme

## MATERIALS NEEDED

0.5% amylase solution*
Beakers (50 and 500 mL)
Distilled water
Funnel
Pipets (1 and 10 mL)
Pipet rubber bulbs
0.5% starch solution
Graduated cylinder (10 mL)
Test tubes
Test-tube clamps

*For Alternative Procedure*:
Small, disposable cups
Distilled water

Wax marker
Iodine-potassium-iodide
    solution
Medicine dropper
Ice
Water bath, 37°C (98.6°F)
Porcelain test plate
Benedict's solution
Hot plates
Test-tube rack
Thermometer

*The amylase should be free of sugar for best results; a low-maltose solution of amylase yields good results.

## ⚠ SAFETY

- Review the Laboratory Safety Guidelines in Appendix 1.
- Use only a mechanical pipetting device (never your mouth). Use pipets with rubber bulbs or dropping pipets.
- Wear safety glasses when working with acids and when heating test tubes.
- Use test-tube clamps when handling hot test tubes.
- If an open flame is used for heating the test solutions, keep clothes and hair away from the flame.
- Dispose of chemicals according to appropriate directions.
- Use appropriate disinfectant to wash laboratory tables before and after the procedures.
- If student saliva is used as a source of amylase, it is important that students wear disposable gloves and handle only their own materials.
- Wash your hands before leaving the laboratory.

## PURPOSE OF THE EXERCISE

To investigate the action of amylase and the effect of heat on its enzymatic activity.

###  LEARNING OUTCOMES AP|R

After completing this exercise, you should be able to

1. Test a solution for the presence of starch or the presence of sugar.
2. Explain the action of amylase.
3. Test the effects of varying temperatures on the activity of amylase.

The digestive enzyme in salivary secretions is called *salivary amylase.* Pancreatic amylase is secreted among several other pancreatic enzymes. A bacterial extraction of amylase is available for laboratory experiments. This enzyme catalyzes the reaction of changing starch molecules into sugar (disaccharide) molecules, which is the first step in the digestion of complex carbohydrates.

As in the case of other enzymes, amylase is a protein catalyst. Its activity is affected by exposure to certain environmental factors, including various temperatures, pH, radiation, and electricity. As temperatures increase, faster chemical reactions occur as the collisions of molecules happen at a greater frequency. Eventually, temperatures increase to a point that the enzyme is denatured and the rate of the enzyme activity rapidly declines. As temperatures decrease, enzyme activity also decreases due to fewer collisions of the molecules; however, the colder temperatures do not denature the enzyme. Normal body temperature provides an environment for enzyme activity near the optimum for enzymatic reactions.

The pH of the environment where enzymes are secreted also has a major influence on the enzyme reactions. The optimum pH for amylase activity is between 6.8 and 7.0, typical of salivary secretions. When salivary amylase arrives in the stomach, hydrochloric acid deactivates the enzyme, diminishing any further chemical digestion of remaining starch in the stomach. Pancreatic amylase is secreted into the small intestine, where an optimum pH for amylase is once more provided. Other enzymes in the digestive system have different optimum pH ranges of activity compared to amylase. For example, pepsin from stomach secretions has an optimum activity around pH 2, whereas trypsin from pancreatic secretions operates best around pH 7–8.

## PRACTICE

### PROCEDURE A—Amylase Activity

1. Study the lock-and-key model of amylase action on starch digestion (fig. 44.1).
2. Examine the digestive locations where starch is converted to glucose (fig. 44.2).
3. Mark three clean test tubes as *tubes 1, 2,* and *3,* and prepare the tubes as follows (fig. 44.3):

   *Tube 1*: **Add 6 mL of amylase solution.**

   *Tube 2*: **Add 6 mL of starch solution.**

   *Tube 3*: **Add 5 mL of starch solution and 1 mL of amylase solution.**

### ALTERNATIVE PROCEDURE

Human saliva can be used as a source of amylase solutions instead of bacterial amylase preparations. Collect about 5 mL of saliva in a small, disposable cup. Add an equal amount of distilled water and mix together for the amylase solutions during the laboratory procedures. Be sure to follow all of the safety guidelines.

4. Shake the tubes well to mix the contents, and place them in a warm water bath, 37°C (98.6°F), for 10 minutes.
5. At the end of the 10 minutes, test the contents of each tube for the presence of starch. To do this, follow these steps:

   a. Place 1 mL of the solution to be tested in a depression of a porcelain test plate.

   b. Next add 1 drop of iodine-potassium-iodide solution, and note the color of the mixture. If the solution becomes blue-black, starch is present.

   c. Record the results in Part A of Laboratory Report 44.

6. Test the contents of each tube for the presence of sugar (disaccharides, in this instance). To do this, follow these steps:

   a. Place 1 mL of the solution to be tested in a clean test tube.

   b. Add 1 mL of Benedict's solution.

   c. Place the test tube with a test-tube clamp in a beaker of boiling water for 2 minutes.

   d. Note the color of the liquid. If the solution becomes green, yellow, orange, or red, sugar is present. Blue indicates a negative test, whereas green indicates a

**FIGURE 44.1** Lock-and-key model of enzyme action of amylase on starch digestion. AP|R

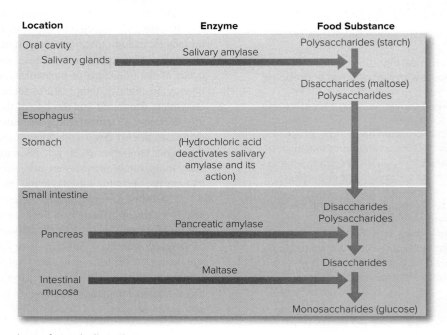

**FIGURE 44.2** Flowchart of starch digestion.

Add 6 mL amylase solution

Add 6 mL starch solution

Add 5 mL starch solution

Add 1 mL amylase solution

Shake the tubes to mix the contents. Transfer the three test tubes from the test-tube rack into the warm water bath (37°C) for 10 minutes.

Tube 1    Tube 2    Tube 3

**FIGURE 44.3**    Test tubes prepared for testing amylase activity.

positive test with the least amount of sugar, and red indicates the greatest amount of sugar present.

   **e.** Record the results in Part A of the laboratory report.

**7.** Complete Part A of the laboratory report.

 **PRACTICE**

### PROCEDURE B—Effect of Heat

**1.** Mark three clean test tubes as *tubes 4, 5,* and *6.*

**2.** Add 1 mL of amylase solution to each of the tubes, and expose each solution to a different test temperature for 3 minutes as follows:

*Tube 4*: **Place in beaker of ice water (about 0°C [32°F]).**

*Tube 5*: **Place in warm water bath (about 37°C [98.6°F]).**

*Tube 6*: **Place in beaker of boiling water (about 100°C [212°F]). Use a test-tube clamp.**

**3.** *It is important that the 5 mL of starch solution added to tube 4 be at ice-water temperature before it is added to the 1 mL of amylase solution.* Add 5 mL of starch solution to each tube, shake to mix the contents, and

return the tubes to their respective test temperatures for 10 minutes.

**4.** At the end of the 10 minutes, test the contents of each tube for the presence of starch and the presence of sugar by following the same directions as in steps 5 and 6 of Procedure A.

**5.** Complete Part B of the laboratory report.

 **LEARN: LAB IN MOTION**

Devise an experiment to test the effect of some other environmental factor on amylase activity. For example, you might test the effect of a strong acid by adding a few drops of concentrated hydrochloric acid to a mixture of starch and amylase solutions. Be sure to include a control in your experimental plan. That is, include a tube containing everything except the factor you are testing. Then you will have something with which to compare your results. *Carry out your experiment only if it has been approved by the laboratory instructor.*

The Ⓐ corresponds to the indicated Learning Outcome(s) found at the beginning of the laboratory exercise.

# Action of a Digestive Enzyme

## PART A ASSESSMENTS

1. Test results: Ⓐ **1**

| Tube | Starch | Sugar |
|---|---|---|
| 1 Amylase solution | | |
| 2 Starch solution | | |
| 3 Starch-amylase solution | | |

2. Complete the following:

   **a.** Explain the reason for including tube 1 in this experiment. Ⓐ **2** _____

   _____

   _____

   _____

   **b.** What is the importance of tube 2? Ⓐ **2** _____

   _____

   _____

   _____

   **c.** What do you conclude from the results of this experiment? Ⓐ **2** _____

   _____

   _____

   _____

**PART B ASSESSMENTS**

1. Test results: **3**

| Tube | Starch | Sugar |
|---|---|---|
| 4  0°C (32°F) | | |
| 5  37°C (98.6°F) | | |
| 6  100°C (212°F) | | |

2. Complete the following:

   **a.** What do you conclude from the results of this experiment? **3** _____

   _____

   _____

   _____

   _____

   **b.** If digestion failed to occur in one of the tubes in this experiment, how can you tell whether the amylase was destroyed by the factor being tested or the amylase activity was simply inhibited by the test treatment? **3** _____

   _____

   _____

   _____

   _____

**ASSESS**

**CRITICAL THINKING**

What test result would occur if the amylase you used contained sugar? _____
Would your results be as valid? Explain your answer. **2**

_____

_____

_____

_____

# 45

# Respiratory Organs

## MATERIALS NEEDED

Textbook
Human skull (sagittal section)
Human torso model
Larynx model
Thoracic organs model
Compound light microscope
Prepared microscope slides of the following:
    Trachea (cross section)
    Lung, human (normal)
Animal lung with trachea (fresh or preserved)
Bicycle pump
Prepared microscope slides of the following:
    Lung tissue (smoker)
    Lung tissue (emphysema)

 **SAFETY**

- Wear disposable gloves when working on the fresh or preserved animal lung demonstration.
- Wash your hands before leaving the laboratory.

## PURPOSE OF THE EXERCISE

To review the structure and function of the respiratory organs and to examine the tissues of some of these organs.

## LEARNING OUTCOMES AP|R

After completing this exercise, you should be able to

1. Locate the major organs and structural features of the respiratory system.
2. Describe the functions of these organs.
3. Sketch and label features of tissue sections of the trachea and lung.

The organs of the respiratory system include the nose, nasal cavity, paranasal sinuses, pharynx, larynx, trachea, bronchial tree, and lungs. They mainly process incoming air

and transport it to and from the atmosphere outside the body and the air sacs of the lungs.

In the alveoli (air sacs), gas exchanges take place between the air and the blood of nearby capillaries. The blood, in turn, transports gases to and from the alveoli and the body cells. This entire process of transporting and exchanging gases between the atmosphere and the body cells is called *respiration*.

##  PRACTICE

### PROCEDURE A—Respiratory Organs

1. Review the section entitled "Organs of the Respiratory System" in chapter 19 of the textbook.
2. As a review activity, label figures 45.1–45.4.
3. Examine the sagittal section of the human skull, and locate the following features:

**nasal cavity**
    nostril (naris)
    nasal septum
    nasal conchae
    meatuses

**paranasal sinuses**
    maxillary sinus
    frontal sinus
    ethmoidal sinus (ethmoidal air cells)
    sphenoidal sinus

4. Observe the larynx model, the thoracic organs model, and the human torso model. Locate the features listed in step 3. Also locate the following features:

**pharynx**
    nasopharynx
    oropharynx
    laryngopharynx
**larynx** (palpate your larynx)
    vocal cords
        false vocal cords (vestibular folds)
        true vocal cords (vocal folds)
    thyroid cartilage ("Adam's apple")
    cricoid cartilage

cricothyroid ligament
epiglottis
epiglottic cartilage
arytenoid cartilages
corniculate cartilages
cuneiform cartilages
glottis
**trachea** (palpate your trachea)
**bronchi**
right and left main (primary) bronchi
lobar (secondary) bronchi
segmental (tertiary) bronchi
**bronchioles**
**lung**
hilum
lobes
superior lobe
middle lobe (right only)
inferior lobe
**visceral pleura**

**parietal pleura**
**pleural cavity**

5. Study figure 45.5 of the bronchial tree.
6. Complete Part A of Laboratory Report 45.

 **LEARN: LAB IN MOTION**

Observe the animal lung and the attached trachea. Identify the larynx, major laryngeal cartilages, trachea, and incomplete cartilaginous rings of the trachea. Open the larynx and locate the vocal folds. Examine the visceral pleura on the surface of a lung. A bicycle pump can be used to demonstrate lung inflation. Section the lung and locate some bronchioles and alveoli. Squeeze a portion of a lung between your fingers. How do you describe the texture of the lung?

_____

_____

_____

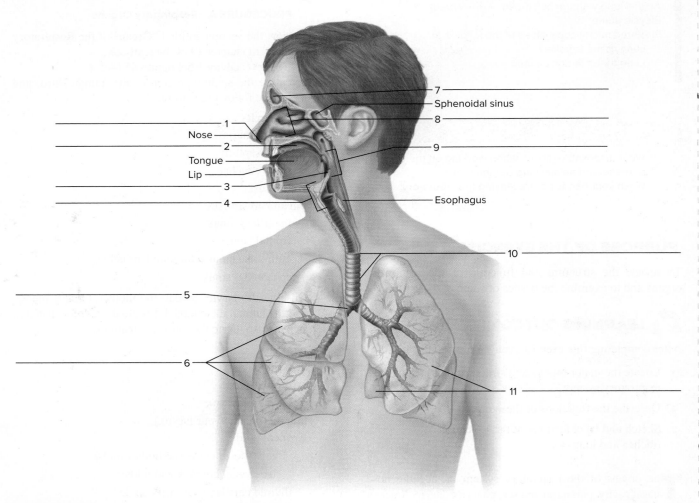

**FIGURE 45.1**   Label the major features of the respiratory system.

**FIGURE 45.2** Label the features of this sagittal section of the upper respiratory tract. **A** **AP|R**

1 _____
2 _____
3 _____
4 _____
5 _____
6 _____
Hyoid bone
7 _____

8 _____
9 _____
10 _____
11 _____
12 _____
13 _____
14 _____
Esophagus

(a)

1 _____
Hyoid bone
2 _____
Cricothyroid ligament
3 _____
Trachea

(b)

Hyoid bone
4 _____
5 _____
6 _____
Trachea

**FIGURE 45.3** Label the major features of the larynx: (*a*) anterior view; (*b*) posterior view. **A** **AP|R**

Anterior

Posterior portion of tongue
1 _____
2 _____
Corniculate cartilage

3 _____
4 _____
Cuneiform cartilage

**FIGURE 45.4** Label the features of the superior aspect of the larynx with the glottis closed. **A**

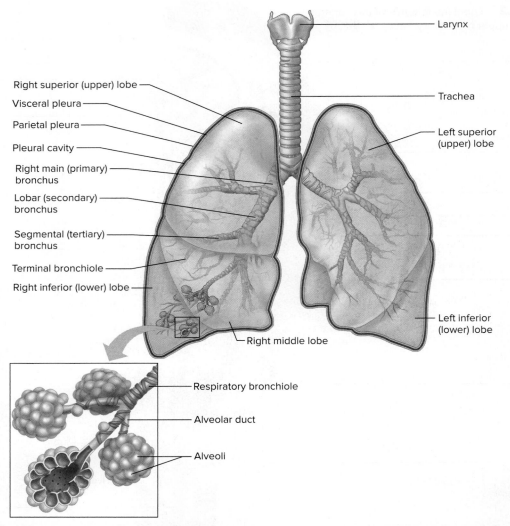

Right superior (upper) lobe

Visceral pleura

Parietal pleura

Pleural cavity

Right main (primary) bronchus

Lobar (secondary) bronchus

Segmental (tertiary) bronchus

Terminal bronchiole

Right inferior (lower) lobe

Larynx

Trachea

Left superior (upper) lobe

Left inferior (lower) lobe

Right middle lobe

Respiratory bronchiole

Alveolar duct

Alveoli

**FIGURE 45.5** The bronchial tree consists of the passageways that connect the trachea and the alveoli. The alveolar ducts and alveoli are enlarged to show their locations. AP|R

 **PRACTICE**

**PROCEDURE B—Respiratory Tissues**

1. Obtain a prepared microscope slide of a trachea, and use low-power magnification to examine it. Notice the inner lining of ciliated pseudostratified columnar epithelium and the deep layer of hyaline cartilage, which represents a portion of an incomplete (C-shaped) tracheal ring (fig. 45.6).

2. Use high-power magnification to observe the cilia on the free surface of the epithelial lining. Locate the wineglass-shaped goblet cells, which secrete a protective mucus, in the epithelium (fig. 45.7).

3. Prepare a labeled sketch of a representative portion of the tracheal wall in Part B of the laboratory report.

4. Study figure 45.8 of the respiratory tubes and the alveoli.

Ciliated epithelium          Lumen of trachea

Hyaline cartilage          Connective tissue

**FIGURE 45.6** Micrograph of a section of the tracheal wall (100×). AP|R (© Biophoto Associates/Science Source)

**FIGURE 45.7** Micrograph of ciliated pseudostratified epithelium in the respiratory tract (275×). **AP|R** (©Biophoto Associates/ Science Source)

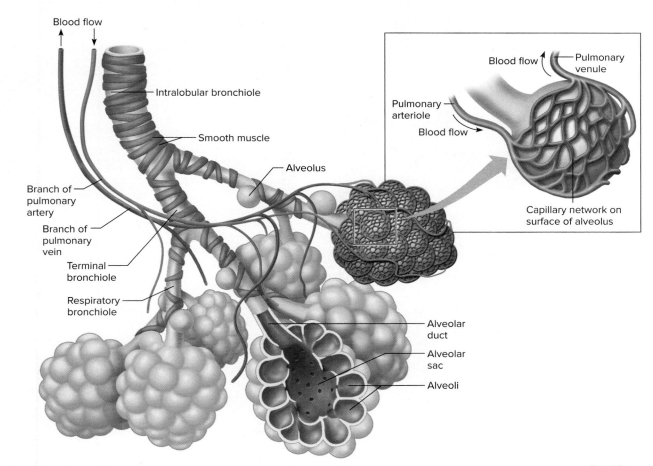

**FIGURE 45.8** The respiratory tubes end in tiny alveoli, each of which is surrounded by a capillary network. **AP|R**

5. Obtain a prepared microscope slide of human lung tissue. Examine it, using low-power magnification, and note the many open spaces of the air sacs (alveoli). Look for a bronchiole—a tube with a relatively thick wall and a wavy inner lining. Locate the smooth muscle tissue in the wall of this tube (fig. 45.9). You also may see a section of cartilage as part of the bronchiole wall.

6. Use high-power magnification to examine the alveoli (fig. 45.10). Their walls are composed of simple squamous epithelium. The thin walls of the alveoli and the associated capillaries allow efficient gas exchange of oxygen and carbon dioxide. You also may see sections of blood vessels containing blood cells.

7. Prepare a labeled sketch of a representative portion of the lung in Part B of the laboratory report.

8. Complete Part C of the laboratory report.

Alveolar sac

Alveolus

Lumen of
bronchiole

Blood vessel

Bronchiole

**FIGURE 45.9**   Micrograph of human lung tissue (70×).
**AP|R**  (©Science Stock Photography/Science Source)

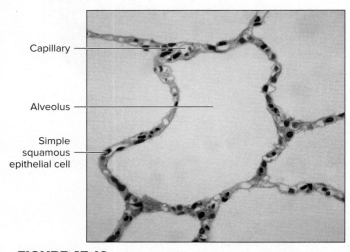

Capillary

Alveolus

Simple
squamous
epithelial cell

**FIGURE 45.10**   Micrograph of human lung tissue
(250×).  **AP|R**   (©Ed Reschke/Getty Images)

Examine the prepared microscope slides of the lung tissue of a smoker and a person with emphysema, using low-power magnification. How does the smoker's lung tissue compare with that of the normal lung tissue that you examined previously?

_____

_____

_____

_____

_____

How does the emphysema patient's lung tissue compare with the normal lung tissue?

_____

_____

_____

_____

_____

Name _____

Date _____

Section _____

The 🅐 corresponds to the indicated Learning Outcome(s) found at the beginning of the laboratory exercise.

LABORATORY
REPORT

**45**

# Respiratory Organs

### ♻ PART A ASSESSMENTS

Match the terms in column A with the descriptions in column B. Place the letter of your choice in the space provided. 🅐 🄰

**Column A**

a. Alveolus
b. Cricoid cartilage
c. Epiglottis
d. Glottis
e. Lung
f. Nasal concha
g. Pharynx
h. Pleural cavity
i. Sinus (paranasal sinus)
j. Vocal cord (true)

**Column B**

_____ **1.** Potential space between visceral and parietal pleurae

_____ **2.** Most inferior portion of larynx

_____ **3.** Air-filled space in skull bone that opens into nasal cavity

_____ **4.** Microscopic air sac for gas exchange

_____ **5.** Consists of large lobes

_____ **6.** Opening between and including true vocal cords

_____ **7.** Fold of mucous membrane containing elastic fibers responsible for sounds

_____ **8.** Increases surface area of nasal mucous membrane

_____ **9.** Passageway for air and food

_____ **10.** Partially covers opening of larynx during swallowing

### ♻ PART B ASSESSMENTS

Sketch and label a portion of the tracheal wall and a portion of lung tissue. 🄱

| | |
|---|---|
| | |
| **Tracheal wall (_____×)** | **Lung tissue (_____×)** |

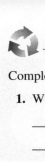

## PART C ASSESSMENTS

Complete the following:

1. What is the function of the mucus secreted by the goblet cells? **2** _____

_____

_____

2. Describe the function of the cilia in the respiratory tubes. **2** _____

_____

_____

3. How is breathing affected if the smooth muscle of the bronchial tree relaxes? **2** _____

_____

_____

4. How is breathing affected if smooth muscle of the bronchial tree contracts? **2** _____

_____

_____

## ASSESS

### CRITICAL THINKING

What is the functional advantage of the alveolar walls being so thin? **2**

_____

_____

_____

_____

# Breathing and Respiratory Volumes

## MATERIALS NEEDED

Textbook
Spirometer, handheld (dry portable)
Alcohol swabs
Disposable mouthpieces
Nose clips
Meterstick
Lung function model

## ⚠ SAFETY

- Clean the spirometer with an alcohol swab before each use.
- Place a new disposable mouthpiece on the stem of the spirometer before each use.
- Dispose of the alcohol swabs and mouthpieces according to directions from your laboratory instructor.

## PURPOSE OF THE EXERCISE

To review the mechanisms of breathing and to measure or calculate certain respiratory air volumes and respiratory capacities.

##  LEARNING OUTCOMES AP|R

After completing this exercise, you should be able to

① Differentiate between the mechanisms responsible for inspiration and expiration.

② Match the respiratory air volumes and respiratory capacities with their definitions.

③ Measure respiratory air volumes using a spirometer.

④ Calculate respiratory capacities using the data obtained from respiratory air volumes.

**B**reathing involves the movement of air from outside the body through the bronchial tree and into the alveoli and the reversal of this air movement to allow gas (oxygen and carbon dioxide) exchange between air and blood. These movements are caused by changes in the size of the thoracic cavity that result from skeletal muscle contractions and from the elastic recoil of stretched tissues.

The volumes of air that move into and out of the lungs during various phases of breathing are called *respiratory air volumes* and *capacities*. A *volume* is a single measurement, whereas a *capacity* represents two or more combined volumes. Respiratory volumes can be measured by using an instrument called a *spirometer*. The values obtained vary with a person's age, sex, height, weight, stress, and physical fitness. Various respiratory capacities can be calculated by combining two or more of the respiratory volumes.

##  PRACTICE

### PROCEDURE A—Breathing Mechanisms AP|R

1. Review the sections entitled "Inspiration" and "Expiration" in chapter 19 of the textbook.
2. Complete Part A of Laboratory Report 46.

Observe the mechanical lung function model (fig. 46.1). It consists of a heavy plastic bell jar with rubber sheeting clamped over its wide open end. Its narrow upper opening is plugged with a rubber stopper through which a Y tube is passed. Small rubber balloons are fastened to the arms of the Y. What happens to the balloons when the rubber sheeting is pulled downward? _____

_____

_____

What happens when the sheeting is pushed upward?

_____

_____

How do you explain these changes?

_____

_____

_____

What part of the respiratory system is represented by the rubber sheeting? _____

The bell jar? _____

The Y tube? _____

The balloons? _____

Air moves into and out of the lungs in much the same way, as can be demonstrated using a syringe (fig. 46.2).

> **WARNING** *If you begin to feel dizzy or light-headed while performing Procedure B, stop the exercise and breathe normally.*

**PROCEDURE B—Respiratory Volumes and Capacities**

1. Review the section entitled "Respiratory Volumes and Capacities" in chapter 19 of the textbook.
2. Complete Part B of the laboratory report.
3. Get a handheld spirometer (fig. 46.3). Point the needle to zero by rotating the adjustable dial. Before using the instrument, clean it with an alcohol swab and place a new disposable mouthpiece over its stem. The instrument should be held with the dial upward, and the air should be blown only into the disposable mouthpiece (fig. 46.4). If air tends to exit the nostrils during exhalation, use a nose clip or pinch your nose as a prevention when exhaling into the spirometer. Movement of the needle indicates the air volume that leaves the lungs. The exhalations should be slowly and forcefully performed. Too rapid an exhalation can result in erroneous data or damage to the spirometer.
4. *Tidal volume (TV)* (about 500 mL) is the volume of air that enters (or leaves) the lungs during a *respiratory cycle* (one inspiration plus the following expiration). *Resting tidal volume* is the volume of air that enters (or leaves) the lungs during normal, quiet breathing (fig. 46.5). To measure this volume, follow these steps:
   a. Sit quietly for a few moments.
   b. Position the spirometer dial so that the needle points to zero.
   c. Place the mouthpiece between your lips and exhale three ordinary expirations into it after inhaling through the nose each time. *Do not force air out of your lungs; exhale normally.*

**FIGURE 46.1** A lung function model. (©J & J Photography)

Rubber stopper
Y tube
Bell jar
Rubber balloons
Clamp
Rubber sheeting

**(a)**           **(b)**

**FIGURE 46.2** Moving the plunger of a syringe as in (*a*) results in air movements into the barrel of the syringe; moving the plunger of the syringe as in (*b*) results in air movements out of the barrel of the syringe. Air movements in and out of the lung function model and the lungs during breathing occur for similar reasons.

**d.** Divide the total value indicated by the needle by 3, and record this amount as your resting tidal volume on the table in Part C of the laboratory report.

**5.** *Expiratory reserve volume (ERV)* (about 1,100 mL) is the volume of air in addition to the tidal volume that leaves the lungs during forced expiration (fig. 46.5). To measure this volume, follow these steps:

    **a.** Breathe normally for a few moments. Set the needle to zero.

    **b.** At the end of an ordinary expiration, place the mouthpiece between your lips and exhale all of the air you can force from your lungs through the spirometer. Use a nose clip or pinch your nose to prevent air from exiting your nostrils.

    **c.** Record the results as your expiratory reserve volume in Part C of the laboratory report.

**6.** *Vital capacity (VC)* (about 4,600 mL) is the maximum volume of air that can be exhaled after taking the deepest breath possible (fig. 46.5). To measure this volume, follow these steps:

    **a.** Breathe normally for a few moments. Set the needle at zero.

    **b.** Breathe in and out deeply a couple of times; then take the deepest breath possible.

    **c.** Place the mouthpiece between your lips and exhale all the air out of your lungs, slowly and forcefully. Use a nose clip or pinch your nose to prevent air from exiting your nostrils.

**FIGURE 46.3** A handheld spirometer can be used to measure respiratory air volumes. (©J & J Photography)

**FIGURE 46.4** Demonstration of a handheld spirometer. Air should be blown slowly and forcefully only into a disposable mouthpiece. *Use a nose clip or pinch your nose when measuring expiratory reserve volume and vital capacity volume if air exits the nostrils.* (©J & J Photography)

**d.** Record the value as your vital capacity in Part C of the laboratory report. Compare your result with that expected for a person of your sex, age, and height listed in tables 46.1 and 46.2. Use the meterstick to determine your height in centimeters if necessary or multiply your height in inches times 2.54 to calculate your height in centimeters. Considerable individual variations from the expected will be noted due to parameters other than sex, age, and height, which can include fitness level, health, medications, and others.

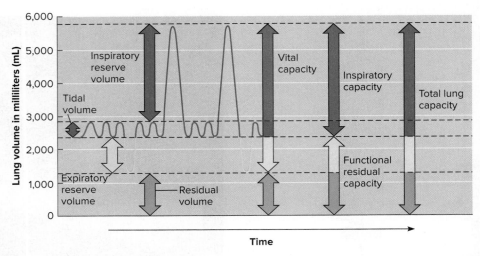

FIGURE 46.5 Graphic representation of respiratory volumes and capacities.

 ## ASSESS

### CRITICAL THINKING

It can be noted from the data in tables 46.1 and 46.2 that vital capacities gradually decrease with age. Propose an explanation for this normal correlation.

_____

_____

_____

_____

_____

7. *Inspiratory reserve volume (IRV)* (about 3,000 mL) is the volume of air in addition to the tidal volume that enters the lungs during forced inspiration (fig. 46.5). Calculate your inspiratory reserve volume by subtracting your tidal volume (TV) and your expiratory reserve volume (ERV) from your vital capacity (VC):

$$IRV = VC - (TV + ERV)$$

8. *Inspiratory capacity (IC)* (about 3,500 mL) is the maximum volume of air a person can inhale following exhalation of the tidal volume (fig. 46.5). Calculate your inspiratory capacity by adding your tidal volume (TV) and your inspiratory reserve volume (IRV):

$$IC = TV + IRV$$

9. *Functional residual capacity (FRC)* (about 2,300 mL) is the volume of air that remains in the lungs following exhalation of the tidal volume (fig. 46.5). Calculate your functional residual capacity (FRC) by adding your expiratory reserve volume (ERV) and your residual volume (RV), which you can assume is 1,200 mL:

$$FRC = ERV + 1,200$$

10. *Residual volume (RV)* (about 1,200 mL) is the volume of air that always remains in the lungs after the most forceful expiration. Although it is part of *total lung capacity* (about 5,800 mL), it cannot be measured with a spirometer. The residual air allows gas exchange and the alveoli to remain open during the respiratory cycle (fig. 46.5).

11. Complete Part C of the laboratory report.

 ## LEARN: LAB IN MOTION

Determine your *minute respiratory volume.* To do this, follow these steps:

1. Sit quietly for a while, and then to establish your breathing rate, count the number of times you breathe in 1 minute. This might be inaccurate because conscious awareness of breathing rate can alter the results. You might ask a laboratory partner to record your breathing rate sometime when you are not expecting it to be recorded. A normal breathing rate is about 12–15 breaths per minute.

2. Calculate your minute respiratory volume by multiplying your breathing rate by your tidal volume:

_____ × _____ × _____
(breathing       (tidal           (minute respiratory
rate)            volume)          volume)

3. This value indicates the total volume of air that moves into your respiratory passages during each minute of ordinary breathing.

**TABLE 46.1  Predicted Vital Capacities (in milliliters) for Females**

**HEIGHT IN CENTIMETERS**

| Age | 146 | 148 | 150 | 152 | 154 | 156 | 158 | 160 | 162 | 164 | 166 | 168 | 170 | 172 | 174 | 176 | 178 | 180 | 182 | 184 | 186 | 188 | 190 | 192 | 194 |
|---|---|---|---|---|---|---|---|---|---|---|---|---|---|---|---|---|---|---|---|---|---|---|---|---|---|
| 16 | 2950 | 2990 | 3030 | 3070 | 3110 | 3150 | 3190 | 3230 | 3270 | 3310 | 3350 | 3390 | 3430 | 3470 | 3510 | 3550 | 3590 | 3630 | 3670 | 3715 | 3755 | 3800 | 3840 | 3880 | 3920 |
| 18 | 2920 | 2960 | 3000 | 3040 | 3080 | 3120 | 3160 | 3200 | 3240 | 3280 | 3320 | 3360 | 3400 | 3440 | 3480 | 3520 | 3560 | 3600 | 3640 | 3680 | 3720 | 3760 | 3800 | 3840 | 3880 |
| 20 | 2890 | 2930 | 2970 | 3010 | 3050 | 3090 | 3130 | 3170 | 3210 | 3250 | 3290 | 3330 | 3370 | 3410 | 3450 | 3490 | 3525 | 3565 | 3605 | 3645 | 3695 | 3720 | 3760 | 3800 | 3840 |
| 22 | 2860 | 2900 | 2940 | 2980 | 3020 | 3060 | 3095 | 3135 | 3175 | 3215 | 3255 | 3290 | 3330 | 3370 | 3410 | 3450 | 3490 | 3530 | 3570 | 3610 | 3650 | 3685 | 3725 | 3765 | 3800 |
| 24 | 2830 | 2870 | 2910 | 2950 | 2985 | 3025 | 3065 | 3100 | 3140 | 3180 | 3220 | 3260 | 3300 | 3335 | 3375 | 3415 | 3455 | 3490 | 3530 | 3570 | 3610 | 3650 | 3685 | 3725 | 3765 |
| 26 | 2800 | 2840 | 2880 | 2920 | 2960 | 3000 | 3035 | 3070 | 3110 | 3150 | 3190 | 3230 | 3265 | 3300 | 3340 | 3380 | 3420 | 3455 | 3495 | 3530 | 3570 | 3610 | 3650 | 3685 | 3725 |
| 28 | 2775 | 2810 | 2850 | 2890 | 2930 | 2965 | 3000 | 3040 | 3070 | 3115 | 3155 | 3190 | 3230 | 3270 | 3305 | 3345 | 3380 | 3420 | 3460 | 3495 | 3535 | 3570 | 3610 | 3650 | 3685 |
| 30 | 2745 | 2780 | 2820 | 2860 | 2895 | 2935 | 2970 | 3010 | 3045 | 3085 | 3120 | 3160 | 3195 | 3235 | 3270 | 3310 | 3345 | 3385 | 3420 | 3460 | 3495 | 3535 | 3570 | 3610 | 3645 |
| 32 | 2715 | 2750 | 2790 | 2825 | 2865 | 2900 | 2940 | 2975 | 3015 | 3050 | 3090 | 3125 | 3160 | 3200 | 3235 | 3275 | 3310 | 3350 | 3385 | 3425 | 3460 | 3495 | 3535 | 3570 | 3610 |
| 34 | 2685 | 2725 | 2760 | 2795 | 2835 | 2870 | 2910 | 2945 | 2980 | 3020 | 3055 | 3090 | 3130 | 3165 | 3200 | 3240 | 3275 | 3310 | 3350 | 3385 | 3425 | 3460 | 3495 | 3535 | 3570 |
| 36 | 2655 | 2695 | 2730 | 2765 | 2805 | 2840 | 2875 | 2910 | 2950 | 2985 | 3020 | 3060 | 3095 | 3130 | 3165 | 3205 | 3240 | 3275 | 3310 | 3350 | 3385 | 3420 | 3460 | 3495 | 3530 |
| 38 | 2630 | 2665 | 2700 | 2735 | 2770 | 2810 | 2845 | 2880 | 2915 | 2950 | 2990 | 3025 | 3060 | 3095 | 3130 | 3170 | 3205 | 3240 | 3275 | 3310 | 3350 | 3385 | 3420 | 3455 | 3490 |
| 40 | 2600 | 2635 | 2670 | 2705 | 2740 | 2775 | 2810 | 2850 | 2885 | 2920 | 2955 | 2990 | 3025 | 3060 | 3095 | 3135 | 3170 | 3205 | 3240 | 3275 | 3310 | 3345 | 3380 | 3420 | 3455 |
| 42 | 2570 | 2605 | 2640 | 2675 | 2710 | 2745 | 2780 | 2815 | 2850 | 2885 | 2920 | 2955 | 2990 | 3025 | 3060 | 3100 | 3135 | 3170 | 3205 | 3240 | 3275 | 3310 | 3345 | 3380 | 3415 |
| 44 | 2540 | 2575 | 2610 | 2645 | 2680 | 2715 | 2750 | 2785 | 2820 | 2855 | 2890 | 2925 | 2960 | 2995 | 3030 | 3060 | 3095 | 3130 | 3165 | 3200 | 3235 | 3270 | 3305 | 3340 | 3375 |
| 46 | 2510 | 2545 | 2580 | 2615 | 2650 | 2685 | 2715 | 2750 | 2785 | 2820 | 2855 | 2890 | 2925 | 2960 | 2995 | 3030 | 3060 | 3095 | 3130 | 3165 | 3200 | 3235 | 3270 | 3305 | 3340 |
| 48 | 2480 | 2515 | 2550 | 2585 | 2620 | 2650 | 2685 | 2715 | 2750 | 2785 | 2820 | 2855 | 2890 | 2925 | 2960 | 2995 | 3030 | 3060 | 3095 | 3130 | 3160 | 3195 | 3230 | 3265 | 3300 |
| 50 | 2455 | 2485 | 2520 | 2555 | 2590 | 2625 | 2655 | 2690 | 2720 | 2755 | 2785 | 2820 | 2855 | 2890 | 2925 | 2955 | 2990 | 3025 | 3060 | 3090 | 3125 | 3155 | 3190 | 3225 | 3260 |
| 52 | 2425 | 2455 | 2490 | 2525 | 2555 | 2590 | 2625 | 2655 | 2690 | 2720 | 2755 | 2790 | 2820 | 2855 | 2890 | 2925 | 2955 | 2990 | 3020 | 3055 | 3090 | 3125 | 3155 | 3190 | 3220 |
| 54 | 2395 | 2425 | 2460 | 2495 | 2530 | 2560 | 2590 | 2625 | 2655 | 2690 | 2720 | 2755 | 2790 | 2820 | 2855 | 2885 | 2920 | 2950 | 2985 | 3020 | 3050 | 3085 | 3115 | 3150 | 3180 |
| 56 | 2365 | 2400 | 2430 | 2460 | 2495 | 2525 | 2560 | 2590 | 2625 | 2655 | 2690 | 2720 | 2755 | 2790 | 2820 | 2855 | 2885 | 2920 | 2950 | 2980 | 3015 | 3045 | 3080 | 3110 | 3145 |
| 58 | 2335 | 2370 | 2400 | 2430 | 2460 | 2495 | 2525 | 2560 | 2590 | 2625 | 2655 | 2690 | 2720 | 2750 | 2785 | 2815 | 2850 | 2880 | 2920 | 2945 | 2975 | 3010 | 3040 | 3075 | 3105 |
| 60 | 2305 | 2340 | 2370 | 2400 | 2430 | 2460 | 2495 | 2525 | 2560 | 2590 | 2625 | 2655 | 2685 | 2720 | 2750 | 2780 | 2810 | 2845 | 2875 | 2915 | 2940 | 2970 | 3000 | 3035 | 3065 |
| 62 | 2280 | 2310 | 2340 | 2370 | 2405 | 2435 | 2465 | 2495 | 2525 | 2560 | 2590 | 2620 | 2655 | 2685 | 2715 | 2745 | 2775 | 2810 | 2840 | 2870 | 2900 | 2935 | 2965 | 2995 | 3025 |
| 64 | 2250 | 2280 | 2310 | 2340 | 2370 | 2400 | 2430 | 2465 | 2495 | 2525 | 2555 | 2585 | 2620 | 2650 | 2680 | 2710 | 2740 | 2770 | 2805 | 2835 | 2865 | 2895 | 2920 | 2955 | 2990 |
| 66 | 2220 | 2250 | 2280 | 2310 | 2340 | 2370 | 2400 | 2430 | 2460 | 2495 | 2525 | 2555 | 2585 | 2615 | 2645 | 2675 | 2705 | 2735 | 2765 | 2800 | 2825 | 2860 | 2890 | 2920 | 2950 |
| 68 | 2190 | 2220 | 2250 | 2280 | 2310 | 2340 | 2370 | 2400 | 2430 | 2460 | 2490 | 2520 | 2550 | 2580 | 2610 | 2640 | 2670 | 2700 | 2730 | 2760 | 2795 | 2820 | 2850 | 2880 | 2910 |
| 70 | 2160 | 2190 | 2220 | 2250 | 2280 | 2310 | 2340 | 2370 | 2400 | 2425 | 2455 | 2485 | 2515 | 2545 | 2575 | 2605 | 2635 | 2665 | 2695 | 2725 | 2755 | 2780 | 2810 | 2840 | 2870 |
| 72 | 2130 | 2160 | 2190 | 2220 | 2250 | 2280 | 2310 | 2335 | 2365 | 2395 | 2425 | 2455 | 2480 | 2510 | 2540 | 2570 | 2600 | 2630 | 2660 | 2685 | 2715 | 2745 | 2775 | 2805 | 2830 |
| 74 | 2100 | 2130 | 2160 | 2190 | 2220 | 2245 | 2275 | 2305 | 2335 | 2360 | 2390 | 2420 | 2450 | 2475 | 2505 | 2535 | 2565 | 2590 | 2620 | 2650 | 2680 | 2710 | 2740 | 2765 | 2795 |

From E. DeF. Baldwin and E. W. Richards, Jr., "Pulmonary Insufficiency 1, Physiologic Classification, Clinical Methods of Analysis, Standard Values in Normal Subjects" in *Medicine 27:243*, © by William & Wilkins. Used by permission.

**TABLE 46.2  Predicted Vital Capacities (in milliliters) for Males**

| | | | | | | | | | | | | HEIGHT IN CENTIMETERS | | | | | | | | | | | | | |
|---|---|---|---|---|---|---|---|---|---|---|---|---|---|---|---|---|---|---|---|---|---|---|---|---|---|
| Age | 146 | 148 | 150 | 152 | 154 | 156 | 158 | 160 | 162 | 164 | 166 | 168 | 170 | 172 | 174 | 176 | 178 | 180 | 182 | 184 | 186 | 188 | 190 | 192 | 194 |
| 16 | 3765 | 3820 | 3870 | 3920 | 3975 | 4025 | 4075 | 4130 | 4180 | 4230 | 4285 | 4335 | 4385 | 4440 | 4490 | 4540 | 4590 | 4645 | 4695 | 4745 | 4800 | 4850 | 4900 | 4955 | 5005 |
| 18 | 3740 | 3790 | 3840 | 3890 | 3940 | 3995 | 4045 | 4095 | 4145 | 4200 | 4250 | 4300 | 4350 | 4405 | 4455 | 4505 | 4555 | 4610 | 4660 | 4710 | 4760 | 4815 | 4865 | 4915 | 4965 |
| 20 | 3710 | 3760 | 3810 | 3860 | 3910 | 3960 | 4015 | 4065 | 4115 | 4165 | 4215 | 4265 | 4320 | 4370 | 4420 | 4470 | 4520 | 4570 | 4625 | 4675 | 4725 | 4775 | 4825 | 4875 | 4930 |
| 22 | 3680 | 3730 | 3780 | 3830 | 3880 | 3930 | 3980 | 4030 | 4080 | 4135 | 4185 | 4235 | 4285 | 4335 | 4385 | 4435 | 4485 | 4535 | 4585 | 4635 | 4685 | 4735 | 4790 | 4840 | 4890 |
| 24 | 3635 | 3685 | 3735 | 3785 | 3835 | 3885 | 3935 | 3985 | 4035 | 4085 | 4135 | 4185 | 4235 | 4285 | 4330 | 4380 | 4430 | 4480 | 4530 | 4580 | 4630 | 4680 | 4730 | 4780 | 4830 |
| 26 | 3605 | 3655 | 3705 | 3755 | 3805 | 3855 | 3905 | 3955 | 4000 | 4050 | 4100 | 4150 | 4200 | 4250 | 4300 | 4350 | 4395 | 4445 | 4495 | 4545 | 4595 | 4645 | 4695 | 4740 | 4790 |
| 28 | 3575 | 3625 | 3675 | 3725 | 3775 | 3820 | 3870 | 3920 | 3970 | 4020 | 4070 | 4115 | 4165 | 4215 | 4265 | 4310 | 4360 | 4410 | 4460 | 4510 | 4555 | 4605 | 4655 | 4705 | 4755 |
| 30 | 3550 | 3595 | 3645 | 3695 | 3740 | 3790 | 3840 | 3890 | 3935 | 3985 | 4035 | 4080 | 4130 | 4180 | 4230 | 4275 | 4325 | 4375 | 4425 | 4470 | 4520 | 4570 | 4615 | 4665 | 4715 |
| 32 | 3520 | 3565 | 3615 | 3665 | 3710 | 3760 | 3810 | 3855 | 3905 | 3950 | 4000 | 4050 | 4095 | 4145 | 4195 | 4240 | 4290 | 4340 | 4385 | 4435 | 4485 | 4530 | 4580 | 4625 | 4675 |
| 34 | 3475 | 3525 | 3570 | 3620 | 3665 | 3715 | 3760 | 3810 | 3855 | 3905 | 3950 | 4000 | 4045 | 4095 | 4140 | 4190 | 4225 | 4285 | 4330 | 4380 | 4425 | 4475 | 4520 | 4570 | 4615 |
| 36 | 3445 | 3495 | 3540 | 3585 | 3635 | 3680 | 3730 | 3775 | 3825 | 3870 | 3920 | 3965 | 4010 | 4060 | 4105 | 4155 | 4200 | 4250 | 4295 | 4340 | 4390 | 4435 | 4485 | 4530 | 4580 |
| 38 | 3415 | 3465 | 3510 | 3555 | 3605 | 3650 | 3695 | 3745 | 3790 | 3840 | 3885 | 3930 | 3980 | 4025 | 4070 | 4120 | 4165 | 4210 | 4260 | 4305 | 4350 | 4400 | 4445 | 4495 | 4540 |
| 40 | 3385 | 3435 | 3480 | 3525 | 3575 | 3620 | 3665 | 3710 | 3760 | 3805 | 3850 | 3900 | 3945 | 3990 | 4035 | 4085 | 4130 | 4175 | 4220 | 4270 | 4315 | 4360 | 4410 | 4455 | 4500 |
| 42 | 3360 | 3405 | 3450 | 3495 | 3540 | 3590 | 3635 | 3680 | 3725 | 3770 | 3820 | 3865 | 3910 | 3955 | 4000 | 4050 | 4095 | 4140 | 4185 | 4230 | 4280 | 4325 | 4370 | 4415 | 4460 |
| 44 | 3315 | 3360 | 3405 | 3450 | 3495 | 3540 | 3585 | 3630 | 3675 | 3725 | 3770 | 3815 | 3860 | 3905 | 3950 | 3995 | 4040 | 4085 | 4130 | 4175 | 4220 | 4270 | 4315 | 4360 | 4405 |
| 46 | 3285 | 3330 | 3375 | 3420 | 3465 | 3510 | 3555 | 3600 | 3645 | 3690 | 3735 | 3780 | 3825 | 3870 | 3915 | 3960 | 4005 | 4050 | 4095 | 4140 | 4185 | 4230 | 4275 | 4320 | 4365 |
| 48 | 3255 | 3300 | 3345 | 3390 | 3435 | 3480 | 3525 | 3570 | 3615 | 3655 | 3700 | 3745 | 3790 | 3835 | 3880 | 3925 | 3970 | 4015 | 4060 | 4105 | 4150 | 4190 | 4235 | 4280 | 4325 |
| 50 | 3210 | 3255 | 3300 | 3345 | 3390 | 3430 | 3475 | 3520 | 3565 | 3610 | 3650 | 3695 | 3740 | 3785 | 3830 | 3870 | 3915 | 3960 | 4005 | 4050 | 4090 | 4135 | 4180 | 4225 | 4270 |
| 52 | 3185 | 3225 | 3270 | 3315 | 3355 | 3400 | 3445 | 3490 | 3530 | 3575 | 3620 | 3660 | 3705 | 3750 | 3795 | 3835 | 3880 | 3925 | 3970 | 4010 | 4055 | 4100 | 4140 | 4185 | 4230 |
| 54 | 3155 | 3195 | 3240 | 3285 | 3325 | 3370 | 3415 | 3455 | 3500 | 3540 | 3585 | 3630 | 3670 | 3715 | 3760 | 3800 | 3845 | 3890 | 3930 | 3975 | 4020 | 4060 | 4105 | 4145 | 4190 |
| 56 | 3125 | 3165 | 3210 | 3255 | 3295 | 3340 | 3380 | 3425 | 3465 | 3510 | 3550 | 3595 | 3640 | 3680 | 3725 | 3765 | 3810 | 3850 | 3895 | 3940 | 3980 | 4025 | 4065 | 4110 | 4150 |
| 58 | 3080 | 3125 | 3165 | 3210 | 3250 | 3290 | 3335 | 3375 | 3420 | 3460 | 3500 | 3545 | 3585 | 3630 | 3670 | 3715 | 3755 | 3800 | 3840 | 3880 | 3925 | 3965 | 4010 | 4050 | 4095 |
| 60 | 3050 | 3095 | 3135 | 3175 | 3220 | 3260 | 3300 | 3345 | 3385 | 3430 | 3470 | 3500 | 3555 | 3595 | 3635 | 3680 | 3720 | 3760 | 3805 | 3845 | 3885 | 3930 | 3970 | 4015 | 4055 |
| 62 | 3020 | 3060 | 3110 | 3150 | 3190 | 3230 | 3270 | 3310 | 3350 | 3390 | 3440 | 3480 | 3520 | 3560 | 3600 | 3640 | 3680 | 3730 | 3770 | 3810 | 3850 | 3890 | 3930 | 3970 | 4020 |
| 64 | 2990 | 3030 | 3080 | 3120 | 3160 | 3200 | 3240 | 3280 | 3320 | 3360 | 3400 | 3440 | 3490 | 3530 | 3570 | 3610 | 3650 | 3690 | 3730 | 3770 | 3810 | 3850 | 3900 | 3940 | 3980 |
| 66 | 2950 | 2990 | 3030 | 3070 | 3110 | 3150 | 3190 | 3230 | 3270 | 3310 | 3350 | 3390 | 3430 | 3470 | 3510 | 3550 | 3600 | 3640 | 3680 | 3720 | 3760 | 3800 | 3840 | 3880 | 3920 |
| 68 | 2920 | 2960 | 3000 | 3040 | 3080 | 3120 | 3160 | 3200 | 3240 | 3280 | 3320 | 3360 | 3400 | 3440 | 3480 | 3520 | 3560 | 3600 | 3640 | 3680 | 3720 | 3760 | 3800 | 3840 | 3880 |
| 70 | 2890 | 2930 | 2970 | 3010 | 3050 | 3090 | 3130 | 3170 | 3210 | 3250 | 3290 | 3330 | 3370 | 3410 | 3450 | 3480 | 3520 | 3560 | 3600 | 3640 | 3680 | 3720 | 3760 | 3800 | 3840 |
| 72 | 2860 | 2900 | 2940 | 2980 | 3020 | 3060 | 3100 | 3140 | 3180 | 3210 | 3250 | 3290 | 3330 | 3370 | 3410 | 3450 | 3490 | 3530 | 3570 | 3610 | 3650 | 3680 | 3720 | 3760 | 3800 |
| 74 | 2820 | 2860 | 2900 | 2930 | 2970 | 3010 | 3050 | 3090 | 3130 | 3170 | 3200 | 3240 | 3280 | 3320 | 3360 | 3400 | 3440 | 3470 | 3510 | 3550 | 3590 | 3630 | 3670 | 3710 | 3740 |

From E. DeF. Baldwin and E. W. Richards, Jr., "Pulmonary Insufficiency 1, Physiologic Classification, Clinical Methods of Analysis, Standard Values in Normal Subjects" in *Medicine* 27:243, © by William & Wilkins. Used by permission.

Name _____

Date _____

Section _____

The 🅐 corresponds to the indicated Learning Outcome(s) found at the beginning of the laboratory exercise.

# Breathing and Respiratory Volumes

## PART A ASSESSMENTS

Complete the following statements:

1. Breathing can also be called _____ .

2. The weight of air causes a force called _____ pressure.

3. The weight of air at sea level is sufficient to support a column of mercury within a tube _____ mm high.

4. If the pressure inside the lungs decreases, outside air is forced into the airways by _____ . 🅐

5. Impulses are conducted to the diaphragm by the _____ nerves. 🅐

6. When the diaphragm contracts, the size of the thoracic cavity _____ . 🅐

7. The ribs are raised by contraction of the _____ muscles, which increases the size of the thoracic cavity. 🅐

8. Only a thin film of lubricating serous fluid separates the parietal pleura from the _____ attached to the surface of a lung.

9. A mixture of lipoproteins, called _____ , acts to reduce the tendency of alveoli to collapse.

10. The force responsible for normal resting expiration comes from _____ of tissues and from surface tension. 🅐

11. Muscles that help to force out more than the normal volume of air by pulling the ribs downward and inward include the _____ . 🅐

12. The diaphragm can be forced to move higher than normal by contraction of the _____ wall muscles. 🅐

## PART B ASSESSMENTS

Match the air volumes in column A with the definitions in column B. Place the letter of your choice in the space provided. 🄶

| Column A | Column B |
|---|---|
| **a.** Expiratory reserve volume | _____ 1. Volume of air in addition to tidal volume that leaves the lungs during forced expiration |
| **b.** Functional residual capacity | |
| **c.** Inspiratory capacity | _____ 2. Vital capacity plus residual volume |
| **d.** Inspiratory reserve volume | _____ 3. Volume of air that remains in lungs after the most forceful expiration |
| **e.** Residual volume | _____ 4. Volume of air that enters or leaves lungs during a respiratory cycle |
| **f.** Tidal volume | |
| **g.** Total lung capacity | _____ 5. Volume of air in addition to tidal volume that enters lungs during forced inspiration |
| **h.** Vital capacity | |
| | _____ 6. Maximum volume of air a person can exhale after taking the deepest possible breath |
| | _____ 7. Maximum volume of air a person can inhale following exhalation of the tidal volume |
| | _____ 8. Volume of air remaining in the lungs following exhalation of the tidal volume |

1. Test results for respiratory air volumes and capacities:

| Respiratory Volume or Capacity | Expected Value* (approximate) | Test Results | Percent of Expected Value (test result/expected value × 100) |
|---|---|---|---|
| Tidal volume (resting) (TV) | 500 mL | | |
| Expiratory reserve volume (ERV) | 1,100 mL | | |
| Vital capacity (VC) | 4,600 mL | | |
| | (enter yours from table 46.1 or 46.2) | | |
| Inspiratory reserve volume (IRV) | 3,000 mL | | |
| Inspiratory capacity (IC) | 3,500 mL | | |
| Functional residual capacity (FRC) | 2,300 mL | | |

*The values listed are most characteristic for a healthy, tall, young adult male. In general, adult females have smaller bodies and therefore smaller lung volumes and capacities. If your expected value for vital capacity is considerably different than 4,600 mL, your other values will vary accordingly.

2. Complete the following:

   **a.** How do your test results compare with the expected values?

   _____

   _____

   _____

   **b.** How does your vital capacity compare with the average value for a person of your sex, age, and height?

   _____

   _____

   _____

   **c.** What measurement in addition to vital capacity is needed before you can calculate your total lung capacity?

   _____

3. If your experimental results are considerably different than the predicted vital capacities, propose reasons for the differences. As you write this paragraph, consider factors such as smoking, physical fitness, respiratory disorders, stress, and medications. (Your instructor might have you make some class correlations from class data.)

   _____

   _____

   _____

   _____

   _____

# 47

# Control of Breathing

## MATERIALS NEEDED

Textbook
Clock with second hand
Paper bags, small
Flasks
Glass tubing
Rubber stoppers, two-hole
Calcium hydroxide solution (limewater)
Pneumograph
Physiological recording apparatus

## PURPOSE OF THE EXERCISE

To review the muscles that control breathing, and to describe the mechanisms that regulate the rate and depth of breathing.

##  LEARNING OUTCOMES AP|R

After completing this exercise, you should be able to

1. Locate the respiratory areas in the brainstem.
2. Describe the mechanisms that control and influence breathing.
3. Select the respiratory muscles involved in inspiration and forced expiration.
4. Test and record the effect of various factors on the rate and depth of breathing.

Several skeletal muscles contract during breathing. The principal muscles involved during inspiration are the diaphragm and external intercostals. The sternocleidomastoid, scalenes, and pectoralis minor are synergistic during forceful inhalation. During quiet respiration, there is minimal involvement of the expiratory muscles. During forced expiration, the principal muscles are the internal intercostals, but the abdominal wall muscles can provide extra force.

Breathing is controlled from regions of the brainstem called the *respiratory areas,* which control both inspiration and expiration. These areas initiate and regulate impulses that travel to various breathing muscles, causing rhythmic breathing movements and adjustments to the rate and depth of breathing to meet various cellular needs of oxygen supply and carbon dioxide removal (fig. 47.1). The *medullary respiratory center* is composed of two bilateral groups of neurons in the medulla oblongata. They are called the *ventral respiratory group* and the *dorsal respiratory group.* The ventral respiratory group involves regulation of the basic rhythm of breathing. The dorsal respiratory group primarily stimulates the diaphragm contraction, resulting in inspiration, and helps process respiratory sensory information involving the cardiovascular system. Neurons in another part of the brainstem, the pons, compose the bilateral *pontine respiratory group* (formerly the pneumotaxic center). These neurons may contribute to the rhythm of breathing by modifying the respirations during such situations as exercise and sleep (fig. 47.2).

Various factors can influence the respiratory areas and thus affect the rate and depth of breathing. These factors include stretch of the lung tissues, emotional state, and presence in the blood of certain chemicals, such as carbon dioxide, hydrogen ions, and oxygen. The breathing rate increases primarily as the blood concentration of carbon dioxide or hydrogen ions increases, but much less often as the concentration of oxygen decreases.

##  PRACTICE

### PROCEDURE A—Control of Breathing

1. Review the section entitled "Control of Breathing" in chapter 19 of the textbook.
2. Respiratory areas in the brainstem control the cycle of breathing. The dorsal respiratory group of neurons is located in the medulla oblongata. Impulses from the dorsal respiratory group stimulate the muscles of inspiration, especially the diaphragm (see fig. 47.1). This results in a normal respiratory rate of 12–15 breaths per minute. The ventral respiratory group of neurons will fire during inspiration and expiration to control the appropriate muscles of inspiration and expiration and the basic rhythm of breathing (fig. 47.2).
3. Complete Part A of Laboratory Report 47.

Midbrain

Fourth ventricle

Pontine respiratory group

Pons

Medulla oblongata

Ventral respiratory group

Dorsal respiratory group

Medullary respiratory center

Internal (expiratory) intercostal muscles

External (inspiratory) intercostal muscles

Diaphragm

**FIGURE 47.1** The respiratory areas are located in the pons and the medulla oblongata. **AP|R**

## LEARN: ACTIVITY

When a solution of calcium hydroxide is exposed to carbon dioxide, a chemical reaction occurs and a white precipitate of calcium carbonate is formed, as indicated by the following reaction:

$$Ca(OH)_2 + CO_2 \longrightarrow CaCO_3 + H_2O$$

Thus, a clear water solution of calcium hydroxide (lime-water) can be used to detect the presence of carbon dioxide because the solution becomes cloudy if this gas is bubbled through it.

The laboratory instructor will demonstrate this test for carbon dioxide by drawing some air through lime-water in an apparatus such as that shown in figure 47.3. Then the instructor will blow an equal volume of expired air through a similar apparatus. (*Note*: A new sterile mouth-piece should be used each time the apparatus is demonstrated.) Watch for the appearance of a precipitate that causes the limewater to become cloudy. Was there any carbon dioxide in the atmospheric air drawn through the limewater?

_____

If so, how did the amount of carbon dioxide in the atmospheric air compare with the amount in the expired air?

_____

_____

_____

Respiratory areas

Pontine respiratory group

Medullary respiratory center

Ventral respiratory group

Dorsal respiratory group

Respiratory muscles

Basic rhythm of breathing

Changes in rate and depth of breathing

**FIGURE 47.2** The medullary respiratory center and the pontine respiratory group control breathing.

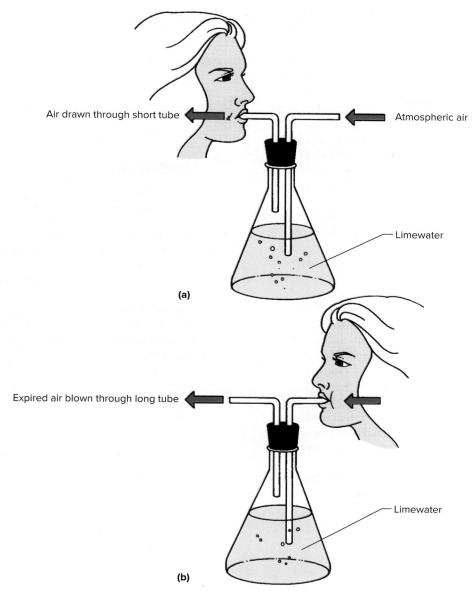

Air drawn through short tube ← → Atmospheric air

Limewater

(a)

Expired air blown through long tube ← → 

Limewater

(b)

**FIGURE 47.3** Apparatus used to demonstrate the presence of carbon dioxide in air: (*a*) atmospheric air is drawn through limewater; (*b*) expired air is blown through limewater.

 **PRACTICE**

**PROCEDURE B—Factors Affecting Breathing**

Perform each of the following tests, using your laboratory partner as a test subject.

1. *Normal breathing.* To determine the subject's normal breathing rate and depth, follow these steps:
   a. Have the subject sit quietly for a few minutes.
   b. After the rest period, ask the subject to count backward mentally, beginning with 500.
   c. While the subject is distracted by counting, watch the subject's chest movements, and count the breaths taken in a minute. Use this value as the normal breathing rate (breaths per minute).

   d. Note the relative depth of the breathing movements.
   e. Record your observations in the table in Part B of the laboratory report.
2. *Effect of hyperventilation.* To test the effect of hyperventilation on breathing, follow these steps:
   a. Seat the subject and *guard to prevent the possibility of the subject falling over.*
   b. Have the subject breathe rapidly and deeply for a maximum of 20 seconds. *If the subject begins to feel dizzy, the hyperventilation should be halted immediately to prevent the subject from fainting from complications of alkalosis. The increased blood pH causes vasoconstriction of cerebral arterioles, which decreases circulation and oxygen to the brain.*

**c.** After the period of hyperventilation, determine the subject's breathing rate, and judge the breathing depth as before.

**d.** Record the results in Part B of the laboratory report.

**3.** *Effect of rebreathing air.* To test the effect of rebreathing air on breathing, follow these steps:

   **a.** Have the subject sit quietly (approximately 5 minutes) until the breathing rate returns to normal.

   **b.** Have the subject breathe deeply into a small paper bag that is held tightly over the nose and mouth. *If the subject begins to feel light-headed or like fainting, the rebreathing air should be halted immediately to prevent further acidosis and fainting.*

   **c.** After 2 minutes of rebreathing air, determine the subject's breathing rate, and judge the depth of breathing.

   **d.** Record the results in Part B of the laboratory report.

**4.** *Effect of breath holding.* To test the effect of breath holding on breathing, follow these steps:

   **a.** Have the subject sit quietly (approximately 5 minutes) until the breathing rate returns to normal.

   **b.** Have the subject hold his/her breath as long as possible. *If the subject begins to feel light-headed or like fainting, breath holding should be halted immediately to prevent further acidosis and fainting.*

   **c.** As the subject begins to breathe again, determine the rate of breathing, and judge the depth of breathing.

   **d.** Record the results in Part B of the laboratory report.

**5.** *Effect of exercise.* To test the effect of exercise on breathing, follow these steps:

   **a.** Have the subject sit quietly (approximately 5 minutes) until breathing rate returns to normal.

   **b.** Have the subject exercise by moderately running in place for 3–5 minutes. *This exercise should be avoided by anyone with health risks.*

   **c.** After the exercise, determine the breathing rate, and judge the depth of breathing.

   **d.** Record the results in Part B of the laboratory report.

**6.** Complete Part B of the laboratory report.

 **LEARN: LAB IN MOTION**

A *pneumograph* is a device that can be used together with some type of recording apparatus to record breathing movements. The laboratory instructor will demonstrate the use of this equipment to record various movements, such as those that accompany coughing, laughing, yawning, and speaking.

Devise an experiment to test the effect of some factor, such as hyperventilation, rebreathing of air, or exercise, on the length of time a person can hold his/her breath. *After the laboratory instructor has approved your plan,* carry out the experiment, using the pneumograph and recording equipment. What conclusion can you draw from the results of your experiment?

_____

_____

_____

_____

_____

The Ⓐ corresponds to the indicated Learning Outcome(s) found at the beginning of the laboratory exercise.

# Control of Breathing

## ♻ PART A ASSESSMENTS

Complete the following statements:

1. The respiratory areas are widely scattered throughout the pons and _____ of the brainstem. Ⓐ₁

2. The _____ respiratory group within the medulla oblongata regulates the basic rhythm of breathing. Ⓐ₁

3. The _____ respiratory group within the medulla oblongata primarily stimulates the diaphragm. Ⓐ₁

4. Chemoreceptors are sensitive to changes in the blood concentrations of carbon dioxide and _____ ions. Ⓐ₂

5. As the blood concentration of carbon dioxide increases, the breathing rate _____ . Ⓐ₂

6. As a result of increased breathing, the blood concentration of carbon dioxide is _____ . Ⓐ₂

7. As a result of hyperventilation, breath-holding time is _____ . Ⓐ₂

8. The principal muscles of forced expiration are the _____ and the abdominal wall muscles. Ⓐ₃

9. The principal muscles of inspiration are the _____ and the external intercostal muscles. Ⓐ₃

## ♻ PART B ASSESSMENTS

1. Record the results of your breathing tests in the table: Ⓐ₄

| Factor Tested | Breathing Rate (breaths/min) | Breathing Depth (+, + +, + + +) |
|---|---|---|
| Normal | | |
| Hyperventilation (shortly after) | | |
| Rebreathing air (shortly after) | | |
| Breath holding (shortly after) | | |
| Exercise (shortly after) | | |

**2.** Explain the reason for the changes in breathing that occurred in each of the following cases: **2**

    **a.** Hyperventilation

_____

_____

_____

    **b.** Rebreathing air

_____

_____

_____

    **c.** Breath holding

_____

_____

_____

_____

    **d.** Exercise

_____

_____

_____

_____

**3.** Complete the following:

    **a.** Why is it important to distract a person when you are determining the normal rate of breathing?

_____

_____

    **b.** How can the depth of breathing be measured accurately?

_____

_____

_____

## ASSESS

**CRITICAL THINKING**

Why is it dangerous for a swimmer to hyperventilate in order to hold the breath for a longer period of time? **2**

_____

_____

_____

_____

_____

# 48

# Urinary Organs

## MATERIALS NEEDED

Textbook
Human torso model
Kidney model
Preserved pig (or sheep) kidney
Dissecting tray
Dissecting instruments
Long knife
Compound light microscope
Prepared microscope slides of the following:
    Kidney section
    Ureter
    Urinary bladder

## ⚠ SAFETY

- Wear disposable gloves when working on the kidney dissection.
- Dispose of the kidney and gloves as directed by your laboratory instructor.
- Wash the dissecting tray and instruments as instructed.
- Wash your laboratory table.
- Wash your hands before leaving the laboratory.

## PURPOSE OF THE EXERCISE

To review the structure of the urinary organs, to dissect a kidney, and to observe the major structures of a nephron.

##  LEARNING OUTCOMES AP|R

After completing this exercise, you should be able to

1. Locate and identify the major organs of the urinary system.
2. Locate and identify the major structures of a kidney.
3. Identify and sketch the structures of a nephron.
4. Trace the path of filtrate through a nephron.
5. Trace the path of blood and blood pressure changes through the renal blood vessels.
6. Identify and sketch the structures of a ureter and the urinary bladder.

The two kidneys are the primary organs of the urinary system. They are located in the abdominal cavity, against the posterior wall and behind the parietal peritoneum (retroperitoneal). Masses of adipose tissue associated with the kidneys hold them in place at a vertebral level between T12 and L3. The right kidney is slightly more inferior due to the large mass of the liver near its superior border. The movements of the muscular walls of the ureters force urine by means of peristaltic waves into the urinary bladder, which temporarily stores urine. The urethra conveys urine to the outside of the body.

Each kidney contains about 1 million nephrons, which serve as the basic structural and functional units of the kidney. A glomerular capsule, proximal tubule, nephron loop, and distal tubule compose the microscopic, multicellular structure of a relatively long nephron tubule. Several nephrons merge and drain into a common collecting duct. Approximately 80% of the nephrons are cortical nephrons with short nephron loops, whereas the remaining represent juxtamedullary nephrons, with long nephron loops extending deeper into the renal medulla. An elaborate network of blood vessels surrounds the entire nephron. Glomerular filtration, tubular reabsorption, and tubular secretion represent three processes resulting in urine as the final product.

A variety of functions occur in the kidneys. They remove metabolic wastes from the blood; help regulate blood volume, blood pressure, and pH of blood; control water and electrolyte concentrations; and secrete renin and erythropoietin.

##  PRACTICE

### PROCEDURE A—Kidney

1. Review the section entitled "Kidney Structure" in chapter 20 of the textbook.
2. As a review activity, label figures 48.1 and 48.2.

Labels visible on figure: Renal vein, Adrenal gland, Renal artery, Inferior vena cava, Abdominal aorta

1

2

3

4

**FIGURE 48.1**   Label the major organs of the urinary system.  A **AP|R**

Labels visible on figure: Renal artery, Renal vein, Fat in renal sinus

1

2

3

4

5

6

7

8

9

10

**FIGURE 48.2**   Label the major structures in the longitudinal section of a kidney.  A **AP|R**

3. Complete Part A of Laboratory Report 48.
4. Observe the human torso model and the kidney model. Locate the following:

**kidneys**
**ureters**
**urinary bladder**
**urethra**
**renal sinus**
**renal pelvis**
    major calyces
    minor calyces
**renal medulla**
    renal pyramids
    renal papillae
**renal cortex**
**renal columns**
**nephrons**
    cortical nephrons (80% of nephrons)
    juxtamedullary nephrons (20% of nephrons)

5. Obtain a pig or sheep kidney along with a dissecting tray, dissecting instruments, and disposable gloves, and follow these steps:
  **a.** Rinse the kidney with water to remove as much of the preserving fluid as possible and place it in a dissecting tray.

**b.** Carefully remove any adipose tissue from the surface of the specimen.
**c.** Locate the following external anatomical features:

**renal (fibrous) capsule**
**hilum of kidney**
**renal artery**
**renal vein**
**ureter**

**d.** Use a long knife to cut the kidney in half longitudinally along the frontal plane, beginning on the convex border.
**e.** Rinse the interior of the kidney with water, and using figure 48.3 as a reference, locate the following:

**renal pelvis**
    major calyces
    minor calyces
**renal cortex**
**renal columns** (extensions of renal cortical tissue between renal pyramids)
**renal medulla**
    renal pyramids
    renal papillae

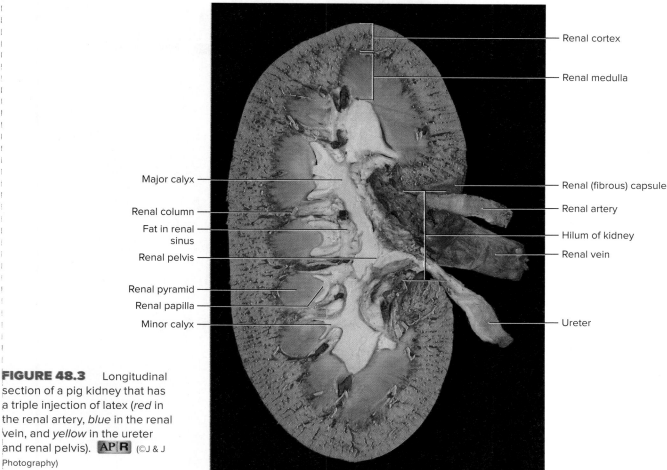

Renal cortex
Renal medulla
Major calyx
Renal column
Fat in renal sinus
Renal pelvis
Renal pyramid
Renal papilla
Minor calyx
Renal (fibrous) capsule
Renal artery
Hilum of kidney
Renal vein
Ureter

**FIGURE 48.3** Longitudinal section of a pig kidney that has a triple injection of latex (*red* in the renal artery, *blue* in the renal vein, and *yellow* in the ureter and renal pelvis). **AP|R** (©J & J Photography)

## PRACTICE

### PROCEDURE B—Renal Blood Vessels and Nephrons

1. Review the sections entitled "Renal Blood Vessels" and "Nephrons" in chapter 20 of the textbook.
2. As a review activity, label figure 48.4.
3. Complete Part B of the laboratory report.
4. Obtain a microscope slide of a kidney section, and examine it using low-power magnification. Locate the *renal capsule,* the *renal cortex* (which appears somewhat granular and may be more darkly stained than the other renal tissues), and the *renal medulla* (fig. 48.5).
5. Examine the renal cortex, using high-power magnification. Locate a *renal corpuscle.* These structures appear as isolated circular areas. Identify the *glomerulus,*

the capillary cluster inside the corpuscle, and the *glomerular capsule,* which appears as a clear area surrounding the glomerulus. A glomerulus and a glomerular capsule compose a *renal corpuscle.* Also note the numerous sections of renal tubules that occupy the spaces between renal corpuscles (fig. 48.5*a*). The renal cortex contains the proximal and distal convoluted tubules.

6. Prepare a labeled sketch of a representative section of the renal cortex in Part C of the laboratory report.
7. Examine the renal medulla, using high-power magnification. Identify longitudinal sections of various collecting ducts. These ducts are lined with simple epithelial cells, which vary in shape from squamous to cuboidal (fig. 48.5*b*).
8. Prepare a labeled sketch of a representative section of the renal medulla in Part C of the laboratory report.

**FIGURE 48.4** Label the major structures of the nephron and associated structures. **3** **AP|R**

(a)

(b)

**FIGURE 48.5** (*a*) Micrograph of a section of the renal cortex (220×). (*b*) Micrograph of a section of the renal medulla (80×). (*Note*: With about a million nephrons in a kidney, there are many different orientations visible in micrograph sections of renal cortex and renal medulla.) **AP|R** (*(a)* ©Biophoto Associates/Science Source; *(b)* ©McGraw-Hill Education/Al Tesler, photographer)

**PROCEDURE C—Ureter and Urinary Bladder**

1. Review the section entitled "Ureters" in chapter 20 of the textbook.
2. Obtain a microscope slide of a cross section of a ureter, and examine it using low-power magnification. Locate the *mucous coat* layer next to the lumen. Examine the middle *muscular coat* composed of longitudinal and circular smooth muscle cells, responsible for the peristaltic waves that propel urine from the kidneys to the urinary bladder. The outer *fibrous coat,* composed of connective tissue, secures the ureter in the retroperitoneal position (fig. 48.6).
3. Examine the mucous coat using high-power magnification. The specialized tissue is transitional epithelium, which allows changes in its thickness when unstretched and stretched.
4. Prepare a labeled sketch of a ureter in Part D of the laboratory report.

5. Review the section entitled "Urinary Bladder" in chapter 20 of the textbook.
6. Obtain a microscope slide of a segment of the wall of a urinary bladder, and examine it using low-power magnification. Examine the *mucous coat* next to the lumen and the *submucous coat* composed of connective tissue just beneath the mucous coat. Examine the *muscular coat* composed of bundles of smooth muscle cells interlaced in many directions. This thick, muscular layer is called the *detrusor muscle* and functions in the elimination of urine. An outer *serous coat* is composed of connective tissue (fig. 48.7). The serous coat on the upper surface of the urinary bladder consists of the parietal peritoneum.
7. Examine the mucous coat using high-power magnification. The tissue is transitional epithelium, which allows changes in its thickness from unstretched when the bladder is empty to stretched when the bladder distends with urine.
8. Prepare a labeled sketch of a segment of the urinary bladder wall in Part D of the laboratory report.

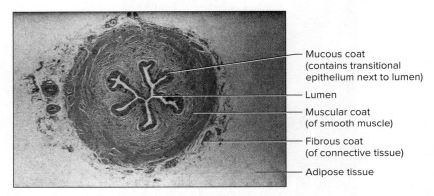

Mucous coat
(contains transitional
epithelium next to lumen)

Lumen

Muscular coat
(of smooth muscle)

Fibrous coat
(of connective tissue)

Adipose tissue

**FIGURE 48.6** Micrograph of a cross section of a ureter (75×). AP|R (©Biophoto Associates/Science Source)

Lumen

Mucous coat
(contains transitional
epithelium next to lumen)

Submucous coat
(of connective tissue)

Muscular coat
(of smooth muscle)
(detrusor muscle)

**FIGURE 48.7** Micrograph of a segment of the human urinary bladder wall (6×). AP|R (©McGraw-Hill Education/Al Tesler, photographer)

Name _____

Date _____

Section _____

LABORATORY
REPORT

The **A** corresponds to the indicated Learning Outcome(s) found at the beginning of the laboratory exercise.

# Urinary Organs

## PART A ASSESSMENTS

Match the terms in column A with the descriptions in column B. Place the letter of your choice in the space provided. **2**

| Column A | Column B |
|---|---|
| **a.** Calyces | _____ **1.** Superficial region around the renal medulla |
| **b.** Hilum of kidney | _____ **2.** Extensions of renal pelvis to renal papillae containing urine |
| **c.** Nephron | _____ **3.** Conical mass of tissue within renal medulla |
| **d.** Renal column | _____ **4.** Projection with tiny openings into a minor calyx |
| **e.** Renal cortex | _____ **5.** Hollow chamber within kidney |
| **f.** Renal papilla | _____ **6.** Microscopic functional subunit of kidney |
| **g.** Renal pelvis | _____ **7.** Cortical tissue extensions between renal pyramids |
| **h.** Renal pyramid | _____ **8.** Superior, funnel-shaped sac at end of ureter inside the renal sinus |
| **i.** Renal sinus | _____ **9.** Medial depression for blood vessels and ureter to enter kidney chamber |

## PART B ASSESSMENTS

Complete the following:

**1.** Distinguish between a renal corpuscle and a renal tubule. **3** _____

_____

_____

**2.** Number the following structures to indicate their respective positions in relation to the nephron. Assign the number 1 to the structure nearest the glomerulus. **4**

_____ Ascending limb of nephron loop

_____ Collecting duct

_____ Descending limb of nephron loop

_____ Distal tubule

_____ Glomerular capsule

_____ Proximal tubule

_____ Renal papilla

**3.** Number the following structures to indicate their respective positions in the blood pathway within the kidney. Assign the number 1 to the vessel nearest the renal artery. (Not every renal blood vessel is represented in the choices provided.) △5

_____ Afferent arteriole

_____ Efferent arteriole

_____ Glomerulus (glomerular capillaries)

_____ Peritubular capillaries

_____ Renal vein

## PART C ASSESSMENTS

Sketch a representative section of the renal cortex and the renal medulla. Label the glomerulus, the glomerular capsule, and sections of renal tubules in the renal cortex and a longitudinal section and cross section of a collecting duct in the renal medulla. △3

| | |
|---|---|
| Renal cortex (_____×) | Renal medulla (_____×) |

## PART D ASSESSMENTS

Sketch a cross section of a ureter. Label the three layers and the lumen. Sketch a segment of a urinary bladder and label the four layers and the lumen. △6

| | |
|---|---|
| Ureter (_____×) | Urinary bladder (_____×) |

# 49

# Urinalysis

## MATERIALS NEEDED

Normal and abnormal simulated urine specimens can be substituted for collected urine.
Disposable urine-collecting container
Paper towel
Urinometer cylinder (jar)
Urinometer hydrometer
Laboratory thermometer
pH test paper
Reagent strips (individual or combination test strips such as Chemstrip or Multistix) to test for the presence of the following:
    Glucose
    Protein
    Ketones
    Bilirubin
    Hemoglobin/occult blood
Compound light microscope
Microscope slide
Coverslip
Centrifuge
Centrifuge tube
Graduated cylinder, 10 mL
Medicine dropper
Sedi-stain

## ⚠ SAFETY

- Consider using normal and abnormal simulated urine samples available from various laboratory supply houses.
- Use safety glasses, laboratory coats, and disposable gloves when working with body fluids.
- Work only with your own urine sample.
- Use an appropriate disinfectant to wash the laboratory table before and after the procedures.
- Place glassware in a disinfectant when finished.
- Dispose of contaminated items as directed by your laboratory instructor.
- Wash your hands before leaving the laboratory.

## PURPOSE OF THE EXERCISE

To perform the observations and tests commonly used to analyze the characteristics and composition of urine.

## LEARNING OUTCOMES

After completing this exercise, you should be able to

1. Evaluate the color, transparency, and specific gravity of a urine sample.
2. Measure the pH of a urine sample.
3. Test a urine sample for the presence of glucose, protein, ketones, bilirubin, and hemoglobin.
4. Perform a microscopic study of urine sediment.
5. Summarize the results of these observations and tests.

Urine is the product of three processes of the nephrons within the kidneys: glomerular filtration, tubular reabsorption, and tubular secretion. As a result of these processes, various waste substances are removed from the blood, and body fluid and electrolyte balance are maintained. Consequently, the composition of urine varies considerably because of differences in dietary intake and physical activity from day to day and person to person. The normal urinary output ranges from 1.0 to 1.8 liters per day. Typically, urine consists of 95% water and 5% solutes. The volume of urine produced by the kidneys varies with such factors as fluid intake, environmental temperature, relative humidity, respiratory rate, and body temperature.

An analysis of urine composition and volume often is used to evaluate the functions of the kidneys and other organs. This procedure, called *urinalysis,* is a clinical assessment and a diagnostic tool for certain pathological conditions and general overall health. A urinalysis and a complete blood analysis complement each other for an evaluation of certain diseases such as diabetes mellitus, and general health.

A urinalysis involves three aspects: *physical characteristics, chemical analysis,* and a *microscopic examination.* Physical characteristics of urine that are noted include volume, color, transparency, and odor. The chemical analysis of solutes in urine addresses urea and other nitrogenous wastes; electrolytes; pigments; and possible glucose, protein, ketones, bilirubin, and hemoglobin. The specific gravity and pH of urine are greatly influenced by the components and amounts of solutes. An examination of microscopic solids, including cells, casts, and crystals, assists in the diagnosis of injury, various diseases, and urinary infections.

> **WARNING** *While performing the following tests, you should wear disposable latex gloves so that skin contact with urine is avoided. Observe all safety procedures listed for this laboratory exercise. (Normal and abnormal simulated urine specimens can be used instead of real urine.)*

### PRACTICE

#### PROCEDURE A—Physical and Chemical Analysis

1. Proceed to the restroom with a clean, disposable container if collected urine is to be used for this exercise. The first small volume of urine should not be collected because it contains abnormally high levels of microorganisms from the urethra. Collect a midstream sample of about 50 mL of urine. The best collections are the first specimen in the morning or one taken 3 hours after a meal. Refrigerate samples if they are not used immediately.

2. Place a sample of urine in a clean, transparent container. Describe the *color* of the urine. Normal urine varies from light yellow to amber, depending on the presence of urochromes, end-product pigments produced during the decomposition of hemoglobin. Dark urine indicates a high concentration of pigments.

   Abnormal urine colors include yellow-brown or green, due to elevated concentrations of bile pigments, and red to dark brown, due to the presence of blood. Certain foods, such as beets or carrots, and various drug substances also can cause color changes in urine, but in such cases the colors have no clinical significance. Enter the results of this and the following tests in Part A of Laboratory Report 49.

3. Evaluate the *transparency* of the urine sample (judge whether the urine is clear, slightly cloudy, or very cloudy). Normal urine is clear enough to see through. You can read newsprint through slightly cloudy urine; you can no longer read newsprint through cloudy urine. Cloudy urine indicates the presence of various substances, including mucus, bacteria, epithelial cells, fat droplets, and inorganic salts.

4. Gently wave your hand over the urine sample toward your nose to detect the *odor*. Normal urine should have a slight ammonia-like odor due to the nitrogenous wastes. Some vegetables (such as asparagus), drugs (such as

certain vitamins), and diseases (such as diabetes mellitus) will influence the odor of urine.

5. Determine the *specific gravity* of the urine sample, which indicates the solute concentration. Specific gravity is the ratio of the weight of something to the weight of an equal volume of pure water. For example, mercury (at 15°C) weighs 13.6 times as much as an equal volume of water; thus, it has a specific gravity of 13.6. Although urine is mostly water, it has substances dissolved in it and is slightly heavier than an equal volume of water. The specific gravity of pure (distilled) water is 1.000, while the specific gravity of urine is higher, varying from 1.003 to 1.035 under normal circumstances. If the specific gravity is too low, the urine contains few solutes and represents dilute urine, a likely result of excessive fluid intake or the use of diuretics. A specific gravity above the normal range represents a higher concentration of solutes, likely from a limited fluid intake. Concentrated urine over an extended time increases the risk of the formation of kidney stones (renal calculi).

   To determine the specific gravity of a urine sample, follow these steps:

   **a.** Pour enough urine into a clean urinometer cylinder to fill it about three-fourths full. Any foam that appears should be removed with a paper towel.

   **b.** Use a laboratory thermometer to measure the temperature of the urine.

   **c.** Gently place the urinometer hydrometer into the urine, and *make sure that the float is not touching the sides or the bottom of the cylinder* (fig. 49.1).

Hydrometer

Urine Specimen

Urine in urinometer cylinder

**FIGURE 49.1** Float the hydrometer in the urine, making sure that it does not touch the sides or the bottom of the cylinder.

**d.** Position your eye at the level of the urine surface. Determine which line on the stem of the hydrometer intersects the lowest level of the concave surface (meniscus) of the urine.

**e.** Liquids tend to contract and become denser as they are cooled, or to expand and become less dense as they are heated, so it may be necessary to make a temperature correction to obtain an accurate specific gravity measurement. To do this, add 0.001 to the hydrometer reading for each 3 degrees of urine temperature above 25°C or subtract 0.001 for each 3 degrees below 25°C. Enter this calculated value in the table in Part A of the laboratory report as the test result.

**6.** Individual or combination reagent strips can be used to perform a variety of urine tests (fig. 49.2). In each case, directions for using the strips are found on the strip container. *Be sure to read them.*

To perform each test, follow these steps:

**a.** Obtain a urine sample and the proper reagent test strip.

**b.** Read the directions on the strip container.

**c.** Dip the strip in the urine sample as directed on the container.

**d.** Remove the strip at an angle and let it touch the inside rim of the urine container to remove any excess liquid.

**e.** Wait for the length of time indicated by the directions on the container before you compare the color of the test strip with the standard color scale provided with the container. The value or amount represented by the matching color should be used as the test result for the following tests performed and recorded in Part A of the laboratory report.

**7.** Perform the *pH test*. The pH of normal urine varies from 4.6 to 8.0, but most commonly, it is near 6.0 (slightly acidic). The pH of urine may decrease as a result of a diet high in protein, or it may increase with a vegetarian diet. Significant daily variations within the broad normal range are results of concentrations of excesses from variable diets.

**8.** Perform the *glucose test*. Normally, there is no glucose in urine. However, glucose may appear in the urine

temporarily following a meal high in carbohydrates. Glucose also may appear in the urine as a result of uncontrolled diabetes mellitus.

**9.** Perform the *protein test*. Normally, proteins of large molecular size are not present in urine. However, those of small molecular sizes, such as albumins, may appear in trace amounts, particularly following strenuous exercise. Increased amounts of proteins also may appear as a result of kidney diseases in which the glomeruli are damaged or as a result of high blood pressure.

**10.** Perform the *ketone test*. Ketones are products of fat metabolism. Usually, they are not present in urine. However, they may appear in the urine if the diet fails to provide adequate carbohydrate, as in the case of prolonged fasting or starvation, or as a result of insulin deficiency (diabetes mellitus).

**11.** Perform the *bilirubin test*. Bilirubin, which results from hemoglobin decomposition in the liver, normally is absent in urine. It may appear, however, as a result of liver disorders that cause obstructions of the biliary tract. Urochrome, a normal yellow component of urine, is a result of additional breakdown of bilirubin.

**12.** Perform the *hemoglobin/occult blood test*. Hemoglobin occurs in the red blood cells, and because such cells normally do not pass into the renal tubules, hemoglobin is not found in normal urine. Its presence in urine usually indicates a disease process, a transfusion reaction, an injury to the urinary organs, or menstrual blood.

**13.** Complete Part A of the laboratory report.

 **PRACTICE**

**PROCEDURE B—Microscopic Sediment Analysis**

**1.** A urinalysis usually includes an analysis of urine sediment—the microscopic solids present in a urine sample. This sediment normally includes mucus; certain crystals; and a variety of cells, such as the epithelial cells that line the urinary tubes and an occasional white blood cell. Other types of solids, such as casts or red blood cells, may indicate a disease or injury if they are present in excess. (Casts are cylindrical masses of cells or other substances that form in the renal tubules and are flushed out by the flow of urine.)

To observe urine sediment, follow these steps:

**a.** Thoroughly stir or shake your urine sample to suspend the sediment, which tends to settle to the bottom of the container.

**b.** Pour 10 mL of urine into a clean centrifuge tube and centrifuge it for 5 minutes at slow speed (1,500 rpm). Be sure to balance the centrifuge with an even number of tubes filled to the same levels.

**c.** Carefully decant 9 mL (leave 1 mL) of the liquid from the sediment in the bottom of the centrifuge tube, as directed by your laboratory instructor. Resuspend the 1 mL of sediment.

**d.** Use a medicine dropper to remove some of the sediment and place it on a clean microscope slide.

**FIGURE 49.2** An example of a reagent test strip dipped into urine to determine a variety of urine components.
(©McGraw-Hill Education)

e. Add a drop of Sedi-stain to the sample, and add a coverslip.

f. Examine the sediment with low-power (reduce the light when using low power) and high-power magnifications.

g. With the aid of figure 49.3, identify the types of solids present.

2. In Part B of the laboratory report, make a sketch of each type of sediment that you observed.

3. Complete Part B of the laboratory report.

**Crystals**

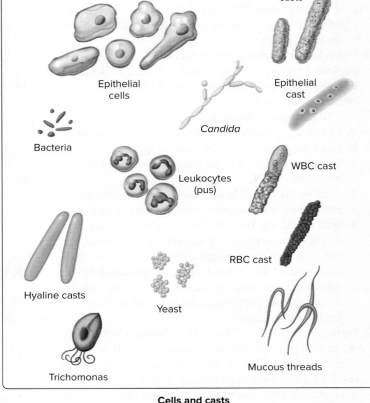

**Cells and casts**

**FIGURE 49.3** Types of urine sediment. Healthy individuals lack many of these sediments and possess only occasional to trace amounts of others. *(Note:* Shades of white to purple sediments are most characteristic when using Sedi-stain.)

The  corresponds to the indicated Learning Outcome(s) found at the beginning of the laboratory exercise.

# Urinalysis

### ♻ PART A ASSESSMENTS

1. Enter your observations, test results, and evaluations in the following table: ⒜ ⒝ ⒞

| Urine Characteristics | Observations and Test Results | Normal Values | Evaluations |
|---|---|---|---|
| Color | | Light yellow to amber | |
| Transparency | | Clear | |
| Odor | | Slight ammonia-like | |
| Specific gravity (corrected for temperature) | | 1.003–1.035 | |
| pH | | 4.6–8.0 | |
| Glucose | | Negative (0) | |
| Protein | | Negative to trace | |
| Ketones | | Negative (0) | |
| Bilirubin | | Negative (0) | |
| Hemoglobin/occult blood | | Negative (0) | |
| (Other) | | | |

2. Summarize the results of the physical and chemical analyses of urine. ⒠ _____

_____

_____

_____

_____

_____

## ASSESS

Why do you think it is important to refrigerate a urine sample if an analysis cannot be performed immediately after collecting it?

_____

_____

_____

_____

## PART B ASSESSMENTS

1. Make a sketch for each type of sediment you observed. Label any from those shown in figure 49.3.  **4**

2. Summarize the results of the microscopic sediment analysis of urine.  **5** _____

_____

_____

_____

_____

_____

# 50

# Male Reproductive System

### MATERIALS NEEDED

Textbook
Human torso model
Model of the male reproductive system
Anatomical chart of the male reproductive system
Compound light microscope
Prepared microscope slides of the following:
    Testis section
    Epididymis, cross section
    Ductus deferens, cross section
    Penis, cross section

## PURPOSE OF THE EXERCISE  AP|R

To review the structures and functions of the male reproductive organs and to examine some of their features.

 ## LEARNING OUTCOMES  AP|R

After completing this exercise, you should be able to

(1) Locate and identify the organs of the male reproductive system.

(2) Describe the functions of these organs.

(3) Sketch and label the major features from microscopic sections of the testis, epididymis, ductus deferens, and penis.

**T**he organs of the male reproductive system are specialized to produce and maintain the male sex cells, to transport these cells together with supporting fluids to the female reproductive tract, and to produce and secrete male sex hormones.

    These organs include the testes, in which sperm and male sex hormones are produced, and sets of internal and external accessory structures. The internal structures include various tubes and glands, whereas the external structures are the scrotum and the penis.

 ## PRACTICE

### PROCEDURE A—Male Reproductive Organs

1. Review the sections entitled "Testes," "Male Internal Accessory Reproductive Organs," and "Male External Accessory Reproductive Organs" in chapter 22 of the textbook.
2. As a review activity, label figures 50.1 and 50.2.
3. Observe the human torso model, the model of the male reproductive system, and the anatomical chart of the male reproductive system. Locate the following features:

**testes**
    seminiferous tubules
    interstitial cells (cells of Leydig)
**inguinal canal**
**spermatic cord**
**epididymis**
**ductus deferens (vas deferens)**
**ejaculatory duct**
**seminal vesicles**
**prostate gland**
**bulbourethral glands**
**scrotum**
**penis**
    corpora cavernosa
    corpus spongiosum
    tunica albuginea
    glans penis
    external urethral orifice
    prepuce (foreskin)

4. Complete Part A of Laboratory Report 50.

**FIGURE 50.1** Label the major structures of the male reproductive system in this sagittal view. **1** **AP|R**

Labels around Figure 50.1:

- Urinary bladder
- Superior pubic ramus (cut)
- 1
- 2
- Corpus cavernosum
- Corpus spongiosum
- 3
- 4
- 5
- 6
- 7
- Ureter
- Large intestine
- 8
- Ejaculatory duct
- 9
- 10
- Anus
- 11

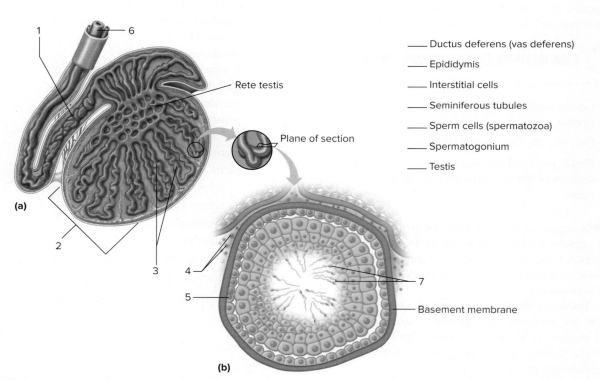

Labels around Figure 50.2:

- 1
- 6
- Rete testis
- Plane of section
- (a)
- 2
- 3
- 4
- 5
- 7
- Basement membrane
- (b)

_____ Ductus deferens (vas deferens)

_____ Epididymis

_____ Interstitial cells

_____ Seminiferous tubules

_____ Sperm cells (spermatozoa)

_____ Spermatogonium

_____ Testis

**FIGURE 50.2** Label the diagram of (*a*) the sagittal section of a testis and (*b*) a cross section of a seminiferous tubule by placing the correct numbers in the spaces provided. **1** **AP|R**

### PROCEDURE B—Microscopic Anatomy

1. Obtain a microscope slide of human testis section and examine it, using low-power magnification (fig. 50.3). Locate the thick *fibrous capsule* (tunica albuginea) on the surface and the numerous sections of *seminiferous tubules* inside.

2. Focus on some of the seminiferous tubules, using high-power magnification (fig. 50.4). Locate the *basement membrane* and the layer of *spermatogonia* (undifferentiated spermatogenic cells) just beneath the basement membrane. Identify some *sustentacular cells* (supporting cells; Sertoli cells), which have pale, oval-shaped nuclei. Spermatogonia give rise to *spermatogenic cells* that are in various stages of spermatogenesis as they are forced toward the lumen. Near the lumen of the tube, find some darkly stained, elongated heads of developing sperm cells. In the spaces between adjacent seminiferous tubules, locate some isolated *interstitial cells* (cells of Leydig) of the endocrine system. Interstitial cells produce the hormone testosterone, transported by the blood.

3. Prepare a labeled sketch of a representative section of the testis in Part B of the laboratory report.

4. Obtain a microscope slide of a cross section of *epididymis* (fig. 50.5). Examine its wall, using high-power magnification. Note the elongated, *pseudostratified columnar epithelial cells* that compose most of the inner lining. These cells have nonmotile stereocilia (elongated microvilli) on their free surfaces that absorb excess fluid secreted by the testes. Also note the thin layer of smooth muscle and connective tissue surrounding the tube.

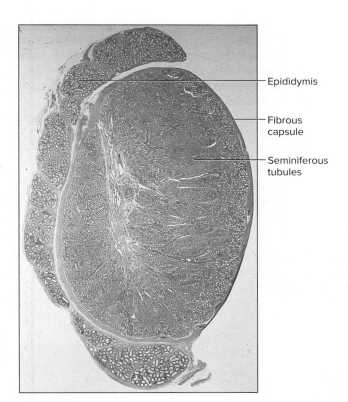

**FIGURE 50.3** Micrograph of a human testis (1.7×).
(©Biophoto Associates/Science Source)

Epididymis

Fibrous capsule

Seminiferous tubules

Basement membrane

Sustentacular cells (Sertoli cells)

Spermatogonia

Seminiferous tubule

Sperm cells

Interstitial cells (cells of Leydig)

**FIGURE 50.4** Micrograph of a seminiferous tubule (250×). **AP|R** (©Ed Reschke)

Pseudostratified columnar epithelial cells

Nonmotile cilia

Sperm cells

Smooth muscle cells

Lumen

**FIGURE 50.5** Micrograph of a cross section of a human epididymis (200×). (©Ed Reschke)

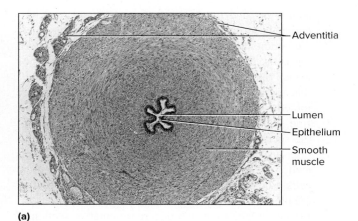

Adventitia
Lumen
Epithelium
Smooth muscle

**(a)**

Sperm in lumen of ductus deferens

Smooth muscle

Pseudostratified columnar epithelium

**(b)**

**FIGURE 50.6** Ductus deferens. (*a*) Micrograph of a cross section of the ductus deferens (40×). (*b*) Micrograph of the wall of the ductus deferens (400×). **AP|R**

((*a,b*) ©McGraw-Hill Education/Al Tesler, photographer)

5. Prepare a labeled sketch of the epididymis wall in Part B of the laboratory report.

6. Obtain a microscope slide of a cross section of the *ductus deferens,* and examine it with scan and low-power magnification (fig. 50.6*a*). Note the small lumen with an epithelium and the thick *smooth muscle* layer composing most of the structure of the duct. Three layers of smooth muscle fibers are distinctive: a middle circular layer between two layers of longitudinal muscle fibers. The peristaltic contractions of the muscular layer transport sperm along the passageway toward the ejaculatory duct and the urethra. An outer, connective tissue covering of the ductus deferens is called the *adventitia*. Use high-power magnification and examine the epithelium layer composed of pseudostratified columnar epithelial cells (fig. 50.6*b*).

7. Prepare a labeled sketch of a representative section of a ductus deferens in Part B of the laboratory report.

8. Obtain a microscope slide of a *penis* cross section, and examine it with low-power magnification (fig. 50.7). Identify the following features:

**corpora cavernosa**
**corpus spongiosum**
**tunica albuginea**
**urethra**
**skin**

9. Prepare a labeled sketch of a penis cross section in Part B of the laboratory report.

10. Complete Part B of the laboratory report.

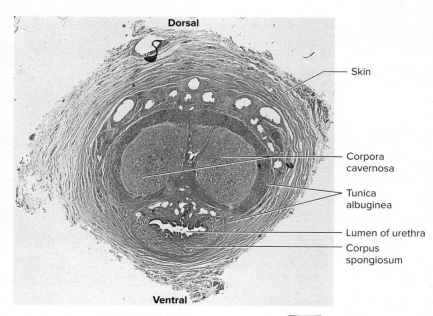

Dorsal

Skin

Corpora cavernosa

Tunica albuginea

Lumen of urethra

Corpus spongiosum

Ventral

**FIGURE 50.7** Micrograph of a cross section of the body of the penis (5×). **AP|R** (©Ed Reschke)

Name _____

Date _____

Section _____

The  corresponds to the indicated Learning Outcome(s) found at the beginning of the laboratory exercise.

LABORATORY
REPORT

50

# Male Reproductive System

## PART A ASSESSMENTS

Complete the following statements:

1. Each testis is suspended by a(an) _____ within the scrotum. **1**

2. As the testes descend, they pass through the _____ of the abdominal wall into the scrotum. **1**

3. Connective tissue subdivides a testis into many _____ , which contain seminiferous tubules. **1**

4. The _____ is a highly coiled tube on the surface of the testis. **1**

5. The _____ cells of the epithelium that line the seminiferous tubule give rise to sperm cells. **2**

6. _____ is the process by which sperm cells are formed. **2**

7. The number of chromosomes normally present in a human sperm cell is _____ . **1**

8. Sperm cells undergo maturation while they are stored in the _____ . **2**

9. The secretion of the seminal vesicles is rich in the monosaccharide called _____ . **2**

10. The single _____ gland surrounds the urethra near the urinary bladder. **1**

11. The secretion of the _____ glands lubricates the end of the penis in preparation for sexual intercourse. **2**

12. The sensitive, cone-shaped end of the penis is called the _____ . **1**

13. _____ is the movement of sperm cells and various glandular secretions into the urethra. **2**

14. _____ is the process by which semen is forced out through the urethra. **2**

1. In the space that follows, sketch a representative area of the organ indicated. Label any of the structures that were observed, and indicate the magnification used for each sketch. **3**

| | |
|---|---|
| Testis (_____×) | Epididymis (_____×) |
| Ductus deferens (_____×) | Penis (_____×) |

2. Describe the function of each of the following: **2**

   **a.** Scrotum

   _____

   _____

   **b.** Spermatogenic cell

   _____

   _____

   **c.** Interstitial cells (cells of Leydig)

   _____

   _____

   **d.** Epididymis

   _____

   _____

   **e.** Corpora cavernosa and corpus spongiosum

   _____

   _____

# 51

# Female Reproductive System

## MATERIALS NEEDED

Textbook
Human torso model
Model of the female reproductive system
Anatomical chart of the female reproductive system
Compound light microscope
Prepared microscope slides of the following:
  Ovary section with maturing follicles
  Uterine tube, cross section
  Uterine wall section
  Uterine wall, early proliferative phase
  Uterine wall, secretory phase
  Uterine wall, early menstrual phase

## PURPOSE OF THE EXERCISE AP|R

To review the structures and functions of the female reproductive organs and to examine some of their features.

##  LEARNING OUTCOMES AP|R

After completing this exercise, you should be able to

① Locate and identify the organs of the female reproductive system.

② Describe the functions of these organs.

③ Sketch and label the major features from microscopic sections of the ovary, uterine tube, and uterine wall.

The organs of the female reproductive system are specialized to produce and maintain the female sex cells, to transport these cells to the site of fertilization, to provide a favorable environment for a developing offspring, to move the offspring to the outside, and to produce female sex hormones.

These organs include the ovaries, which produce the oocytes (egg cells) and female sex hormones, and sets of internal and external accessory structures. The internal accessory structures include the uterine tubes, uterus, and vagina. The external accessory structures are the labia majora, labia minora, clitoris, and vestibular glands.

##  PRACTICE

### PROCEDURE A—Female Reproductive Organs

1. Review the sections entitled "Ovaries," "Female Internal Accessory Reproductive Organs," "Female External Accessory Reproductive Organs," and "Mammary Glands" in chapter 22 of the textbook.
2. As a review activity, label figures 51.1–51.5.
3. Observe the human torso model, the model of the female reproductive system, and the anatomical chart of the female reproductive system. Locate the following features:

**ovaries**
  medulla
  cortex
**ligaments**
  broad ligament
  suspensory ligament
  ovarian ligament
  round ligament
**uterine tubes** (Fallopian tubes; oviducts)
  infundibulum
  fimbriae
**uterus**
  fundus
  body
  cervix
  cervical orifice
  uterine wall
    endometrium
    myometrium
    perimetrium
**rectouterine pouch**
**vagina**
  fornices
  vaginal orifice
  hymen
  mucosal layer

**vulva** (external accessory structures)
  mons pubis
  labia majora
  labia minora
  vestibule
  vestibular glands
  clitoris

**breasts**
  nipple
  areola
  alveolar glands (compose milk-producing parts of mammary glands)
  lactiferous ducts
  adipose tissue

**4.** Complete Part A of Laboratory Report 51.

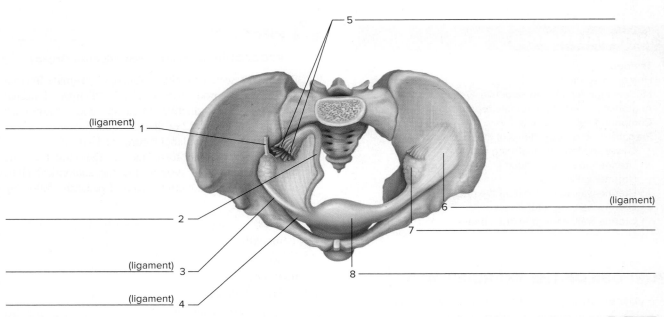

**FIGURE 51.1**    Label the ligaments and other structures associated with the female internal reproductive organs.  A  AP|R

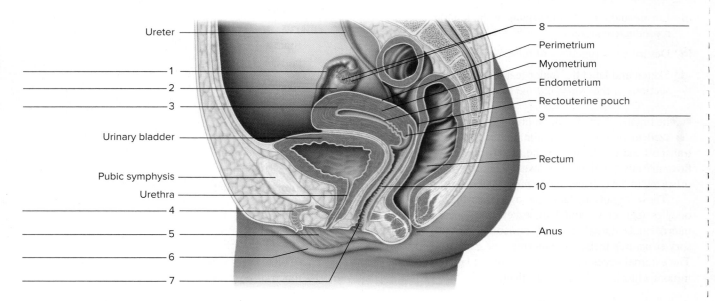

**FIGURE 51.2**    Label the structures of the female reproductive system in this sagittal view.  A  AP|R

**FIGURE 51.3** Label the female external reproductive organs and associated structures. ⒜ AP|R

Ovarian ligament

Ovary

Uterine tube

10

9

8

*Time*

*Time*

*Time*

*Time*

*Time*

Primordial follicles

1

2

3

7

6

5

4

Fimbriae

Infundibulum

_____ Corona radiata          _____ Ovulation site

_____ Corpus albicans        _____ Primary follicle

_____ Corpus luteum          _____ Primary oocyte

_____ First polar body          _____ Secondary oocyte

_____ Mature antral follicle    _____ Zona pellucida

**FIGURE 51.4** Label this ovary, illustrating various stages of follicle maturation, ovulation, and degeneration, by placing the correct numbers in the spaces provided. (*Note*: Although the various stages of ovarian development are depicted in a single illustration, the stages do not migrate around the ovary, but occur in a particular location during a particular ovarian cycle.) ⒜ AP|R

### PROCEDURE B—Microscopic Anatomy

1. Obtain a microscope slide of an ovary section with maturing follicles, and examine it with low-power magnification (fig. 51.6). Locate the outer layer, or *cortex,*

composed of densely packed cells with developing follicles, and the inner layer, or *medulla,* which largely consists of loose connective tissue.

2. Focus on the cortex of the ovary, using high-power magnification (fig. 51.7). Note the thin layer of small cuboidal cells on the free surface. These cells constitute

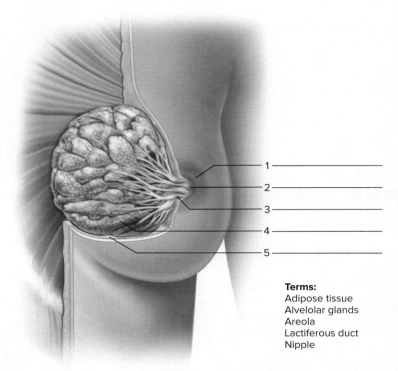

1 —————————
2 —————————
3 —————————
4 —————————
5 —————————

**Terms:**
Adipose tissue
Alvelolar glands
Areola
Lactiferous duct
Nipple

**FIGURE 51.5**   Label the structures of the breast (anterior view), using the terms provided.  🅐 AP|R

**FIGURE 51.6**
Micrograph of the ovary showing various stages of follicular development in the cortex (250×).
(©Biophoto Associates/ Science Source)

Oocyte    Maturing antral follicle         Pre-antral follicle    Primordial follicle the *germinal epithelium*. Also locate some *primordial follicles* just beneath the germinal epithelium. Each follicle consists of a single, relatively large *primary oocyte* with a prominent nucleus and a covering of *follicular cells*.

3. Use low-power magnification to search the ovarian cortex for maturing follicles in various stages of development (fig. 51.6). Locate and compare primordial follicles and *primary follicles*. A certain primary follicle can develop into a *pre-antral follicle* and then into a *mature antral follicle* just before ovulation (figs. 51.4 and 51.7).

4. Prepare two labeled sketches in Part B of the laboratory report to illustrate the changes that occur in a follicle as it matures.

5. Obtain a microscope slide of a cross section of a uterine tube. Examine it, using low-power magnification. The shape of the lumen is very irregular because of folds of the mucosa layer.

6. Focus on the inner lining of the uterine tube, using high-power magnification (fig. 51.8). The lining is composed of *simple columnar epithelium*, and some of the epithelial cells are ciliated on their free surfaces.

**FIGURE 51.7** Micrograph of a mature antral follicle (250×). AP|R (©McGraw-Hill Education/Al Tesler, photographer)

Labels: Fluid-filled antrum, Corona radiata, Zona pellucida, Secondary oocyte

**FIGURE 51.8** Micrograph of a cross section of the uterine tube (800×). AP|R (©Ed Reschke/Photolibrary/Getty Images)

Labels: Lumen, Cilia, Nucleus, Cytoplasm, Basement membrane, Connective tissue layer

7. Prepare a labeled sketch of a representative region of the wall of the uterine tube in Part B of the laboratory report.

8. Obtain a microscope slide of the uterine wall section (fig. 51.9). Examine it, using low-power magnification, and locate the following:

   **endometrium** (inner mucosal layer)
   **myometrium** (middle thick, muscular layer)
   **perimetrium** (outer serosal layer)

9. Prepare a labeled sketch of a representative section of the uterine wall in Part B of the laboratory report.

10. Complete Part B of the laboratory report.

 **LEARN: ACTIVITY**

Observe the slides in the demonstration microscopes. Each slide contains a section of uterine mucosa taken during a different phase in the reproductive cycle. In the *early proliferative phase,* note the simple columnar epithelium on the free surface of the mucosa and the many sections of tubular uterine glands in the tissues beneath the epithelium. In the *secretory phase,* the endometrium is thicker and the uterine glands appear more extensive and they are coiled. In the *early menstrual phase,* the endometrium is thinner because its surface functional layer has been shed, while retaining the basal layer. Also, the uterine glands are less apparent, and the spaces between the glands contain many leukocytes. The basal layer of the endometrium renews the functional layer after menstruation ends.

Lumen
Endometrium
Myometrium
Perimetrium

**FIGURE 51.9** Micrograph of the uterine wall (10×).
**AP|R** (©McGraw-Hill Education/Carol D. Jacobson, Ph.D., Iowa State University, photographer)

Name _____

Date _____

Section _____

The  corresponds to the indicated Learning Outcome(s) found at the beginning of the laboratory exercise.

# Female Reproductive System

## PART A ASSESSMENTS

Complete the following statements:

1. The ovaries are located in the lateral wall of the _____ cavity. Ⓐ

2. The largest of the ovarian attachments is called the _____ ligament. Ⓐ

3. The ovarian cortex appears granular because of the presence of _____. Ⓐ

4. The meiosis of egg formation is called _____. ②

5. A primary oocyte is closely surrounded by epithelial cells called _____ cells. Ⓐ

6. When a primary oocyte divides, a secondary oocyte and a(an) _____ are produced. ②

7. Primordial follicles are stimulated to develop into primary follicles by the hormone called _____. ②

8. _____ is the process by which a secondary oocyte is released from the ovary. ②

9. Uterine tubes are also called oviducts or _____. Ⓐ

10. The _____ is the funnel-shaped expansion at the end of a uterine tube. Ⓐ

11. A portion of the uterus called the _____ extends downward into the upper portion of the vagina. Ⓐ

12. The inner mucosal layer of the uterus is called the _____. Ⓐ

13. The myometrium is largely composed of _____ tissue. Ⓐ

14. The vaginal orifice is partially closed by a thin membrane called the _____. Ⓐ

15. The group of external accessory organs that surround the openings of the urethra and vagina comprise the _____. Ⓐ

16. The rounded mass of fatty tissue overlying the pubic symphysis of the female is called the _____. Ⓐ

17. The female organ that corresponds to the male penis is the _____. Ⓐ

18. The _____ of the female corresponds to the bulbourethral glands of the male. Ⓐ

1. In the space that follows, sketch a representative area of the organ indicated. Label any of the structures observed, and indicate the magnification used for each sketch. **3**

| | |
|---|---|
| Primordial or primary follicle (_____×) | Mature antral follicle (_____×) |
| Uterine tube (_____×) | Wall of uterus (_____×) |

2. Complete the following:

   **a.** Describe the fate of a mature antral follicle. **2** _____

   _____

   _____

   **b.** Describe the function of the cilia in the lining of the uterine tube. **2** _____

   _____

   _____

   **c.** Briefly describe the changes that occur in the uterine lining during a reproductive cycle. **2** _____

   _____

   _____

# 52

# Fertilization and Early Development

## MATERIALS NEEDED

Textbook
Sea urchin egg suspension*
Sea urchin sperm suspension*
Compound light microscope
Depression microscope slide
Coverslip
Medicine droppers
Prepared microscope slide of the following:
    Sea urchin embryos (cleavage, blastula, and
      gastrula stages)
Models of human development
Vaseline
Toothpick
Preserved mammalian embryos

*See the Instructor's Manual for a source of materials.

## PURPOSE OF THE EXERCISE

To review the process of fertilization, to observe sea urchin eggs being fertilized, and to examine embryos in early stages of development.

 **LEARNING OUTCOMES** AP|R

After completing this exercise, you should be able to

1. Describe the process and structures of fertilization.

2. Describe and distinguish the early developmental stages of a human.

3. Identify and sketch the early developmental stages of a sea urchin.

Fertilization is the process by which the nuclei of a secondary oocyte and a sperm cell come together and combine their chromosomes, forming a single cell called a zygote.

Ordinarily, before fertilization can occur in a human female, an oocyte (egg cell) must be released from the ovary and must be carried into a uterine tube. Also, semen containing sperm cells must be deposited in the vagina; some of these sperm cells must travel through the uterus and into the uterine tubes. Although many sperm cells may reach an oocyte (egg cell), only one will participate in the fertilization of the secondary oocyte.

Shortly after fertilization, the zygote undergoes division (mitosis) to form two cells. These two cells become four, they in turn divide into eight, and so forth. The resulting mass of cells continues to grow and undergoes developmental changes and growth that give rise to an offspring.

 **PRACTICE**

### PROCEDURE A—Fertilization

1. Review the sections entitled "Fertilization" and "Period of Cleavage" in chapter 23 of the textbook.
2. As a review activity, label figure 52.1.
3. Complete Part A of Laboratory Report 52.
4. Although it is difficult to observe fertilization in mammals since the process occurs internally, it is possible to view forms of external fertilization. For example, egg and sperm cells can be collected from sea urchins, and the process of fertilization can be observed microscopically. To make this observation, follow these steps:
   a. Place a drop of sea urchin egg-cell suspension in the chamber of a depression slide, and add a coverslip.
   b. Examine the egg cells, using low-power magnification.
   c. Focus on a single unfertilized egg cell with high-power magnification, and sketch the cell in Part B of the laboratory report.
   d. Remove the coverslip and add a drop of sea urchin sperm-cell suspension to the depression slide. Replace the coverslip, and observe the sperm cells with high-power magnification as they cluster around the egg cells. This attraction is stimulated by gamete secretions.
   e. Observe the egg cells with low-power magnification once again, and watch for the appearance of *fertilization membranes*. Such a membrane forms as soon as an egg cell is penetrated by a sperm cell; it looks like a clear halo surrounding the egg cell.
   f. Focus on a single fertilized egg cell, and sketch it in Part B of the laboratory report.

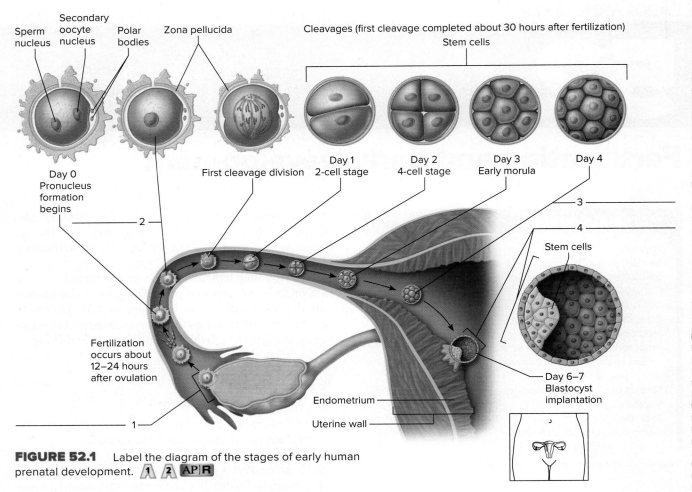

**FIGURE 52.1** Label the diagram of the stages of early human prenatal development. 1️⃣ 2️⃣ AP|R

## LEARN: LAB IN MOTION

Use a toothpick to draw a thin line of Vaseline around the chamber of the depression slide containing the fertilized sea urchin egg cells. Place a coverslip over the chamber, and gently press it into the Vaseline to seal the chamber and prevent the liquid inside from evaporating. Keep the slide in a cool place so that the temperature never exceeds 22°C (72°F). Using low-power magnification, examine the slide every 30 minutes, and look for the appearance of two-, four-, and eight-cell stages of developing sea urchin embryos.

## PRACTICE

**PROCEDURE B—Sea Urchin Early Development**

1. Review the sections entitled "Period of Cleavage" and "Gastrulation and Organogenesis" in chapter 23 of the textbook.

2. Obtain a prepared microscope slide of developing sea urchin embryos. This slide contains embryos in various stages of cleavage. Search the slide, using low-power magnification, and locate embryos in two-, four-, and eight-cell and morula stages. Observe that cleavage results in an increase of cell numbers; however, the cells get progressively smaller.

3. Using low-power magnification, locate the blastula, and identify its blastocoel (hollow central cavity). Then locate the gastrula, and identify ectoderm (outermost germ layer) and endoderm (innermost germ layer).

4. Prepare a sketch of each stage in Part C of the laboratory report.

## PRACTICE

**PROCEDURE C—Human Early Development**

1. Review the sections entitled "Embryonic Stage" and "Extraembryonic Membrane Formation/Placentation" in chapter 23 of the textbook.

2. As a review activity, study and label figure 52.2.

**3.** Study figures 52.3 and 52.4.

**4.** Observe the models of human development, and identify the following features:

**blastocyst**
    inner cell mass (embryoblast)
    trophoblast
    blastocyst cavity (blastocoel)

**primary germ layers**
    ectoderm
    endoderm
    mesoderm

**chorion**
**chorionic villi**
**amnion**
**amniotic fluid**
**yolk sac**
**allantois**
**umbilical cord**
    umbilical arteries (2)
    umbilical vein (1)
**placenta**

**5.** Complete Part D of the laboratory report.

## LEARN: ACTIVITY

Observe the preserved mammalian embryos that are on display. In addition to observing the developing external body structures, identify such features as the chorion, chorionic villi, amnion, yolk sac, umbilical cord, and placenta. What special features provide clues to the types of mammals these embryos represent?

_____

_____

_____

_____

_____

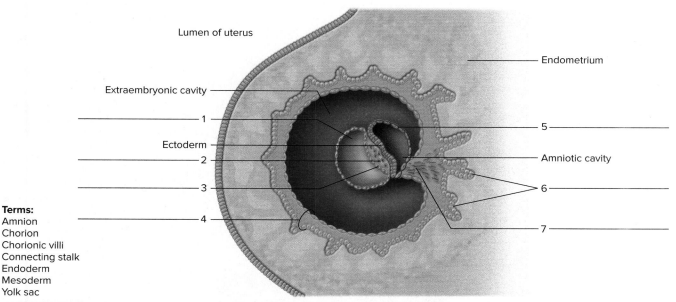

**Terms:**
Amnion
Chorion
Chorionic villi
Connecting stalk
Endoderm
Mesoderm
Yolk sac

**FIGURE 52.2**  Label the major features of the early embryo and the structures associated with it, using the terms provided. **2**

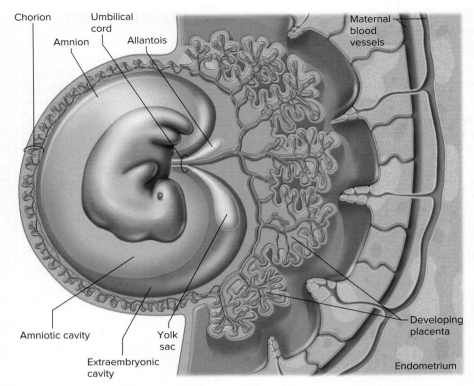

**FIGURE 52.3** Human embryo with extraembryonic membranes and developing umbilical cord and placenta.

**FIGURE 52.4** The umbilical cord, containing two oxygen-poor arteries and a single oxygen-rich vein, is connected to the placenta. The placenta, composed of an embryonic and maternal portion, serves as an exchange site of nutrients, gases, and wastes between embryonic and maternal blood.

Name _____

Date _____

Section _____

The  corresponds to the indicated Learning Outcome(s) found at the beginning of the laboratory exercise.

LABORATORY
REPORT

**52**

# Fertilization and Early Development

## ↻ PART A ASSESSMENTS

Complete the following statements:

1. The zona pellucida surrounds the cell membrane of a(an) _____ . **A**

2. Enzymes secreted by the _____ of a sperm cell help it to penetrate the zona pellucida. **A**

3. The cell resulting from fertilization is called a(an) _____ . **A**

4. The cell resulting from fertilization divides by the process of _____ . **A**

5. _____ is the phase of development during which cellular divisions result in smaller and smaller cells. **2**

6. _____ on the inner lining of the uterine tube aid in moving a developing embryo. **A**

7. The trip from the site of fertilization to the uterine cavity takes about _____ days. **A**

8. The hollow ball of cells formed early in development is called a(an) _____ . **2**

9. A human offspring is called a(an) _____ until the end of the eighth week of development. **2**

10. After the eighth week, a developing human is called a(an) _____ until the time of birth. **2**

## ↻ PART B ASSESSMENTS

Prepare sketches of the following: **3**

| | |
|---|---|
| Single unfertilized sea urchin egg (_____×) | Single fertilized sea urchin egg (_____×) |

Prepare sketches of the following: **3**

| | |
|---|---|
| Two-cell sea urchin embryo (_____×) | Four-cell sea urchin embryo (_____×) |
| Eight-cell sea urchin embryo (_____×) | Morula sea urchin embryo (_____×) |
| Blastula of sea urchin (_____×) | Gastrula of sea urchin (_____×) |

### PART D ASSESSMENTS

Match the terms in column A with the descriptions in column B. Place the letter of your choice in the space provided. **2**

| Column A | Column B |
|---|---|
| **a.** Amniotic fluid | _____ **1.** Germ layer that gives rise to muscle and bone tissues |
| **b.** Blastocyst | _____ **2.** Composed of parts of chorion and uterine wall |
| **c.** Chorionic villi | |
| **d.** Ectoderm | _____ **3.** Protects the embryo from being jarred |
| **e.** Endoderm | _____ **4.** Hollow ball of cells |
| **f.** Mesoderm | _____ **5.** Cells forming wall of blastocyst |
| **g.** Morula | _____ **6.** Solid ball of about sixteen cells |
| **h.** Placenta | _____ **7.** Inner germ layer of embryonic disc |
| **i.** Trophoblast | |
| **j.** Umbilical cord | _____ **8.** Slender extensions that grow out from the trophoblast |
| | _____ **9.** Contains two arteries and one vein |
| | _____ **10.** Outer germ layer of embryonic disc |

# 53

# Genetics

## MATERIALS NEEDED

Textbook
Pennies (or other coins)
Dice
PTC paper
Astigmatism chart
Ichikawa's or Ishihara's color plates for
    colorblindness test

## PURPOSE OF THE EXERCISE

To observe some selected human traits, to use pennies and dice to demonstrate laws of probability, and to solve some genetic problems using a Punnett square.

## LEARNING OUTCOMES

After completing this exercise, you should be able to

1. Examine and record twelve genotypes and phenotypes of selected human traits.

2. Demonstrate the laws of probability using tossed pennies and dice and interpret the results.

3. Predict genotypes and phenotypes of complete dominance, codominance, and sex-linked problems using Punnett squares.

Many of human genetics in this laboratory exercise are basic, external features to observe. Some traits are based upon simple Mendelian genetics. As our knowledge of genetics continues to develop, traits such as tongue roller and free earlobe may no longer be considered examples of simple Mendelian genetics. We may need to continue to abandon some simple Mendelian models. New evidence may involve polygenic inheritance, effects of other genes, or environmental factors as more appropriate explanations. Therefore, the analysis of family genetics is not always feasible or meaningful.

Genetics is the study of the inheritance of characteristics. The genes that transmit this information are coded in segments of DNA in chromosomes. Homologous chromosomes possess the same gene at the same *locus.* These genes may exist in variant forms, called *alleles.* If a person possesses two identical alleles, the person is said to be *homozygous* for that particular trait. If a person possesses two different alleles, the person is said to be *heterozygous* for that particular trait. The particular combination of these gene variants (alleles) represents the person's *genotype;* the appearance or physical manifestation of the individual that develops as a result of the way the genes are expressed represents the person's *phenotype.*

If one allele determines the phenotype by masking the expression of the other allele in a heterozygous individual, the allele is termed *dominant.* The allele whose expression is masked is termed *recessive.* If the heterozygous condition determines an intermediate phenotype, the inheritance represents *incomplete dominance.* However, different alleles are *codominant* if both are expressed in the heterozygous condition. Some characteristics inherited on the sex chromosomes result in phenotype frequencies that might be more prevalent in males or females. Such characteristics are called *sex-linked (X-linked or Y-linked) characteristics.*

As a result of meiosis during the formation of eggs and sperm, a mother and father each transmit an equal number of chromosomes (the haploid number 23) to form the zygote (diploid number 46). An offspring will receive one allele from each parent. These gametes combine randomly in the formation of each offspring. Hence, the *laws of probability* can be used to predict possible genotypes and phenotypes of offspring. A genetic tool called a *Punnett square* simulates all possible combinations (probabilities) in offspring genotypes and resulting phenotypes.

## PRACTICE

### PROCEDURE A—Human Genotypes and Phenotypes

A complete set of genetic instructions in one human cell constitutes one's *genome.* The human genome contains about 2.9 billion base pairs representing approximately

**TABLE 53.1   Examples of Some Common Human Phenotypes**

| Dominant Traits and Genotypes | Recessive Traits and Genotypes |
|---|---|
| Tongue roller (*R___*) | Nonroller (*rr*) |
| Freckles (*F___*) | No freckles (*ff*) |
| Widow's peak (*W___*) | Straight hairline (*ww*) |
| Dimples (*D___*) | No dimples (*dd*) |
| Free earlobe (*E___*) | Attached earlobe (*ee*) |
| Normal skin coloration (*M___*) | Albinism (*mm*) |
| Astigmatism (*A___*) | Normal vision (*aa*) |
| Curly hair (*C___*) | Straight hair (*cc*) |
| PTC taster (*T___*) | Nontaster (*tt*) |
| Blood type A (*I^A___*), B (*I^B___*), or AB (*I^A I^B*) | Blood type O (*ii*) |
| Normal color vision (X^C X^C), (X^C X^c), or (X^C Y) | Red-green colorblindness (X^c X^c) or (X^c Y) |

20,500 protein-encoding genes. These instructions represent our genotypes and are expressed as phenotypes sometimes clearly observable on our bodies. Some of these traits are listed in table 53.1 and are discernible in figure 53.1.

A dominant trait might be homozygous or heterozygous, so only one capital letter is used along with a blank for the possible second dominant or recessive allele. For a recessive trait, two lowercase letters represent the homozygous recessive genotype for that characteristic. Dominant does not always correlate with the predominance of the allele in the gene pool; dominant means one allele will determine the appearance of the phenotype.

1. **Tongue roller/nonroller:** The dominant allele (*R*) determines the person's ability to roll the tongue into a U-shaped trough. The homozygous recessive condition (*rr*) prevents this tongue rolling (fig. 53.1). Record your results in the table in Part A of Laboratory Report 53.
2. **Freckles/no freckles:** The dominant allele (*F*) determines the appearance of freckles. The homozygous recessive condition (*ff*) does not produce freckles (fig. 53.1). Record your results in the table in Part A of the laboratory report.
3. **Widow's peak/straight hairline:** The dominant allele (*W*) determines the appearance of a hairline above the forehead that has a distinct downward point in the center, called a widow's peak. The homozygous recessive condition (*ww*) produces a straight hairline (fig. 53.1). A receding hairline would prevent this phenotype determination. Record your results in the table in Part A of the laboratory report.
4. **Dimples/no dimples:** The dominant allele (*D*) determines the appearance of a distinct dimple in one or both cheeks upon smiling. The homozygous recessive condition (*dd*) results in the absence of dimples (fig. 53.1).

Record your results in the table in Part A of the laboratory report.

5. **Free earlobe/attached earlobe:** The dominant allele (*E*) codes for the appearance of an inferior earlobe that hangs freely below the attachment to the head. The homozygous recessive condition (*ee*) determines the earlobe attaching directly to the head at its inferior border (fig. 53.1). Record your results in the table in Part A of the laboratory report.
6. **Normal skin coloration/albinism:** The dominant allele (*M*) determines the production of some melanin, producing normal skin coloration. The homozygous recessive condition (*mm*) determines albinism due to the inability to produce or use the enzyme tyrosinase in pigment cells. An albino does not produce melanin in the skin, hair, or the middle tunic (choroid coat, ciliary body, and iris) of the eye. The absence of melanin in the middle tunic allows the pupil to appear slightly red to nearly black. Remember that the pupil is an opening in the iris filled with transparent aqueous humor. An albino human has pale white skin, flax-white hair, and a pale blue iris. Record your results in the table in Part A of the laboratory report.
7. **Astigmatism/normal vision:** The dominant allele (*A*) results in an abnormal curvature to the cornea or the lens. As a consequence, some portions of the image projected on the retina are sharply focused, and other portions are blurred. The homozygous recessive condition (*aa*) generates normal cornea and lens shapes and normal vision. Use the astigmatism chart and directions to assess this possible defect described in Laboratory Exercise 35. Other eye defects, such as nearsightedness (myopia) and farsightedness (hyperopia), are different genetic traits due to genes at other locations. Record your results in the table in Part A of the laboratory report.

**Dominant Traits**

**Recessive Traits**

**(a)** Tongue roller

**(b)** Nonroller

**(c)** Freckles

**(d)** No freckles

**(e)** Widow's peak

**(f)** Straight hairline

**FIGURE 53.1** Representative genetic traits comparing dominant and recessive phenotypes: (*a*) tongue roller; (*b*) nonroller; (*c*) freckles; (*d*) no freckles; (*e*) widow's peak; (*f*) straight hairline; (*g*) dimples; (*h*) no dimples; (*i*) free earlobe; (*j*) attached earlobe.

((*a–j*) ©J & J Photography)

**Dominant Traits**

**(g)** Dimples

**Recessive Traits**

**(h)** No dimples

**(i)** Free earlobe

**(j)** Attached earlobe

**FIGURE 53.1** *Continued.*

8. **Curly hair/straight hair:** The dominant allele (*C*) determines the appearance of curly hair. Curly hair is somewhat flattened in cross section, as the hair follicle of a similar shape served as a mold for the root of the hair during its formation. The homozygous recessive condition (*cc*) produces straight hair. Straight hair is nearly round in cross section from being molded into this shape in the hair follicle. In some populations (Caucasians), the heterozygous condition (*Cc*) expresses the intermediate wavy hair phenotype (incomplete dominance). This trait determination assumes no permanents or hair straightening procedures have been performed. Such hair alterations do not change the hair follicle shape, and future hair growth results in original genetic hair conditions. Record your unaltered hair appearance in the table in Part A of the laboratory report.

9. **PTC taster/nontaster:** The dominant allele (*T*) determines the ability to experience a bitter sensation when PTC paper is placed on the tongue. About 70% of people possess this dominant gene. The homozygous recessive condition (*tt*) makes a person unable to notice the substance. Place a piece of PTC (phenylthiocarbamide) paper on the upper tongue surface and chew it slightly to see if you notice a bitter sensation from this harmless chemical. The nontaster of the PTC paper does not detect any taste at all from this substance. Record your results in the table in Part A of the laboratory report.

10. **Blood type A, B, or AB/blood type O:** There are three alleles ($I^A$, $I^B$, and $i$) in the human population affecting RBC membrane structure. These alleles are located on a single pair of homologous chromosomes, so a person possesses either two of the three alleles or two of the same allele. All of the possible combinations of these alleles of genotypes and the resulting phenotypes are depicted in table 53.2. The expression of the blood type AB is a result of both codominant alleles located in the same individual. Possibly you have already determined your blood type in Laboratory Exercise 37 or have it recorded on a blood donor card. (If simulated blood-typing kits were used for Laboratory Exercise 37, those results would not be valid for your genetic factors.) Record your results in the table in Part A of the laboratory report.

11. **Sex determination:** A person with sex chromosomes XX displays a female phenotype. A person with sex chromosomes XY displays a male phenotype. Record your results in the table in Part A of the laboratory report.

12. **Normal color vision/red-green colorblindness:** This condition is a sex-linked (X-linked) characteristic. The alleles for color vision are also on the X chromosome, but they are absent on the Y chromosome. As a result, a female might possess both alleles ($C$ and $c$), one on each of the X chromosomes. The dominant allele ($C$) determines normal color vision; the homozygous recessive condition ($cc$) results in red-green colorblindness. However, a male would possess only one of the two alleles for color vision because there is only a single X chromosome in a male. Hence, a male with even a single recessive gene for colorblindness possesses the defect. Note all the possible genotypes and phenotypes for this condition (table 53.1). Review the color vision test in Laboratory Exercise 35 using the color plate in figure 35.4 and color plates in Ichikawa's or Ishihara's book. Record your results in the table in Part A of the laboratory report.

13. Complete Part A of the laboratory report.

**TABLE 53.2   Genotypes and Phenotypes (Blood Types)**

| Genotypes | Phenotypes (Blood Types) |
|---|---|
| $I^A I^A$ or $I^A i$ | A |
| $I^B I^B$ or $I^B i$ | B |
| $I^A I^B$ (codominant) | AB |
| $ii$ | O |

## PRACTICE

### PROCEDURE B—Laws of Probability

The laws of probability provide a mathematical way to determine the likelihood of events occurring by chance. This prediction is often expressed as a ratio of the number of results from experimental events to the number of results considered possible. For example, when tossing a coin there is an equal chance of the results displaying heads or tails. Hence, the probability is one-half of obtaining either a heads or a tails (there are two possibilities for each toss). When all of the probabilities of all possible outcomes are considered for the result, they will always add up to 1. To predict the probability of two or more events occurring in succession, multiply the probabilities of each individual event. For example, the probability of tossing a die and displaying a 4 twice in a row is $1/6 \times 1/6 = 1/36$ (there are six possibilities for each toss). Each toss in a sequence is an *independent event* (chance has no memory). The same laws apply when parents have multiple children (each fertilization is an independent event). Perform the following experiments to demonstrate the laws of probability:

1. Use a single penny (or other coin) and toss it 20 times. Predict the number of heads and tails that would occur from the 20 tosses. Record your prediction and the actual results observed in Part B of the laboratory report.

2. Use a single die (*pl.* dice) and toss it 24 times. Predict the number of times a number below 3 (numbers 1 and 2) would occur from the 24 tosses. Record your prediction and the actual results in Part B of the laboratory report.

3. Use two pennies and toss them simultaneously 32 times. Predict the number of times two heads, a heads and a tails, and two tails would occur. Record your prediction and the actual results in Part B of the laboratory report.

4. Use a pair of dice and toss them simultaneously 32 times. Predict the number of times for both dice coming up with odd numbers, one die an odd and the other an even number, and both dice coming up with even numbers. Record your prediction and the actual results in Part B of the laboratory report.

5. Obtain class totals for all of the coins and dice tossed by adding your individual results to a class tally location, as on the blackboard.

6. A Punnett square can be used to visually demonstrate the probable results for two pennies tossed simultaneously. For the purpose of a genetic comparison, an *h* (heads) will represent one "allele" on the coin; a *t* (tails) will represent a different "allele" on the coin.

**Coin #1 Possibilities**

|              |   | h  | t  |
|--------------|---|----|----|
| Coin #2      | h | hh | ht |
| Possibilities| t | ht | tt |

Possible Combinations of Two Tossed Pennies (in boxes)

7. Complete Part B of the laboratory report.

 **PRACTICE**

### PROCEDURE C—Genetic Problems

1. A Punnett square can be constructed to demonstrate a visual display of the predicted offspring from parents with known genotypes. Recall that, in complete dominance, a dominant allele is expressed in the phenotype, as it can mask the other recessive allele on the homologous chromosome pair. During meiosis, the homologous chromosomes with their alleles separate (Mendel's Law of Segregation) into different gametes. An example of such a cross might be a homozygous dominant mother for dimples ($DD$) who has offspring with a father homozygous recessive ($dd$) for the same trait. The results of such a cross, according to the laws of probability, would be represented by the following Punnett square:

**Female Gametes**

|         |   | D  | D  |
|---------|---|----|----|
| Male    | d | Dd | Dd |
| Gametes | d | Dd | Dd |

Possible Genotypes of Offspring (in boxes)

*Results:* Genotypes: 100% $Dd$ (all heterozygous)

Phenotypes: 100% dimples

In another example, assume that both parents are heterozygous ($Dd$) for dimples. The results of such a cross, according to the laws of probability, would be represented by the following Punnett square:

**Female Gametes**

|         |   | D  | d  |
|---------|---|----|----|
| Male    | D | DD | Dd |
| Gametes | d | Dd | dd |

Possible Genotypes of Offspring (in boxes)

*Results:* Genotypes: 25% DD (homozygous dominant); 50% Dd (heterozygous); 25% dd (homozygous recessive) (1:2:1 genotypic ratio)

Phenotypes: 75% dimples; 25% no dimples (3:1 phenotypic ratio)

2. Work genetic problems 1 and 2 in Part C of the laboratory report.

3. The ABO blood type inheritance represents an example of codominance. Review table 53.2 for the genotypes and phenotypes for the expression of this trait. A Punnett square can be constructed to predict the offspring of parents of known genotypes. In this example, assume the genotype of the mother is $I^A I^B$, and the father is *ii*. The results of such a cross would be represented by the following Punnett square:

**Female Gametes**

|         |   | $I^A$   | $I^B$   |
|---------|---|---------|---------|
| Male    | i | $I^A i$ | $I^B i$ |
| Gametes | i | $I^A i$ | $I^B i$ |

Possible Genotypes of Offspring (in boxes)

*Results:* Genotypes: 50% $I^A i$ (heterozygous for A); 50% $I^B i$ (heterozygous for B) (1:1 genotypic ratio)

Phenotypes: 50% blood type A; 50% blood type B (1:1 phenotypic ratio)

*Note:* In this cross, all of the children would have blood types unlike either parent.

4. Work genetic problems 3 and 4 in Part C of the laboratory report.

5. Review the inheritance of red-green colorblindness, an X-linked characteristic, in table 53.1. A Punnett square can be constructed to predict the offspring of parents of known genotypes. In this example, assume the genotype of the mother is heterozygous $X^C X^c$ (normal color vision but a carrier for the colorblindness defect), and the father is $X^C Y$ (normal color vision; no allele on the Y chromosome). The results of such a cross would be represented by the following Punnett square:

**Female Gametes**

|         |       | $X^C$       | $X^c$       |
|---------|-------|-------------|-------------|
| Male    | $X^C$ | $X^C X^C$   | $X^C X^c$   |
| Gametes | Y     | $X^C Y$     | $X^c Y$     |

Possible Genotypes of Offspring (in boxes)

*Results:* Genotypes: 25% $X^C X^C$; 25% $X^C X^c$; 25% $X^C Y$; 25% $X^c Y$ (1:1:1:1 genotypic ratio)

Phenotypes for sex determination: 50% females; 50% males (1:1 phenotypic ratio)

Phenotypes for color vision: Females 100% normal color vision (however, 50% are heterozygous carriers for colorblindness)

Phenotypes for color vision: Males 50% normal; 50% with red-green colorblindness (1:1 phenotypic ratio)

*Note:* In X-linked inheritance, the colorblind males received the recessive gene from their mothers.

6. Complete Part C of the laboratory report.

The Ⓐ corresponds to the indicated Learning Outcome(s) found at the beginning of the laboratory exercise.

# Genetics

## 🔄 PART A ASSESSMENTS

1. Enter your test results for genotypes and phenotypes in the table. Circle your phenotype and genotype for each of the twelve traits. Ⓐ

| Trait | Dominant Phenotype | Genotype | Recessive Phenotype | Genotype |
|---|---|---|---|---|
| Tongue movement | Roller | $R\_\_$ | Nonroller | $rr$ |
| Freckles | Freckles | $F\_\_$ | No freckles | $ff$ |
| Hairline | Widow's peak | $W\_\_$ | Straight | $ww$ |
| Dimples | Dimples | $D\_\_$ | No dimples | $dd$ |
| Earlobe | Free | $E\_\_$ | Attached | $ee$ |
| Skin coloration | Normal (some melanin) | $M\_\_$ | Albinism | $mm$ |
| Vision | Astigmatism | $A\_\_$ | Normal | $aa$ |
| Hair shape* | Curly | $C\_\_$ | Straight | $cc$ |
| Taste | PTC taster | $T\_\_$ | Nontaster for PTC | $tt$ |
| Blood type | A, B, or AB | $I^A\_\_; I^B\_\_;$ or $I^A I^B$ | O | $ii$ |
| Sex | | XX or XY | | |
| Color vision | Normal | $X^C X\_\_$ or $X^C Y$ | Red-green colorblindness | $X^c X^c$ or $X^c Y$ |

*In some populations (Caucasians) the heterozygous condition (*Cc*) results in the appearance of wavy hair, which actually represents an example of incomplete dominance for this trait.

2. Choose at least three dominant phenotypes that you circled in question 1, and analyze the genotypes for those traits. If it is feasible to observe your biological parents and siblings for any of these traits, are you able to determine if any of your dominant genotypes are homozygous dominant or heterozygous? _____ If so, which ones? _____ _____ Explain the rationale for your response.

_____

_____

_____

_____

_____

_____

_____

_____

1. Single penny tossed 20 times and counting heads and tails:

   Probability (prediction): _____/20 heads _____/20 tails

   *(Note: Traditionally, probabilities are converted to the lowest fractional representation.)*

   Actual results: _____ heads _____ tails

   Class totals: _____ heads _____ tails

2. Single die tossed 24 times and counting the number of times a number below 3 occurs:

   Probability: _____/24 number below 3 (numbers 1 and 2)

   Actual results: _____ number below 3

   Class totals: _____ number below 3 _____ total tosses by class members

3. Two pennies tossed simultaneously 32 times and counting the number of two heads, a heads and a tails, and two tails:

   Probability: _____/32 of two heads _____/32 of a heads and a tails _____/32 of two tails

   Actual results: _____ two heads _____ heads and tails _____ two tails

   Class totals: _____ two heads _____ heads and tails _____ two tails

4. Two dice tossed simultaneously 32 times and counting the number of two odd numbers, an odd and an even number, and two even numbers:

   Probability: _____/32 of two odd numbers _____/32 of an odd and an even number _____/32 of two even numbers

   Actual results: _____ two odd numbers _____ an odd and an even number _____ two even numbers

   Class totals: _____ two odd numbers _____ an odd and an even number _____ two even numbers

5. Use the example of the two dice tossed 32 times, and construct a Punnett square to represent the possible combinations that could be used to determine the probability (prediction) of odd and even numbers for the resulting tosses. Your construction should be similar to the Punnett square for the two coins tossed that is depicted in Procedure B of the laboratory exercise.

6. Complete the following:

   **a.** Are the class totals closer to the predicted probabilities than your results? _____ Explain your response.

   _____

   _____

   **b.** Does the first toss of the penny or the first toss of the die have any influence on the next toss? _____ Explain your response.

   _____

**c.** Assume a family has two boys or two girls. They wish to have one more child but hope for the child to be of the opposite sex from the two they already have. What is the probability that the third child will be of the opposite sex? _____ Explain your response.

_____

**d.** What is the probability (prediction) that a couple without children will eventually have four children, all girls? _____ Explain your response.

_____

_____

## PART C ASSESSMENTS

For each of the genetic problems, (a) determine the parents' genotypes, (b) determine the possible gametes for each parent, (c) construct a Punnett square, and (d) record the resulting genotypes and phenotypes as ratios from the cross. Problems 1 and 2 involve examples of complete dominance; problems 3 and 4 are examples of codominance; problem 5 is an example of sex-linked (X-linked) inheritance.

**1.** Determine the results from a cross of a mother who is heterozygous (*Rr*) for tongue rolling with a father who is homozygous recessive (*rr*). **3**

**2.** Determine the results from a cross of a mother and a father who are both heterozygous for freckles. **3**

**3.** Determine the results from a mother who is heterozygous for blood type B and a father who is homozygous dominant for blood type A. **3**

**4.** Determine the results from a mother who is heterozygous for blood type A and a father who is heterozygous for blood type B.

**5.** Colorblindness is an example of X-linked inheritance. Hemophilia is another example of X-linked inheritance, also from a recessive allele (*h*). The dominant allele (*H*) determines whether the person possesses normal blood clotting. A person with hemophilia has a permanent tendency for hemorrhaging due to a deficiency of one of the clotting factors (VIII—antihemophilic factor). Determine the offspring from a cross of a mother who is a carrier (heterozygous) for the disease and a father with normal blood coagulation.

## ASSESS

### CRITICAL THINKING

Assume that the genes for hairline and earlobes are on different pairs of homologous chromosomes. Determine the genotypes and phenotypes of the offspring from a cross if both parents are heterozygous for both traits. (1) First determine the genotypes for each parent. (2) Determine the gametes, but remember each gamete has one allele for each trait (gametes are haploid). (3) Construct a Punnett square, with 16 boxes, that has four different gametes from each parent along the top and the left edges. (This is an application to demonstrate Mendel's Law of Independent Assortment.) (4) List the results of genotypes and phenotypes as ratios.

# Cat Dissection: Musculature

## MATERIALS NEEDED

Textbook
Preserved cat (double injection)
Dissecting tray
Dissecting instruments
Large plastic bag
Identification tag
Disposable gloves
Bone shears
Human torso model
Human upper and lower limb models
Cat skeleton

 **SAFETY**

- Wear disposable gloves when working on the cat dissection.
- Dispose of tissue remnants and gloves as instructed.
- Wash the dissecting tray and instruments as instructed.
- Wash your laboratory table.
- Wash your hands before leaving the laboratory.

## PURPOSE OF THE EXERCISE

To examine and identify the musculature of the cat and to compare it with that of the human.

 **LEARNING OUTCOMES** AP|R

After completing this exercise, you should be able to

1. Name and locate the major skeletal muscles of the cat.
2. Name the origins, insertions, and actions of the skeletal muscles designated by the laboratory instructor.
3. Name and locate the corresponding skeletal muscles of the human.

Although the aim of this exercise is to become more familiar with the human musculature, human cadavers are not always available for dissection. Instead, preserved cats are often used for dissection because they are relatively small and can be purchased from biological suppliers. Because they are mammals, cats have many features in common with humans, including similar skeletal muscles (with similar names).

On the other hand, cats use four limbs for locomotion, whereas humans use only two limbs. Because the musculature of each type of organism is adapted to provide for its special needs, comparisons of the muscles of cats and humans may not be precise.

As you continue your dissection of various systems of the cat, many anatomical similarities between cats and humans will be observed. These fundamental similarities are called homologous structures. Although homologous structures have a similar structure and embryological origin, the functions are sometimes different.

 **MUSCLE DISSECTION TECHNIQUES**

*Dissect*: to expose the entire length of the muscle from origin to insertion. Most of the procedures are accomplished with blunt probes used to separate various connective tissues that hold adjacent structures together. It does not mean to remove or to cut into the muscles or other organs.

*Transect*: to cut through the muscle near its midpoint. The cut is perpendicular to the length of the muscle fibers.

*Reflect*: to lift a transected muscle aside to expose deeper muscles or other organs.

 **PRACTICE**

### PROCEDURE A—Skin Removal

1. Cats usually are preserved with a solution that prevents microorganisms from causing the tissues to decompose. However, because fumes from this preserving

fluid may be annoying and may irritate skin, be sure to work in a well-ventilated room and wear disposable gloves to protect your hands.

2. Obtain a preserved cat, a dissecting tray, a set of dissecting instruments, a large plastic bag, and an identification tag.

3. If the cat is in a storage bag, dispose of any excess preserving fluid in the bag, as directed by the laboratory instructor. However, you may want to save some of the preserving fluid to keep the cat moist until you have completed your work later in the year.

4. If the cat is too wet with preserving fluid, blot the cat dry with paper towels.

5. Remove the skin from the cat. To do this, follow these steps:

   a. Place the cat in the dissecting tray with its ventral surface down.

   b. Use a sharp scalpel to make a short, shallow incision through the skin in the dorsal midline at the base of the cat's neck.

   c. Insert the pointed blade of a dissecting scissors through the opening in the skin and cut along the dorsal midline, following the tips of the vertebral spinous processes to the base of the tail (fig. 54.1).

   d. Use the scissors to make an incision encircling the region of the tail, anus, and genital organs. Do not remove the skin from this area.

**FIGURE 54.1** Incisions to be made for removing the skin of the cat (dorsal view).

Anterior (cranial)

5. Incision around neck

1. Initial incision

6. Lateral incision of forelimb

2. Dorsal midline incision

4. Lateral incision on hindlimb

3. Incision encircling tail

Posterior (caudal)

**e.** Use bone shears or a bone saw to sever the tail at its base and discard it (optional procedure).

**f.** From the incision at the base of the tail, cut along the lateral surface of each hindlimb to the paw, and encircle the ankle. Also clip off the claws from each paw to prevent them from tearing the plastic storage bag and scratching your skin (optional procedure).

**g.** Make an incision from the initial cut at the base of the neck to encircle the neck.

**h.** From the incision around the neck, cut along the lateral surface of each forelimb to the paw and encircle the wrist (fig. 54.1).

**i.** Grasp the skin on either side of the dorsal midline incision, and carefully pull the skin laterally away from the body. At the same time, use your fingers or a blunt probe to help remove the skin from the underlying muscles by separating the loose connective tissue (superficial fascia). As you pull the skin away, you may note a thin sheet of muscle tissue attached to it. This muscle is called the *cutaneous maximus,* and it functions to move the cat's skin. Humans lack the cutaneous maximus, but a similar sheet of muscle (platysma) is present in the neck of a human.

**j.** As you remove the skin, work toward the ventral surface, then work toward the head, and finally work toward the tail. Pull the skin over each limb as if you were removing a glove.

**k.** If the cat is a female, note the mammary glands on the ventral surface of the thorax and abdomen (fig. 54.2). These glands will adhere to the skin and can be removed from it and discarded.

**l.** After the skin has been pulled away, carefully remove as much of the remaining connective tissue as possible to expose the underlying skeletal muscles. The muscles should appear light brown and fibrous.

**6.** After skinning the cat, follow these steps:

**a.** Discard the tissues you have removed as directed by the laboratory instructor.

**b.** Wrap the skin around the cat to help keep its body moist, and place it in a plastic storage bag.

**c.** Write your name in pencil on an identification tag, and tie the tag to the storage bag so that you can identify your specimen.

**7.** Observe the recommended safety procedures for the conclusion of a laboratory session.

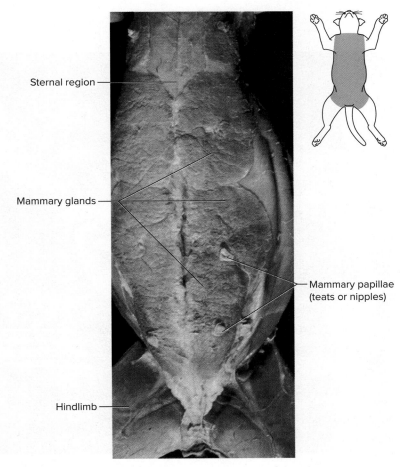

Sternal region

Mammary glands

Mammary papillae (teats or nipples)

Hindlimb

**FIGURE 54.2** Mammary glands on the ventral surface of the cat's thorax and abdomen. (©Dr. Sheril Burton)

### PROCEDURE B—Skeletal Muscle Dissection

1. The purpose of a skeletal muscle dissection is to separate the individual muscles from any surrounding tissues and thus expose the muscles for observation. To do a muscle dissection, follow these steps:
   a. Use the appropriate figure as a guide and locate the muscle to be dissected in the specimen.
   b. Use a blunt probe or a finger to separate the muscle from the surrounding connective tissue along its natural borders. The muscle should separate smoothly. If the border appears ragged, you probably have torn the muscle fibers.
2. If it is necessary to cut through (transect) a superficial muscle to observe a deeper one, use scissors to transect the muscle about halfway between its origin and insertion. Then, lift aside (reflect) the cut ends, leaving their attachments intact.
3. The following procedures will instruct you to dissect some of the larger and more easily identified muscles of the cat. In each case, the procedure will include the names of the muscles to be dissected, figures illustrating their locations, and tables listing the origins, insertions, and actions of the muscles. Compare the muscle origins and insertions to the cat skeleton (fig. 54.3).

(More detailed dissection instructions can be obtained by consulting an additional guide or atlas for cat dissection.) As you dissect each muscle, study the figures and tables in chapter 9 of the textbook and identify any corresponding (homologous) muscles of the human body. Also locate these homologous muscles in the human torso model or models of the human upper and lower limbs.

 **PRACTICE**

### PROCEDURE C—Muscles of the Head and Neck

1. Place the cat in the dissecting tray with its ventral surface up.
2. Remove the skin and underlying connective tissues from each side of the neck, forward to the chin, and up to each ear.
3. Study figures 54.4 and 54.5, and then locate and dissect the following muscles:

   **sternomastoid**
   **mylohyoid**
   **sternohyoid**
   **masseter**
   **digastric**

**FIGURE 54.3** Skeleton of a cat, lateral view. (©McGraw-Hill Education/Christine Eckel, photographer)

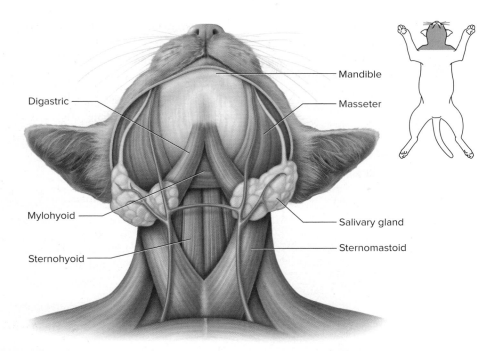

**FIGURE 54.4**   Muscles on the ventral surface of the cat's head and neck.   **AP|R**   (©McGraw-Hill Education/APR)

**FIGURE 54.5**   Muscles of the ventral surface of the cat's head, neck, thorax, and forelimb.   (©McGraw-Hill Education/APR)

4. To find the deep muscles of the cat's neck and chin, carefully remove the sternomastoid, sternohyoid, and mylohyoid muscles from the right side.
5. Study figure 54.6, and locate the following muscles:

   **hyoglossus**
   **cleidomastoid**
   **styloglossus**
   **thyrohyoid**
   **geniohyoid**
   **sternothyroid**

6. See table 54.1 for the origins, insertions, and actions of these head and neck muscles.
7. Complete Part A of Laboratory Report 54.

**TABLE 54.1  Muscles of the Head and Neck**

| Muscle | Origin | Insertion | Action |
|---|---|---|---|
| Sternomastoid | Manubrium of sternum | Mastoid process of temporal bone | Turns and depresses head |
| Sternohyoid | Costal cartilage | Hyoid bone | Depresses hyoid bone |
| Digastric | Mastoid process and occipital bone | Ventral border of mandible | Depresses mandible |
| Mylohyoid | Inner surface of mandible | Hyoid bone | Raises floor of mouth |
| Masseter | Zygomatic arch | Coronoid fossa of mandible | Elevates mandible |
| Hyoglossus | Hyoid bone | Tongue | Pulls tongue back |
| Styloglossus | Styloid process | Tip of tongue | Pulls tongue back |
| Geniohyoid | Inner surface of mandible | Hyoid bone | Pulls hyoid bone forward |
| Cleidomastoid | Clavicle | Mastoid process | Turns head to side |
| Thyrohyoid | Thyroid cartilage | Hyoid bone | Raises larynx |
| Sternothyroid | Sternum | Thyroid cartilage | Pulls larynx back |

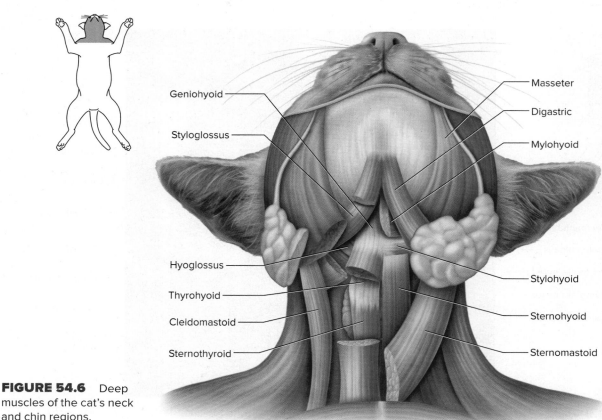

**FIGURE 54.6** Deep muscles of the cat's neck and chin regions.

### PROCEDURE D—Muscles of the Thorax

1. Place the cat in the dissecting tray with its ventral surface up.
2. Remove any remaining fat and connective tissue to expose the muscles in the walls of the thorax and abdomen.
3. Study figures 54.5, 54.7, and 54.8. Transect the superficial pectoantebrachialis to expose the entire pectoralis major. Locate and dissect the following pectoral muscles:

   **pectoantebrachialis** (well developed in cats)
   **pectoralis minor**
   **pectoralis major**
   **xiphihumeralis**

4. To find the deep thoracic muscles, transect the pectoralis minor, pectoralis major, and pectoantebrachialis muscles, and reflect their cut edges to the sides. Remove the underlying fat and connective tissues to expose the following muscles, as shown in figure 54.9:

   **scalenus**
   **levator scapulae**
   **serratus ventralis**
   **transversus costarum**

5. See table 54.2 for the origins, insertions, and actions of these muscles.

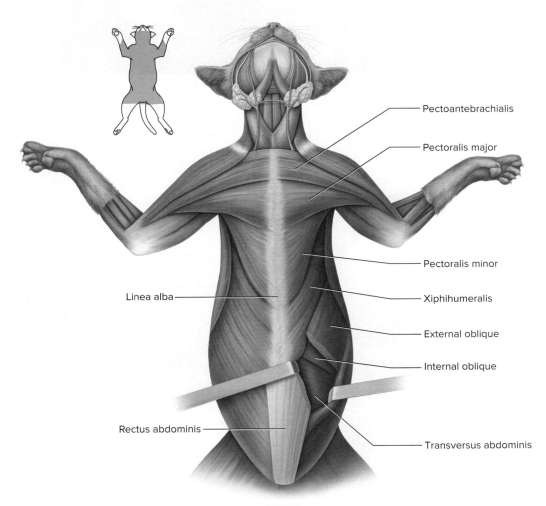

**FIGURE 54.7** Muscles of the cat's thorax and abdomen. AP|R

**FIGURE 54.8** Superficial muscles of the cat's thorax. (©Dr. Sheril Burton)

**FIGURE 54.9** Superficial and deep muscles of the cat's thorax.

**TABLE 54.2  Muscles of the Thorax**

| Muscle | Origin | Insertion | Action |
|--------|--------|-----------|--------|
| Pectoantebrachialis | Manubrium of sternum | Fascia of forelimb | Adducts forelimb |
| Pectoralis major | Manubrium of sternum and costal cartilages | Greater tubercle of humerus | Adducts and rotates forelimb |
| Pectoralis minor | Sternum | Proximal end of humerus | Adducts and rotates forelimb |
| Xiphihumeralis | Xiphoid process of sternum | Proximal end of humerus | Adducts forelimb |
| Scalenus | Ribs | Cervical vertebrae | Flexes neck |
| Serratus ventralis | Ribs | Scapula | Pulls scapula ventrally and posteriorly |
| Levator scapulae | Cervical vertebrae | Scapula | Pulls scapula ventrally and anteriorly |
| Transversus costarum | Sternum | First rib | Pulls sternum toward head |

 **PRACTICE**

**PROCEDURE E—Muscles of the Abdominal Wall**

1. Study figures 54.7 and 54.10.
2. Locate the *external oblique* muscle in the abdominal wall.
3. Make a shallow, longitudinal incision through the external oblique. Reflect the cut edge and expose the *internal oblique* muscle beneath (see fig. 54.7). The fibers of the internal oblique run at a right angle to those of the external oblique.
4. Make a longitudinal incision through the internal oblique. Reflect the cut edge and expose the *transversus abdominis* muscle.
5. Expose the *rectus abdominis* muscle on one side of the midventral line. This muscle lies beneath an aponeurosis.
6. See table 54.3 for the origins, insertions, and actions of these muscles.
7. Complete Part B of the laboratory report.

**TABLE 54.3  Muscles of the Abdominal Wall**

| Muscle | Origin | Insertion | Action |
|--------|--------|-----------|--------|
| External oblique | Ribs and fascia of back | Linea alba | Compresses abdominal wall |
| Internal oblique | Fascia of back | Linea alba | Compresses abdominal wall |
| Transversus abdominis | Lower ribs | Linea alba | Compresses abdominal wall |
| Rectus abdominis | Symphysis pubis | Sternum and costal cartilages | Compresses abdominal wall and flexes trunk |

Serratus ventralis

Latissimus dorsi

Xiphihumeralis

Aponeurosis

Rectus abdominis

Internal oblique

Transversus abdominis

External oblique

**FIGURE 54.10** Muscles of the cat's thorax and abdomen. **AP|R** (©Dr. Sheril Burton)

 **PRACTICE**

**PROCEDURE F—Muscles of the Shoulder and Back**

1. Place the cat in the dissecting tray with its dorsal surface up.
2. Remove any remaining fat and connective tissue to expose the muscles of the shoulder and back.
3. Study figures 54.11 and 54.12, and then locate and dissect the following superficial muscles:

   **clavotrapezius**
   **acromiotrapezius**
   **spinotrapezius**
   **levator scapulae ventralis**
   **clavodeltoid**
   **acromiodeltoid**
   **spinodeltoid**
   **latissimus dorsi**

*Note*: The clavodeltoid (clavobrachialis) appears as a continuation of the clavotrapezius.

4. Using scissors, transect the latissimus dorsi and the group of trapezius muscles. Reflect their cut edges and remove any underlying fat and connective tissue. Study figures 54.11 and 54.13, and then locate and dissect the following deep muscles of the shoulder and back:

   **supraspinatus**
   **infraspinatus**
   **teres major**
   **rhomboid**
   **rhomboid capitis**
   **splenius**

5. See table 54.4 for the origins, insertions, and actions of these muscles in the shoulder and back.
6. Review the locations, origins, insertions, and actions designated by the laboratory instructor without the aid of the figures and tables. **A1 A2**
7. Complete Part C of the laboratory report.

**FIGURE 54.11** Superficial muscles of the cat's shoulder and back (*right side*); deep muscles of the shoulder and back (*left side*). AP|R

**FIGURE 54.12** Muscles of the cat's shoulder, back, forelimb, and hindlimb.
(©McGraw-Hill Education/APR)

**TABLE 54.4   Muscles of the Shoulder and Back**

| Muscle | Origin | Insertion | Action |
|---|---|---|---|
| Clavotrapezius | Occipital bone of skull and spines of cervical vertebrae | Clavicle | Pulls clavicle upward |
| Acromiotrapezius | Spines of cervical and thoracic vertebrae | Spine and acromion process of scapula | Pulls scapula upward |
| Spinotrapezius | Spines of thoracic vertebrae | Spine of scapula | Pulls scapula upward and back |
| Levator scapulae ventralis | Occipital bone of skull | Ventral border of scapula | Pulls scapula toward head |
| Clavodeltoid | Clavicle | Ulna near semilunar notch | Flexes forelimb |
| Acromiodeltoid | Acromion process of scapula | Proximal end of humerus | Flexes and rotates forelimb |
| Spinodeltoid | Spine of scapula | Proximal end of humerus | Flexes and rotates forelimb |
| Latissimus dorsi | Thoracic and lumbar vertebrae | Proximal end of humerus | Pulls forelimb upward and back |
| Supraspinatus | Fossa above spine of scapula | Greater tubercle of humerus | Extends forelimb |
| Infraspinatus | Fossa below spine of scapula | Greater tubercle of humerus | Rotates forelimb |
| Teres major | Posterior border of scapula | Proximal end of humerus | Rotates forelimb |
| Rhomboid | Spines of cervical and thoracic vertebrae | Medial border of scapula | Pulls scapula back |
| Rhomboid capitis | Occipital bone of skull | Medial border of scapula | Pulls scapula forward |
| Splenius | Fascia of neck | Occipital bone | Turns and raises head |

**FIGURE 54.13**   Deep muscles of the cat's back.   (©Dr. Sheril Burton)

**PRACTICE**

### PROCEDURE G—Muscles of the Forelimb

1. Place the cat in the dissecting tray with its ventral surface up.
2. Remove any remaining fat and connective tissue from a forelimb to expose the muscles.
3. Study figures 54.5 and 54.14a, and then locate and dissect the following muscles from the medial surface of the upper forelimb:

   **epitrochlearis** (well developed in cats)
   **triceps brachii**

4. Transect the pectoral muscles (pectoantebrachialis, pectoralis major, and pectoralis minor) and reflect their cut edges.

5. Study figures 54.14b and 54.15, and then locate and dissect the following muscles:

   **coracobrachialis**
   **biceps brachii**
   **long head of the triceps brachii**
   **medial head of the triceps brachii**

6. Study figures 54.16a and 54.17, which show the lateral surface of the upper forelimb.
7. Locate the lateral head of the triceps muscle, transect it, and reflect its cut edges.
8. Study figure 54.16b. Locate and dissect the following deeper muscles:

   **anconeus**
   **brachialis**

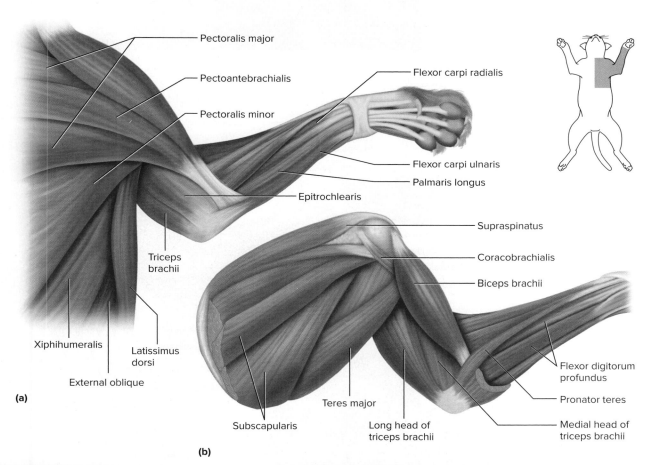

**FIGURE 54.14** (a) Superficial muscles of the cat's left medial shoulder and forelimb. (b) Deep muscles of the left medial shoulder and forelimb. AP|R

Clavodeltoid (reflected)
Epitrochlearis (reflected)
Biceps brachii
Coracobrachialis
Triceps brachii (long head)
Epitrochlearis (reflected)
Latissimus dorsi

**FIGURE 54.15** Deep muscles of the cat's right medial forelimb. (©Dr. Sheril Burton)

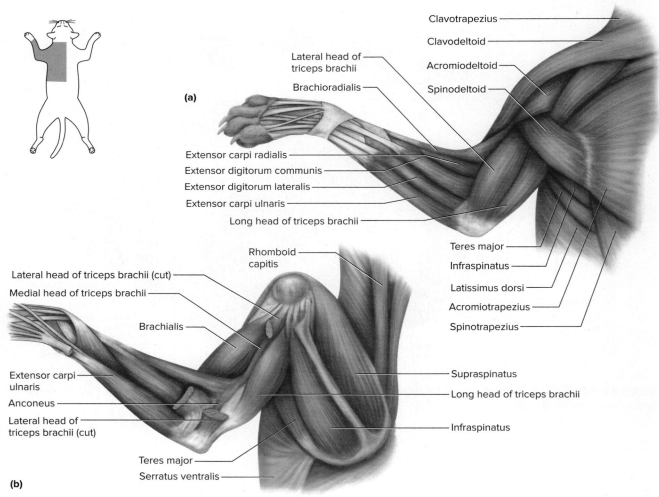

**(a)**

Clavotrapezius
Clavodeltoid
Acromiodeltoid
Lateral head of triceps brachii
Spinodeltoid
Brachioradialis

Extensor carpi radialis
Extensor digitorum communis
Extensor digitorum lateralis
Extensor carpi ulnaris
Long head of triceps brachii

Teres major
Infraspinatus
Latissimus dorsi
Acromiotrapezius
Spinotrapezius

Rhomboid capitis
Lateral head of triceps brachii (cut)
Medial head of triceps brachii
Brachialis
Extensor carpi ulnaris
Anconeus
Lateral head of triceps brachii (cut)
Teres major
Serratus ventralis

Supraspinatus
Long head of triceps brachii
Infraspinatus

**(b)**

**FIGURE 54.16** (*a*) Superficial muscles of the cat's left lateral shoulder and forelimb. (*b*) Deep muscles of the left lateral shoulder and forelimb.

**9.** Separate the muscles in the lower forelimb into an *extensor group* (fig. 54.16*a*) and a *flexor group* (see fig. 54.14*a*). Generally, the extensors are located on the posterior surface of the lower forelimb, and the flexors are located on the anterior surface. A thick, tough layer of fascia (antebrachial fascia) surrounds the forelimb. Remove this fascia to expose and dissect the following muscles:

**brachioradialis**
**extensor carpi radialis**
**extensor digitorum communis**
**extensor digitorum lateralis**

**extensor carpi ulnaris**
**flexor carpi ulnaris**
**palmaris longus**
**flexor carpi radialis**
**pronator teres**

**10.** Transect the flexor carpi ulnaris, palmaris longus, and flexor carpi radialis to expose the flexor digitorum profundus.

**11.** See table 54.5 for the origins, insertions, and actions of these muscles of the forelimb.

**12.** Complete Part D of the laboratory report.

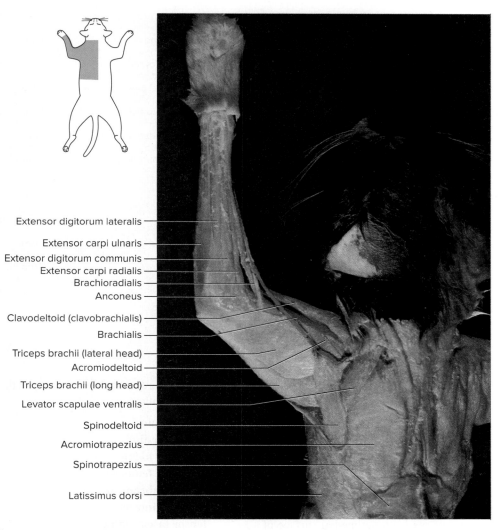

Extensor digitorum lateralis
Extensor carpi ulnaris
Extensor digitorum communis
Extensor carpi radialis
Brachioradialis
Anconeus
Clavodeltoid (clavobrachialis)
Brachialis
Triceps brachii (lateral head)
Acromiodeltoid
Triceps brachii (long head)
Levator scapulae ventralis
Spinodeltoid
Acromiotrapezius
Spinotrapezius
Latissimus dorsi

**FIGURE 54.17** Superficial muscles of the cat's left lateral shoulder and forelimb, dorsal view. (©Dr. Sheril Burton)

TABLE 54.5   Muscles of the Forelimb

| Muscle | Origin | Insertion | Action |
|---|---|---|---|
| Coracobrachialis | Coracoid process of scapula | Proximal end of humerus | Adducts forelimb |
| Epitrochlearis | Fascia of latissimus dorsi | Olecranon process of ulna | Extends forelimb |
| Biceps brachii | Border of glenoid cavity of scapula | Radial tuberosity of radius | Flexes forelimb |
| Triceps brachii | | | |
|   Lateral head | Deltoid tuberosity of humerus | Olecranon process of ulna | Extends forelimb |
|   Long head | Border of glenoid cavity of scapula | Olecranon process of ulna | Extends forelimb |
|   Medial head | Shaft of humerus | Olecranon process of ulna | Extends forelimb |
| Anconeus | Lateral epicondyle of humerus | Olecranon process of ulna | Extends forelimb |
| Brachialis | Lateral surface of humerus | Proximal end of ulna | Flexes forelimb |
| Brachioradialis | Shaft of humerus | Distal end of radius | Supinates lower forelimb |
| Extensor carpi radialis | Lateral surface of humerus | Second and third metacarpals | Extends wrist |
| Extensor digitorum communis | Lateral surface of humerus | Digits two to five | Extends digits |
| Extensor digitorum lateralis | Lateral surface of humerus | Digits two to five | Extends digits |
| Extensor carpi ulnaris | Lateral epicondyle of humerus and ulna | Fifth metacarpal | Extends wrist |
| Flexor carpi ulnaris | Medial epicondyle of humerus | Carpals | Flexes wrist |
| Palmaris longus | Medial epicondyle of humerus | Digits | Flexes digits |
| Flexor carpi radialis | Medial epicondyle of humerus | Second and third metacarpals | Flexes metacarpals |
| Pronator teres | Medial epicondyle of humerus | Radius | Pronates lower forelimb |
| Flexor digitorum profundus | Radius, ulna, and medial epicondyle of humerus | Distal phalanges | Flexes digits |

 **PRACTICE**

**PROCEDURE H—Muscles of the Hip and Hindlimb**

1. Place the cat in the dissecting tray with its ventral surface up.
2. Remove any remaining fat and connective tissue from the hip and hindlimb to expose the muscles.
3. Study figures 54.18*a* and 54.19, and then locate and dissect the following muscles from the medial surface of the thigh:

   **sartorius**
   **gracilis**

4. Using scissors, transect the sartorius and gracilis, and reflect their cut edges to observe the deeper muscles of the thigh.

5. Study figures 54.18*b* and 54.20, and then locate and dissect the following muscles:

   **rectus femoris**
   **vastus medialis**
   **adductor longus**
   **adductor femoris**
   **semimembranosus**
   **semitendinosus**
   **tensor fasciae latae**

6. Transect the tensor fasciae latae and rectus femoris muscles and turn their ends aside. Locate the *vastus intermedius* and *vastus lateralis* muscles beneath (fig. 54.21). (*Note*: In some specimens, the vastus intermedius, vastus lateralis, and vastus medialis muscles are closely united by connective tissue and are difficult to separate.)

**FIGURE 54.18** (*a*) Superficial muscles of the cat's left medial hindlimb. (*b*) Deep muscles of the left medial hindlimb.
**AP|R**

In figure (a), labels include:
- Rectus abdominis
- Adductor longus
- Adductor femoris
- Gracilis
- Semitendinosus
- External oblique
- Vastus medialis
- Sartorius

In figure (b), labels include:
- Adductor femoris
- Semimembranosus
- Semitendinosus
- Tensor fasciae latae
- Rectus femoris
- Adductor longus
- Vastus medialis

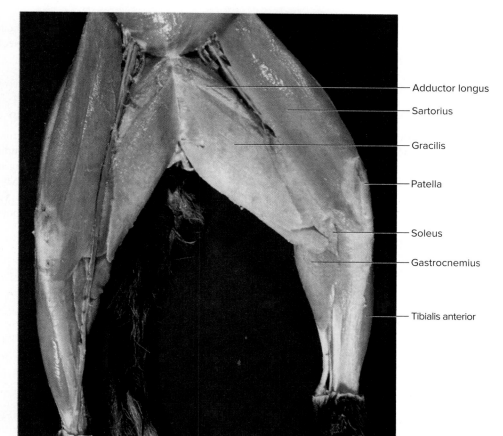

**FIGURE 54.19**
Superficial muscles of the cat's hindlimb, medial view. (©Dr. Sheril Burton)

Labels:
- Adductor longus
- Sartorius
- Gracilis
- Patella
- Soleus
- Gastrocnemius
- Tibialis anterior

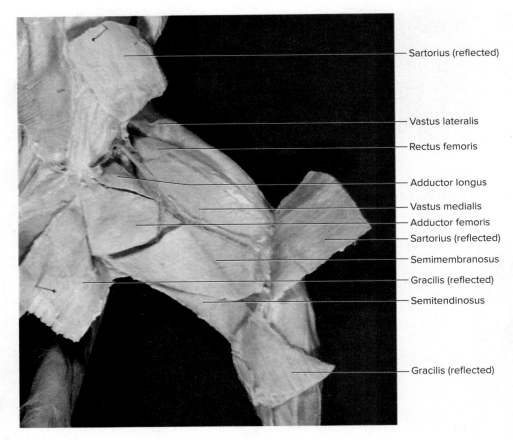

- Sartorius (reflected)
- Vastus lateralis
- Rectus femoris
- Adductor longus
- Vastus medialis
- Adductor femoris
- Sartorius (reflected)
- Semimembranosus
- Gracilis (reflected)
- Semitendinosus
- Gracilis (reflected)

**FIGURE 54.20** Deep muscles of the cat's left hindlimb, medial view. (©Dr. Sheril Burton)

**7.** Study figures 54.12, 54.22*a*, and 54.23, and then locate and dissect the following muscles from the lateral surface of the hip and thigh:

**biceps femoris**
**caudofemoralis**
**gluteus maximus** (small in cats)

**8.** Using scissors, transect the tensor fasciae latae and biceps femoris, and reflect their cut edges to observe the deeper muscles of the thigh.

**9.** Locate and dissect the following muscles:

**tenuissimus**
**gluteus medius**

(*Note:* The tenuissimus muscle and the sciatic nerve may adhere tightly to the dorsal surface of the biceps femoris muscle. Take care to avoid cutting these structures.)

**10.** On the lateral surface of the lower hindlimb (figs. 54.22 and 54.24), locate and dissect the following muscles:

**gastrocnemius**
**soleus**
**tibialis anterior**
**fibularis (peroneus) group**

**11.** See table 54.6 for the origins, insertions, and actions of these muscles of the hip and hindlimb.

**12.** Review the locations, origins, insertions, and actions designated by the laboratory instructor without the aid of the figures and tables. **1** **2**

**13.** Complete Part E of the laboratory report.

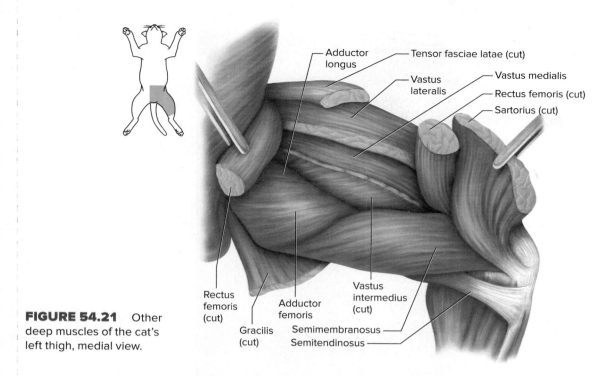

**FIGURE 54.21** Other deep muscles of the cat's left thigh, medial view.

Adductor longus

Tensor fasciae latae (cut)

Vastus lateralis

Vastus medialis

Rectus femoris (cut)

Sartorius (cut)

Rectus femoris (cut)

Gracilis (cut)

Adductor femoris

Vastus intermedius (cut)

Semimembranosus

Semitendinosus

Tensor fasciae latae

Gluteus maximus

Caudofemoralis

Vastus lateralis

Tensor fasciae latae (cut)

Gluteus medius

Gluteus maximus

Sartorius

Biceps femoris

Biceps femoris (cut)

Semitendinosus

Gastrocnemius

Soleus

Tibialis anterior

Extensor digitorum longus

Adductor femoris

Semitendinosus

Semimembranosus

Tenuissimus

Fibularis (peroneus) group

(a)

(b)

**FIGURE 54.22** (a) Superficial muscles of the cat's left lateral hip and hindlimb. (b) Deep muscles of the left lateral hip and hindlimb. **AP|R**

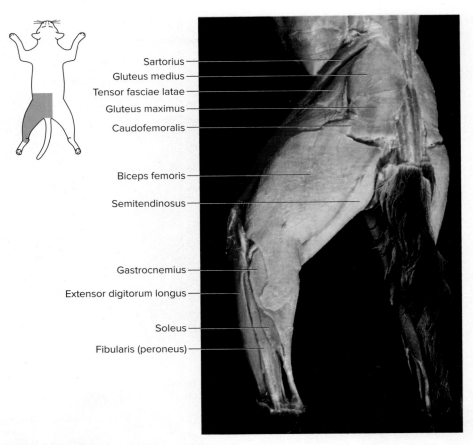

Sartorius

Gluteus medius

Tensor fasciae latae

Gluteus maximus

Caudofemoralis

Biceps femoris

Semitendinosus

Gastrocnemius

Extensor digitorum longus

Soleus

Fibularis (peroneus)

**FIGURE 54.23** Muscles of the cat's left hip and hindlimb, lateral view. (©Dr. Sheril Burton)

Gluteus medius

Tensor fasciae latae

Sartorius

Gluteus maximus

Caudofemoralis (reflected)

Sciatic nerve

Biceps femoris (reflected)

Semimembranosus

Tenuissimus

Semitendinosus

Gastrocnemius

Soleus

Fibularis (peroneus) tertius

Extensor digitorum longus

**FIGURE 54.24** Deep muscles of the cat's left hindlimb, lateral view. (©McGraw-Hill Education/APR)

**TABLE 54.6  Muscles of the Hip and Hindlimb**

| Muscle | Origin | Insertion | Action |
|---|---|---|---|
| Sartorius | Crest of ilium | Fascia of knee and proximal end of tibia | Rotates and extends hindlimb |
| Gracilis | Symphysis pubis and ischium | Proximal end of tibia | Adducts hindlimb |
| Quadriceps femoris | | | |
| Vastus lateralis | Shaft of femur and greater trochanter | Patella | Extends hindlimb |
| Vastus intermedius | Shaft of femur | Patella | Extends hindlimb |
| Rectus femoris | Ilium | Patella | Extends hindlimb |
| Vastus medialis | Shaft of femur | Patella | Extends hindlimb |
| Adductor longus | Pubis | Proximal end of femur | Adducts hindlimb |
| Adductor femoris | Pubis and ischium | Shaft of femur | Adducts hindlimb |
| Semimembranosus | Ischial tuberosity | Medial epicondyle of femur | Extends thigh |
| Tensor fasciae latae | Ilium | Fascia of thigh | Extends thigh |
| Biceps femoris | Ischial tuberosity | Patella and tibia | Abducts thigh and flexes lower hindlimb |
| Tenuissimus | Second caudal vertebra | Tibia and fascia of biceps femoris | Abducts thigh and flexes lower hindlimb |
| Semitendinosus | Ischial tuberosity | Crest of tibia | Flexes lower hindlimb |
| Caudofemoralis | Caudal vertebrae | Patella | Abducts thigh and extends lower hindlimb |
| Gluteus maximus | Sacral and caudal vertebrae | Greater trochanter of femur | Abducts thigh |
| Gluteus medius | Ilium, and sacral and caudal vertebrae | Greater trochanter of femur | Abducts thigh |
| Gastrocnemius | Lateral and medial epicondyles of femur | Calcaneus | Extends foot |
| Soleus | Proximal end of fibula | Calcaneus | Extends foot |
| Tibialis anterior | Proximal end of tibia and fibula | First metatarsal | Flexes foot |
| Fibularis (peroneus) group | Shaft of fibula | Metatarsals | Flexes foot |

# Notes

438

Name _____

Date _____

Section _____

The Ⓐ corresponds to the indicated Learning Outcome(s) found at the beginning of the laboratory exercise.

LABORATORY
REPORT

54

# Cat Dissection: Musculature

## 🔄 PART A ASSESSMENTS

Complete the following statements:

1. The _____ muscle of the human is homologous to the sternomastoid muscle of the cat. Ⓐ

2. The _____ muscle elevates the mandible in the human and in the cat. Ⓐ Ⓐ

3. Two muscles of the cat that are inserted on the hyoid bone are the _____ and the

   _____ . Ⓐ Ⓐ

## 🔄 PART B ASSESSMENTS

Complete the following:

1. Name two muscles that are found in the thoracic wall of the cat but are absent in the human. Ⓐ Ⓐ

   _____        _____

2. Name two muscles that are found in the thoracic wall of the cat and the human. Ⓐ

   _____        _____

3. Name four muscles that are found in the abdominal wall of the cat and the human. Ⓐ

   _____        _____

   _____        _____

## 🔄 PART C ASSESSMENTS

Complete the following:

1. Name three muscles of the cat that together correspond to the trapezius muscle in the human. Ⓐ Ⓐ

   _____

   _____

   _____

2. Name three muscles of the cat that together correspond to the deltoid muscle in the human. Ⓐ Ⓐ

   _____

   _____

   _____

3. Name the muscle in the cat and in the human that occupies the fossa above the spine of the scapula. Ⓐ Ⓐ

   _____

**4.** Name the muscle in the cat and in the human that occupies the fossa below the spine of the scapula.

_____

**5.** Name two muscles found in the cat and in the human that can rotate the forelimb.

_____   _____

## PART D ASSESSMENTS

Complete the following:

**1.** Name two muscles found in the cat and in the human that can flex the forelimb.

_____   _____

**2.** Name a muscle found in the cat but absent in the human that can extend the forelimb.

_____

**3.** Name a muscle that has three heads, can extend the forelimb, and is found in the cat and in the human.

_____

## PART E ASSESSMENTS

Observe the muscles on the human torso model and the upper and lower limb models. Each of the muscles in figure 54.25 is found both in the cat and in the human. Identify each of the numbered muscles in the figure by placing its name in the space next to its number.

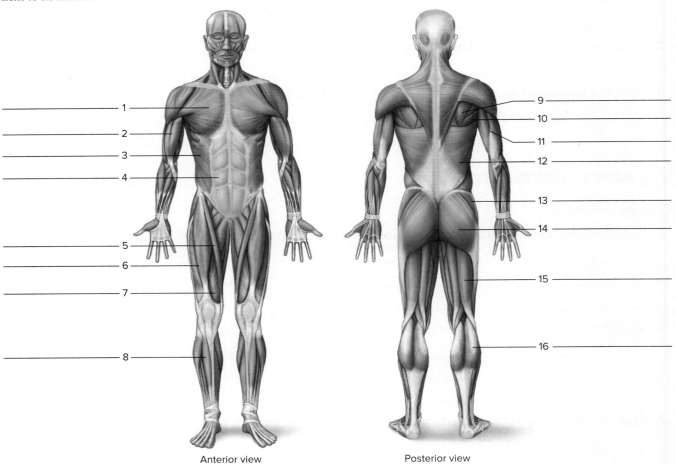

Anterior view          Posterior view

**FIGURE 54.25**  Identify the numbered muscles that occur in both the cat and the human.

# Cat Dissection: Cardiovascular System

## MATERIALS NEEDED

Preserved cat
Dissecting tray
Dissecting instruments
Disposable gloves
Human heart model
Human torso model

## ⚠ SAFETY

- Wear disposable gloves when working on the cat dissection.
- Dispose of tissue remnants and gloves as instructed.
- Wash the dissecting tray and instruments as instructed.
- Wash your laboratory table.
- Wash your hands before leaving the laboratory.

## PURPOSE OF THE EXERCISE

To examine and identify the major organs of the cardiovascular system of the cat and to compare them with the corresponding organs of the human torso model.

##  LEARNING OUTCOMES AP|R

After completing this exercise, you should be able to

1. Locate and identify the major organs of the cardiovascular system of the cat.

2. Compare the features of the cardiovascular system of the cat with those of the human.

3. Identify the corresponding cardiovascular organs of the human.

In this laboratory exercise, you will dissect the major organs of the cardiovascular system of the cat. As before, while you are examining the organs of the cat, compare them with the corresponding organs of the human torso model.

If the cardiovascular system of the cat has been injected, the systemic arteries will be filled with red latex, and the systemic veins will be filled with blue latex. This will make it easier for you to trace the vessels as you dissect them.

##  PRACTICE

### PROCEDURE A—The Arterial System

1. Place the preserved cat in the dissecting tray with its ventral surface up.
2. Open the thoracic cavity, and expose its contents. To do this, follow these steps:
   a. Make a longitudinal incision passing anteriorly from the diaphragm along one side of the sternum. Continue the incision through the neck muscles to the mandible. Try to avoid damaging the internal organs as you cut.
   b. Make a lateral cut on each side along the anterior surface of the diaphragm, and cut the diaphragm loose from the thoracic wall.
   c. Spread the sides of the thoracic wall outward, and use a scalpel to make a longitudinal cut along each side of the inner wall of the rib cage to weaken the ribs. Continue to spread the thoracic wall laterally to break the ribs so that the flaps of the wall will remain open (figs. 55.1 and 55.2).
3. Note the location of the heart and the large blood vessels associated with it. Slit the outer *parietal pericardium* that surrounds the heart by cutting with scissors along the midventral line. Note how this membrane is connected to the *visceral pericardium* that is attached to the surface of the heart. Locate the *pericardial cavity,* the space between the two layers of the pericardium.
4. Examine the heart and closely associated blood vessels (figs. 55.1–55.5). The arrangement of blood vessels coming off the aortic arch is different in the cat than in the human body. Locate the following:

**right atrium**
**left atrium**
**right ventricle**
**left ventricle**
**pulmonary trunk**
**aorta**
**coronary arteries**

**FIGURE 55.1** Arteries of the cat's thorax, neck, and forelimb. **AP|R**

Right common carotid artery

Trachea

Brachial artery

Axillary artery

Right subclavian artery

Brachiocephalic artery

Right atrium

Right ventricle

Left common carotid artery

Vertebral artery

Thyrocervical artery

Left subclavian artery

Aortic arch

Left atrium

Left ventricle

Diaphragm

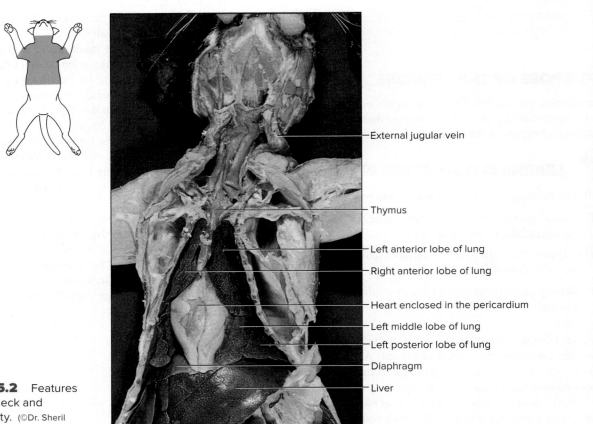

External jugular vein

Thymus

Left anterior lobe of lung

Right anterior lobe of lung

Heart enclosed in the pericardium

Left middle lobe of lung

Left posterior lobe of lung

Diaphragm

Liver

**FIGURE 55.2** Features of the cat's neck and thoracic cavity. (©Dr. Sheril Burton)

External jugular vein

Vagus nerve

Nerves of the
brachial plexus

Brachiocephalic vein

Lung

Right ventricle

Larynx

Trachea

Left common carotid artery

Brachial vein

External jugular vein

Brachial artery

Subscapular vein

Axillary vein

Thymus gland

Pericardium (cut)

Left ventricle

Coronary artery

**FIGURE 55.3**    Blood vessels of the cat's upper forelimb and neck.  (©Dr. Sheril Burton)

5. Use a scalpel to open the heart chambers by making a cut along the frontal plane from its apex to its base. Remove any remaining latex from the chambers. Examine the valves between the chambers, and note the relative thicknesses of the chamber walls (see fig. 55.4). Do not remove the heart, as it is needed for future examination of the relationship between major blood vessels. Compare the heart structures of the cat with the model of the human heart.

6. Using figures 55.1, 55.3–55.5 as guides, locate and dissect the following arteries of the thorax and neck:

**aortic arch**
**brachiocephalic artery** (on right only; first branch of the aortic arch)
**right subclavian artery**

**left subclavian artery**
**right common carotid artery**
**left common carotid artery**

7. Trace the right subclavian artery into the forelimb, and locate the following arteries:

**vertebral artery**
**thyrocervical artery**
**subscapular artery**
**axillary artery**
**brachial artery**

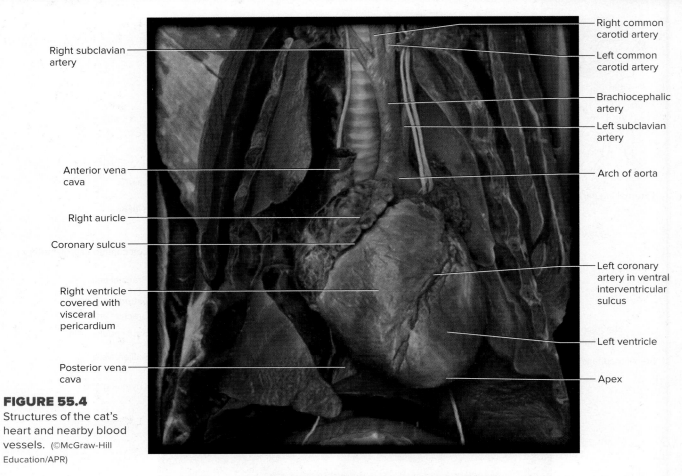

Right subclavian artery

Anterior vena cava

Right auricle

Coronary sulcus

Right ventricle covered with visceral pericardium

Posterior vena cava

Right common carotid artery

Left common carotid artery

Brachiocephalic artery

Left subclavian artery

Arch of aorta

Left coronary artery in ventral interventricular sulcus

Left ventricle

Apex

**FIGURE 55.4**
Structures of the cat's heart and nearby blood vessels. (©McGraw-Hill Education/APR)

Subclavian vein

Axillary vein

Brachiocephalic vein

Anterior vena cava

Azygos vein

Esophagus

Right ventricle

Posterior vena cava

Thoracic aorta

Diaphragm

Common carotid artery

External jugular vein

Subscapular vein

Axillary vein

Left subclavian artery

Pericardium

Left atrium

Left ventricle

Apex of the heart

**FIGURE 55.5**  Blood vessels of the cat's neck and thorax. (©Dr. Sheril Burton)

8. Open the abdominal cavity. To do this, follow these steps:
   a. Use scissors to make an incision through the body wall along the midline from the symphysis pubis to the diaphragm.
   b. Make a lateral incision through the body wall along either side of the inferior border of the diaphragm and along the bases of the thighs.
   c. Reflect the flaps created in the body wall as you would open a book, and expose the contents of the abdominal cavity.
   d. Note the *parietal peritoneum* that forms the inner lining of the abdominal wall. Also note the *greater omentum*, a structure composed of a double layer of peritoneum that hangs from the border of the stomach and covers the lower abdominal organs like a fatty apron (fig. 55.6). Reflect the abdominal organs to one side to expose various blood vessels in the abdominopelvic cavity region.
9. As you expose and dissect blood vessels, try not to destroy other visceral organs needed for future studies.

Using figures 55.7–55.9 as guides, locate and dissect the following arteries of the abdomen:

**abdominal aorta** (unpaired)
**celiac artery** (unpaired)
**hepatic artery** (unpaired)
**gastric artery** (unpaired)
**splenic artery** (unpaired)
**anterior mesenteric artery** (corresponds to superior mesenteric artery; unpaired)
**renal arteries** (paired)
**adrenolumbar arteries** (paired)
**gonadal arteries** (paired)
**posterior mesenteric artery** (corresponds to inferior mesenteric artery; unpaired)
**iliolumbar arteries** (paired)
**external iliac arteries** (paired)
**internal iliac arteries** (paired)
**caudal artery** (unpaired)

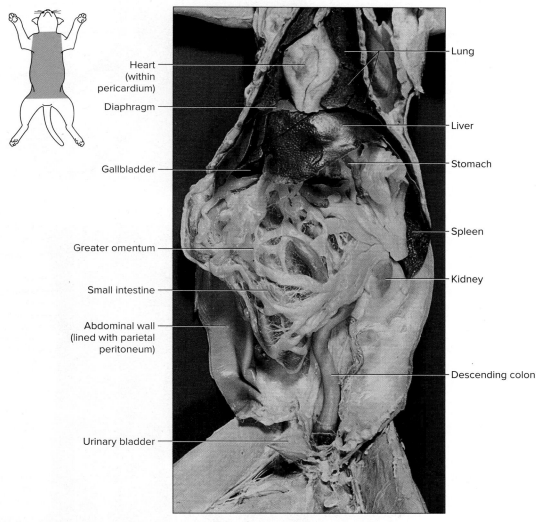

Heart (within pericardium)
Diaphragm
Gallbladder
Greater omentum
Small intestine
Abdominal wall (lined with parietal peritoneum)
Urinary bladder

Lung
Liver
Stomach
Spleen
Kidney
Descending colon

**FIGURE 55.6**  Organs of the cat's thoracic and abdominal cavities. (©Dr. Sheril Burton)

Right adrenolumbar artery

Abdominal aorta

External iliac artery

Femoral artery

Diaphragm

Celiac artery

Anterior mesenteric artery

Renal artery

Gonadal artery

Posterior mesenteric artery

Iliolumbar artery

Internal iliac artery

Caudal artery

Deep femoral artery

**FIGURE 55.7** Arteries of the cat's abdomen and hindlimb. **AP** **R**

Spleen

Pancreas

Small intestine

Large intestine

Diaphragm

Thoracic aorta

Stomach

Gastric artery

Hepatic artery

Celiac artery

Splenic artery

Anterior mesenteric artery

Renal artery

Gonadal artery

Abdominal aorta

Posterior mesenteric artery

External iliac artery

**FIGURE 55.8** Arteries of the cat's abdomen, left lateral view.

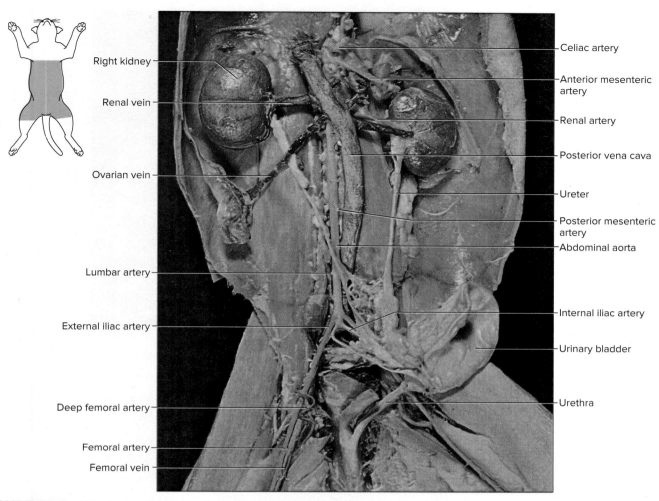

Right kidney
Renal vein
Ovarian vein
Lumbar artery
External iliac artery
Deep femoral artery
Femoral artery
Femoral vein

Celiac artery
Anterior mesenteric artery
Renal artery
Posterior vena cava
Ureter
Posterior mesenteric artery
Abdominal aorta
Internal iliac artery
Urinary bladder
Urethra

**FIGURE 55.9** Blood vessels of the cat's abdominal cavity, with the digestive organs removed. (©Dr. Sheril Burton)

10. Trace the external iliac artery into the right hindlimb (fig. 55.10), and locate the following:

   **femoral artery**
   **deep femoral artery**

11. Review the locations of the heart structures and arteries of the cat without the aid of the figures. **1**

12. Examine the human torso model along with figures 55.1, 55.5, 55.7, and 55.9. Identify the arteries of the cat that correspond with those of the human. **3**

13. Complete Part A of Laboratory Report 55.

 **PRACTICE**

   **PROCEDURE B—The Venous System**

1. Examine the heart again, and locate the following veins:

   **anterior vena cava** (corresponds to superior vena cava)
   **posterior vena cava** (corresponds to inferior vena cava)
   **pulmonary veins**

2. Using figures 55.3, 55.5, and 55.11 as guides, locate and dissect the following veins in the thorax and neck:

   **right brachiocephalic vein**
   **left brachiocephalic vein**
   **right subclavian vein**
   **left subclavian vein**
   **internal jugular vein**
   **external jugular vein**

3. Trace the right subclavian vein into the forelimb, and locate the following veins:

   **axillary vein**
   **subscapular vein**
   **brachial vein**

4. Using figures 55.9, 55.12, and 55.13 as guides, locate and dissect the following veins in the abdomen:

   **posterior vena cava**
   **adrenolumbar vein**
   **anterior mesenteric vein** (corresponds to superior mesenteric vein)

External iliac vein

Femoral vein

External iliac artery

Deep femoral artery

Femoral artery

**FIGURE 55.10** Blood vessels of the cat's upper hindlimb. (©Dr. Sheril Burton)

Right internal jugular vein

Subscapular vein

Brachial vein

Long thoracic vein

Axillary vein

Right subclavian vein

Right brachiocephalic vein

Left external jugular vein

Left subclavian vein

Left brachiocephalic vein

Sternal vein

Anterior vena cava

Posterior vena cava

Diaphragm

**FIGURE 55.11** Veins of the cat's thorax, neck, and forelimb. **AP|R**

**FIGURE 55.12** Veins of the cat's abdomen and hindlimb. AP|R

Labels on the figure:
- Diaphragm
- Left adrenolumbar vein
- Right adrenolumbar vein
- Renal vein
- Posterior vena cava
- Left gonadal vein
- Iliolumbar vein
- Right gonadal vein
- Common iliac vein
- External iliac vein
- Internal iliac vein
- Deep femoral vein
- Femoral vein

**posterior mesenteric vein** (corresponds to inferior mesenteric vein)
**hepatic portal vein**
**splenic vein**
**renal vein**
**gonadal vein**
**iliolumbar vein**
**common iliac vein**
**internal iliac vein**
**external iliac vein**

5. Trace the external iliac vein into the hindlimb (figs. 55.10 and 55.12), and locate the following veins:
   **femoral vein**
   **deep femoral vein**

6. Review the locations of the veins of the cat without the aid of the figures. **1**

7. Examine the human torso model along with figures 55.5, 55.9, 55.11, and 55.12. Identify the veins of the cat that correspond with those of the human. **3**

8. Complete Part B of the laboratory report.

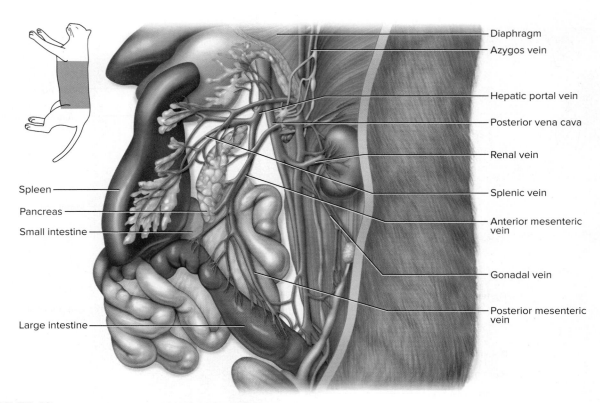

Spleen

Pancreas

Small intestine

Large intestine

Diaphragm

Azygos vein

Hepatic portal vein

Posterior vena cava

Renal vein

Splenic vein

Anterior mesenteric vein

Gonadal vein

Posterior mesenteric vein

**FIGURE 55.13**    Veins of the cat's abdomen and hindlimb.

Name _____

Date _____

Section _____

The  corresponds to the indicated Learning Outcome(s) found at the beginning of the laboratory exercise.

LABORATORY
REPORT

55

# Cat Dissection:
# Cardiovascular System

## PART A ASSESSMENTS

Complete the following:

**1.** Describe the position and attachments of the parietal pericardium of the heart of the cat.  _____

_____

_____

_____

**2.** Describe the relative thicknesses of the walls of the heart chambers of the cat.  _____

_____

_____

_____

**3.** Explain how the wall thicknesses are related to the functions of the chambers.  _____

_____

_____

_____

_____

_____

**4.** Compare the origins of the common carotid arteries of the cat with those of the human.  _____

_____

_____

_____

_____

_____

**5.** Compare the origins of the iliac arteries of the cat with those of the human.  _____

_____

_____

_____

_____

Complete the following:

1. Compare the origins of the brachiocephalic veins of the cat with those of the human. **2** _____

_____

_____

_____

2. Compare the relative sizes of the external and internal jugular veins of the cat with those of the human. **2** _____

_____

_____

_____

_____

_____

3. List twelve veins that cats and humans have in common. **2** **3** _____

_____

_____

_____

_____

_____

# Cat Dissection: Digestive System

## PURPOSE OF THE EXERCISE

To examine and identify the major digestive organs of the cat and to compare these organs with those of the human.

 **LEARNING OUTCOMES** AP|R

After completing this exercise, you should be able to

(1) Locate and identify the major digestive organs of the cat.

(2) Compare the digestive system of the cat with that of the human.

(3) Identify the corresponding organs in the human torso model.

In this laboratory exercise, you will dissect the major digestive organs of the cat. As you observe these organs, compare them with those of the human by observing the parts of the human torso model; however, keep in mind that the cat is a carnivore (flesh-eating mammal) and that the organs of its digestive system are adapted to capturing, holding, eating, and digesting the bodies of other animals. In contrast, humans are omnivores (eat both plant and animal substances). Because humans are adapted to eating and digesting a greater variety of foods, comparisons of the cat and the human digestive organs may not be precise.

 **PRACTICE**

### PROCEDURE—Digestive System Dissection

1. Place the preserved cat in the dissecting tray on its left side.
2. Locate the major salivary glands on one side of the head (fig. 56.1). To do this, follow these steps:
   a. Clear away any remaining fascia and other connective tissue from the region below the ear and near the joint of the mandible.
   b. Identify the *parotid gland,* a relatively large mass of glandular tissue just below the ear. Note the parotid duct (Stensen's duct) that passes over the surface of the masseter muscle and opens into the mouth.
   c. Look for the *submandibular gland* just below the parotid gland, near the angle of the jaw.
   d. Locate the *sublingual gland* that is adjacent and medial to the submandibular gland.
3. Open the oral cavity. To do this, follow these steps:
   a. Use scissors to cut through the soft tissues at the angle of the mouth.
   b. When you reach the bone of the jaw, use a bone cutter to cut through the bone, thus freeing the mandible.

**c.** Open the mouth wide, and locate the following features:

**cheek**

**lip**

**vestibule**

**palate**

hard palate (has transverse ridges to help hold food)

soft palate (lacks a uvula)

**palatine tonsils** (small, rounded masses of glandular tissue in the lateral wall of the soft palate)

**tongue**

frenulum

papillae (examine with a hand lens; spiny projections used to clean fur)

**4.** Examine the teeth of the upper jaw. The adult cat has 6 incisors, 2 canines, 8 premolars, and 2 molars. The lower jaw teeth are similar, except 4 premolars are present.

**5.** Reflect some of the structures of the neck and the thoracic cavity to one side. Look for the *esophagus* in the neck deep to the trachea, or deep to the heart in the thoracic cavity. This ventral view of the cat is also another way to observe the salivary glands (fig. 56.2).

**6.** Complete Part A of Laboratory Report 56.

**7.** Examine organs in the abdominal cavity with the cat positioned with its ventral surface up. Review the normal position and function of the greater omentum.

**8.** Examine the *liver,* which is located just beneath the diaphragm and is attached to the central portion of the diaphragm by the *falciform ligament.* Also, locate the *spleen,* which is posterior to the liver on the left side (fig. 56.3). The liver has five lobes—a right medial, right lateral (which is subdivided into two parts by a deep cleft), left medial, left lateral, and caudate lobe. The caudate lobe is the smallest lobe and is located in the median line, where it projects into the curvature of the stomach. The caudate lobe is covered by a sheet of mesentery, the *lesser omentum,* that connects the liver to the stomach (figs. 56.4 and 56.5). Find the greenish *gallbladder* on the inferior surface of the liver

**FIGURE 56.1** Major salivary glands of the cat, lateral view.

on the right side. Also note the *cystic duct,* by which the gallbladder is attached to the *bile duct,* and the *hepatic duct,* which originates in the liver and attaches to the cystic duct. Trace the bile duct to its connection with the duodenum.

9. Locate the *stomach* on the left side of the abdominal cavity. At its anterior end, note the union with the esophagus, which passes through the diaphragm. Identify the *cardia, fundus, body,* and *pyloric part* (listed from entrance to exit) of the stomach. Use scissors to make an incision along the convex border of the

stomach from the cardia to the pyloric part. The lining of the stomach has numerous *gastric folds (rugae).* Examine the *pyloric sphincter,* which creates a constriction between the stomach and small intestine.

10. Locate the *pancreas.* It appears as a grayish, two-lobed, elongated mass of glandular tissue. One lobe lies dorsal to the stomach and extends across to the duodenum. The other lobe is enclosed by the *mesentery,* which supports the duodenum (figs. 56.4 and 56.5).

11. Trace the *small intestine,* beginning at the pyloric sphincter. The first portion, the *duodenum,* travels

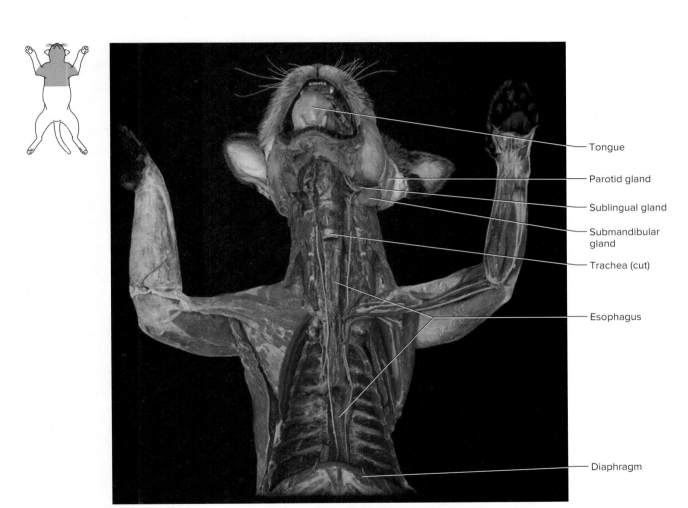

**FIGURE 56.2**  Ventral view of the cat's head, neck, and thoracic cavity with many structures removed in order to observe some deep digestive structures.  (©McGraw-Hill Education/APR)

posteriorly for several centimeters. Then it loops back around a lobe of the pancreas. The proximal half of the remaining portion of the small intestine is the *jejunum,* and the distal half is the *ileum.* Open the duodenum and note the velvety appearance of the villi. Note how the mesentery supports the small intestine from the dorsal body wall. The small intestine terminates on the right side, where it joins the large intestine.

12. Locate the *large intestine,* and identify the *cecum, ascending colon, transverse colon,* and *descending*

*colon.* Also locate the *rectum,* which extends through the pelvic cavity to the *anus.* Make an incision at the junction between the ileum and cecum, and look for the *ileocecal sphincter.* An appendix is absent in cats.

13. Review the locations of the digestive organs without the aid of the figures. 〈1〉

14. Examine the human torso model along with figures 56.1–56.5. For the human digestive organs, identify corresponding digestive organs of the cat. 〈3〉

15. Complete Part B of the laboratory report.

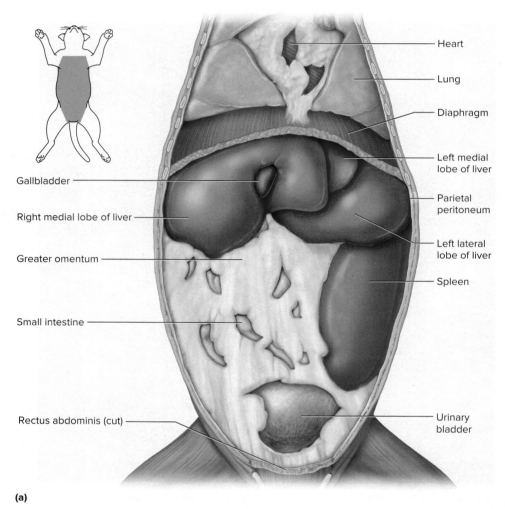

**(a)**

**FIGURE 56.3** Ventral view of abdominal organs of the cat: (*a*) illustration; (*b*) dissected cat with greater omentum removed.
**AP|R** (©McGraw-Hill Education/APR)

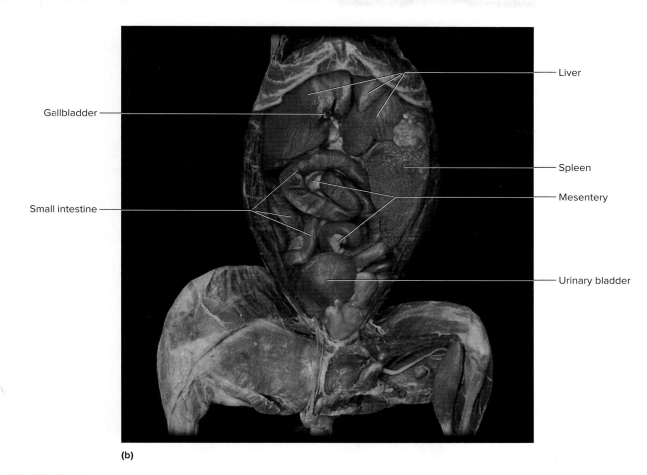

Gallbladder

Small intestine

Liver

Spleen

Mesentery

Urinary bladder

(b)

**FIGURE 56.3** *Continued.*

Right medial lobe of liver

Bile duct

Lesser omentum

Right lateral lobe of liver

Duodenum

Pancreas

Transverse colon

Small intestine

Mesentery

Descending colon

Gallbladder

Diaphragm

Left medial lobe of liver

Left lateral lobe of liver

Esophagus

Cardia

Fundus

Body

Pyloric part

Stomach

Spleen

Kidney

Parietal peritoneum

Rectum

Urinary bladder

**FIGURE 56.4** Ventral view of the cat's abdominal organs with intestines reflected to the right side.

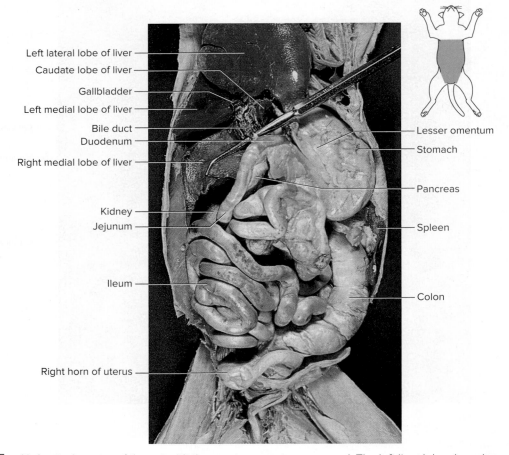

Left lateral lobe of liver

Caudate lobe of liver

Gallbladder

Left medial lobe of liver

Bile duct

Duodenum

Right medial lobe of liver

Kidney

Jejunum

Ileum

Right horn of uterus

Lesser omentum

Stomach

Pancreas

Spleen

Colon

**FIGURE 56.5** Abdominal organs of the cat with the greater omentum removed. The left liver lobes have been reflected to the right side. (©Dr. Sheril Burton)

Name _____

Date _____

Section _____

The 🄰 corresponds to the indicated Learning Outcome(s) found at the beginning of the laboratory exercise.

LABORATORY
REPORT

# Cat Dissection: Digestive System

## 🔄 PART A ASSESSMENTS

Complete the following:

1. Compare the locations of the major salivary glands of the human with those of the cat. 🄰2 _____

_____

_____

_____

_____

2. Compare the types and numbers of teeth present in the cat's upper and lower jaws with those of the human. 🄰2 _____

_____

_____

_____

_____

3. In what ways do the cat's teeth seem to be adapted to a special diet? 🄰1 _____

_____

_____

_____

_____

4. What part of the human soft palate is lacking in the cat? 🄰2 _____

5. What do you think is the function of the transverse ridges (rugae) in the hard palate of the cat? 🄰1 _____

_____

_____

_____

6. How do the papillae on the surface of the human tongue compare with those of the cat? 🄰2 _____

_____

_____

_____

Complete the following:

1. Describe how the peritoneum and mesenteries are associated with the organs in the abdominal cavity.  ⓐ _____

_____

_____

_____

_____

_____

_____

2. Describe the inner lining of the stomach.  ⓐ _____

_____

_____

_____

_____

3. Compare the structure of the human liver with that of the cat.  ② _____

_____

_____

_____

_____

4. Compare the structure and location of the human pancreas with those of the cat.  ② _____

_____

_____

_____

_____

5. What feature of the human cecum is lacking in the cat?  ② _____

_____

# 57

# Cat Dissection: Respiratory System

## MATERIALS NEEDED

Preserved cat
Dissecting tray
Dissecting instruments
Disposable gloves
Human torso model

## ⚠ SAFETY

- Wear disposable gloves when working on the cat dissection.
- Dispose of tissue remnants and gloves as instructed.
- Wash the dissecting tray and instruments as instructed.
- Wash your laboratory table.
- Wash your hands before leaving the laboratory.

## PURPOSE OF THE EXERCISE

To examine and identify the major respiratory organs of the cat and to compare these organs with those of the human.

##  LEARNING OUTCOMES AP|R

After completing this exercise, you should be able to

1. Locate and identify the major respiratory organs of the cat.
2. Compare the respiratory system of the human with that of the cat.
3. Identify the corresponding organs in the human torso model.

In this laboratory exercise, you will dissect the major respiratory organs of a cat. As you observe these structures in the cat, compare them with those of the human torso model.

##  PRACTICE

### PROCEDURE—Respiratory System Dissection

1. Place the preserved cat in a dissecting tray with its ventral surface up.
2. Examine the *nostrils* (*external nares*) and *nasal septum.*
3. Open the oral cavity wide to expose the *hard palate* and the *soft palate* (fig. 57.1). Cut through the tissues of the soft palate, and observe the *nasopharynx* above it. Locate the small openings of the *auditory tubes* in the lateral walls of the nasopharynx. Insert a probe into an opening of an auditory tube.
4. Pull the tongue forward, and locate the *epiglottis* near its base. Also identify the *glottis,* an opening between the *vocal folds* (*vocal cords*) into the larynx. Locate the *esophagus,* which is dorsal to the larynx (figs. 57.1 and 57.2).
5. Open the thoracic cavity, and expose its contents.
6. Dissect the *trachea* in the neck, and expose the *larynx* at the anterior end near the base of the tongue. Note the *tracheal rings,* and locate the lobes of the *thyroid gland* on each side of the trachea, just posterior to the larynx. Also note the *thymus gland,* a rather diffuse mass of glandular tissue extending along the ventral surface of the trachea into the thorax (figs. 57.1 and 57.2).
7. Examine the larynx, and identify the *thyroid cartilage* and *cricoid cartilage* in its wall. Make a longitudinal incision through the ventral wall of the larynx, and locate the vocal folds (vocal cords) inside.
8. Trace the trachea posteriorly to where it divides into *main (primary) bronchi,* which pass into the lungs (fig. 57.3).
9. Examine the *lungs,* each of which is subdivided into three lobes—an *anterior (cranial),* a *middle,* and a *posterior (caudal) lobe.* The right lung of the cat has a small fourth *accessory (mediastinal; cardiac)* lobe associated with the right posterior lobe. Notice the thin membrane, the *visceral pleura,* on the surface of each lung. Also

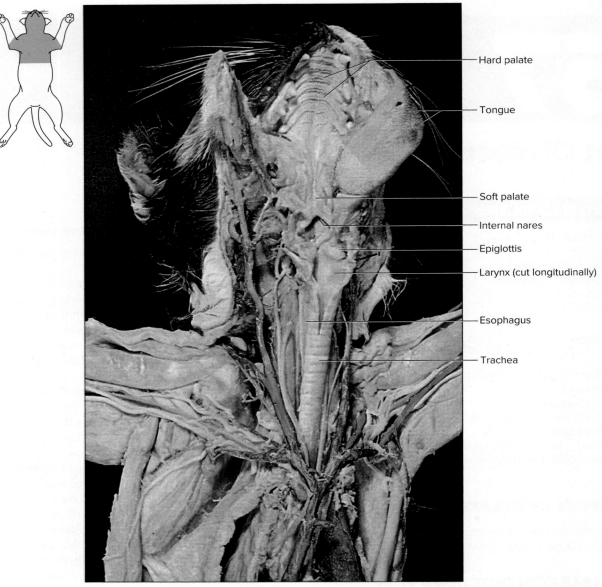

Hard palate

Tongue

Soft palate

Internal nares

Epiglottis

Larynx (cut longitudinally)

Esophagus

Trachea

**FIGURE 57.1** Features of the cat's neck and oral cavity, ventral view. **AP|R** (©Dr. Sheril Burton)

notice the *parietal pleura,* which forms the inner lining of the thoracic wall, and locate the spaces of the *pleural cavities* (see fig. 57.2).

10. Make an incision through a lobe of a lung, and examine its interior. Note the branches of the smaller air passages. The main bronchi branch from the trachea and enter the lungs. Each main bronchus divides into *lobar bronchi*; lobar bronchi divide into even smaller *segmental bronchi* (fig. 57.3).

11. Examine the *diaphragm,* and note that its central portion is tendinous.

12. Locate the *phrenic nerve.* This nerve appears as a white thread passing along the side of the heart to the diaphragm.

13. Examine the organs located in the *mediastinum.* Note that the mediastinum separates the right and left lungs and pleural cavities within the thorax.

14. Review the locations of the respiratory organs without the aid of the figures. **A**

15. Examine the human torso model along with figures 57.1–57.3. For each human respiratory organ, identify the corresponding respiratory organ of the cat. **3**

16. Complete Laboratory Report 57.

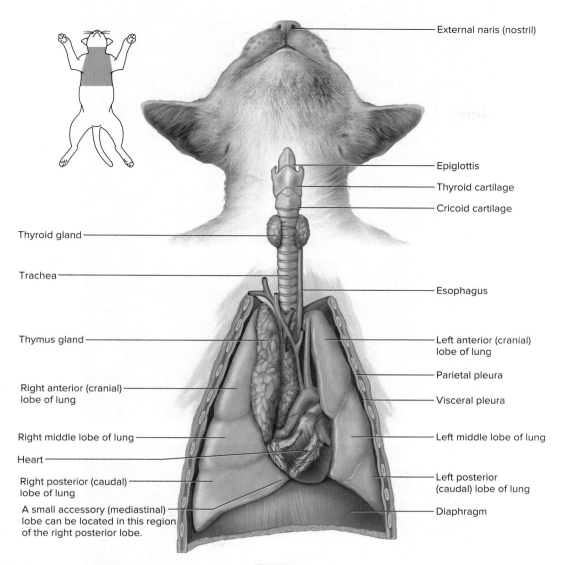

External naris (nostril)

Epiglottis

Thyroid cartilage

Cricoid cartilage

Thyroid gland

Trachea

Esophagus

Thymus gland

Left anterior (cranial) lobe of lung

Parietal pleura

Right anterior (cranial) lobe of lung

Visceral pleura

Right middle lobe of lung

Left middle lobe of lung

Heart

Right posterior (caudal) lobe of lung

Left posterior (caudal) lobe of lung

A small accessory (mediastinal) lobe can be located in this region of the right posterior lobe.

Diaphragm

**FIGURE 57.2**   Ventral view of the cat's thoracic cavity. **AP|R**

**FIGURE 57.3** Ventral view of head, neck, and thoracic cavity of the cat: (*a*) lungs and heart included; (*b*) heart and lungs removed. ((*a,b*) ©McGraw-Hill Education/APR)

The A corresponds to the indicated Learning Outcome(s) found at the beginning of the laboratory exercise.

# Cat Dissection: Respiratory System

Complete the following:

1. Why do the auditory tubes open into the nasopharynx? A _____

_____

_____

_____

2. Distinguish between the glottis and the epiglottis. A _____

_____

_____

_____

3. Are the tracheal rings of the cat complete or incomplete? A _____ How does this

feature compare with that of the human? 2 _____

_____

4. How does the structure of the primary bronchi compare with that of the trachea? A _____

_____

_____

5. Compare the number of lobes in the human lungs with the number of lobes in the cat. 2 _____

_____

_____

_____

_____

6. Describe the attachments of the diaphragm. A _____

_____

_____

7. What major structures are located within the mediastinum of the cat? A _____

_____

_____

8. Compare the organs located within the mediastinum of the cat with those within the human mediastinum. 2 _____

_____

_____

# Notes

# 58

# Cat Dissection: Urinary System

## MATERIALS NEEDED

Preserved cat
Dissecting tray
Dissecting instruments
Disposable gloves
Human torso model

## SAFETY

- Wear disposable gloves when working on the cat dissection.
- Dispose of tissue remnants and gloves as instructed.
- Wash the dissecting tray and instruments as instructed.
- Wash your laboratory table.
- Wash your hands before leaving the laboratory.

## PURPOSE OF THE EXERCISE

To examine and identify the urinary organs of the cat and to compare them with those of the human.

 **LEARNING OUTCOMES** AP|R

After completing this exercise, you should be able to

① Locate and identify the urinary organs of the cat.

② Compare the urinary organs of the cat with those of the human.

③ Identify the corresponding organs in the human torso model.

In this laboratory exercise, you will dissect the urinary organs of the cat. As you observe these structures, compare them with the corresponding human organs by observing the parts of the human torso model.

 **PRACTICE**

### PROCEDURE—Urinary System Dissection

1. Place the preserved cat in a dissecting tray with its ventral surface up.
2. Open the abdominal cavity, and remove the liver, stomach, and spleen (figs. 58.1 and 58.2).
3. Push the intestines to one side, and locate the *kidneys* in the dorsal abdominal wall on either side of the vertebral column. The kidneys are located dorsal to the *parietal peritoneum* (retroperitoneal).
4. Carefully remove the parietal peritoneum and the adipose tissue surrounding the kidneys. Locate the following, using figures 58.1 and 58.2 as guides:

   **ureters**
   **renal arteries**
   **renal veins**

5. Locate the *adrenal glands,* which lie medially and anteriorly to the kidneys and are surrounded by connective tissues. The adrenal glands are attached to blood vessels and are separated by the abdominal aorta and posterior vena cava.
6. Expose the ureters by cleaning away the connective tissues along their lengths. They enter the *urinary bladder* on the dorsal surface.
7. Examine the urinary bladder, attached to the abdominal wall by folds of peritoneum that form ligaments. Locate the *medial ligament* on the ventral surface of the bladder that connects it with the linea alba and the *two lateral ligaments* on the dorsal surface.
8. Remove the connective tissue from around the urinary bladder. Use a sharp scalpel to open the bladder, and examine its interior. Locate the openings of the ureters and the urethra on the inside.
9. Expose the *urethra* at the posterior of the urinary bladder. The urethra terminates with the vagina in a urogenital sinus in the female cat; the urethra enters the penis in the male cat (figs. 58.1–58.3).

**Renal artery**

**Adipose tissue**

**Left kidney**

**Ureter**

**Abdominal aorta**

**Urinary bladder (cut)**

**Urethra**

Renal vein

Posterior vena cava

Ureter

Penis

**FIGURE 58.1** Urinary system of the male cat's abdominal cavity, with most of the digestive and some reproductive organs removed. (©McGraw-Hill Education/APR)

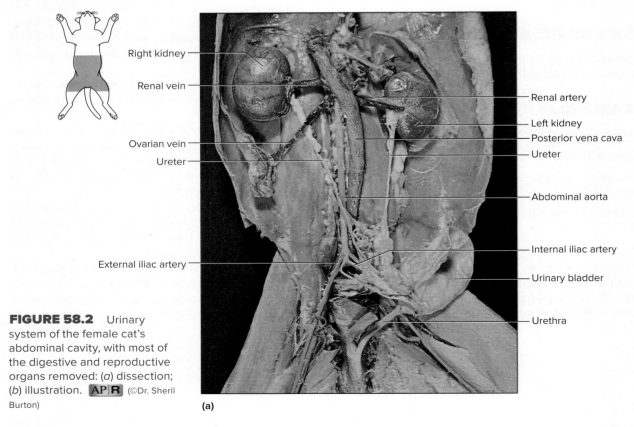

Right kidney

Renal vein

Ovarian vein

Ureter

External iliac artery

**Renal artery**

**Left kidney**

**Posterior vena cava**

**Ureter**

**Abdominal aorta**

**Internal iliac artery**

**Urinary bladder**

**Urethra**

**FIGURE 58.2** Urinary system of the female cat's abdominal cavity, with most of the digestive and reproductive organs removed: (*a*) dissection; (*b*) illustration. **AP|R** (©Dr. Sheril Burton)

(a)

Adrenal glands

Renal artery

Kidney

Renal vein

Posterior vena cava

Abdominal aorta

Ureters

Urinary bladder

Rectum

Urethra

**FIGURE 58.2** *Continued.* **(b)**

Kidney

Renal pelvis

Renal cortex

Renal medulla

Ovary

Ureter

Left uterine horn
(contains embryo)

Body of uterus

Ureter

Urinary bladder

Vagina

Urethra

**FIGURE 58.3** Female urinary and reproductive systems of a pregnant cat. AP|R (©McGraw-Hill Education/Ralph W. Stevens III, Ph.d., photographer)

**10.** Remove one kidney, and section it longitudinally (figs. 58.3 and 58.4). Identify the following features:

**fibrous capsule**
**renal cortex** (superficial, lighter region)
**renal medulla** (deeper, darker region)
**renal pyramid** (single in cat)
**renal papilla** (single in cat)
**renal sinus**
**renal pelvis**
**hilum of kidney**

**11.** Review the locations of the urinary organs without the aid of the figures. ◢**1**

**12.** Discard the organs and tissues that were removed from the cat, as directed by the laboratory instructor.

**13.** Examine the human torso model along with figures 58.1–58.4. Identify urinary organs of the cat that correspond with those of the human. ◢**3**

**14.** Complete Laboratory Report 58.

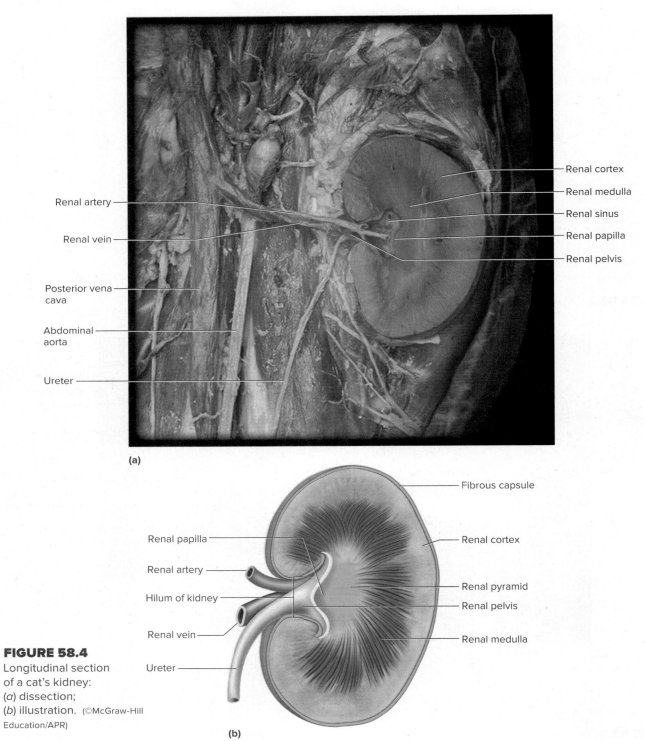

(a)

**FIGURE 58.4**
Longitudinal section of a cat's kidney:
(*a*) dissection;
(*b*) illustration. (©McGraw-Hill Education/APR)

(b)

The Ⓐ corresponds to the indicated Learning Outcome(s) found at the beginning of the laboratory exercise.

# Cat Dissection: Urinary System

Complete the following:

1. Compare the positions of the kidneys in the cat with those in the human. ②  _____
   _____
   _____
   _____
   _____
   _____

2. Compare the locations of the adrenal glands in the cat with those in the human. ②  _____
   _____
   _____
   _____

3. What structures of the cat urinary system are retroperitoneal? ①  _____
   _____
   _____

4. Describe the wall of the urinary bladder of the cat. ①  _____
   _____
   _____
   _____

5. Compare the renal pyramids and renal papillae of the human kidney with those of the cat. ②  _____
   _____
   _____
   _____
   _____

# Notes

472

# Cat Dissection: Reproductive Systems

## MATERIALS NEEDED

Preserved cat
Dissecting tray
Dissecting instruments
Bone cutters
Disposable gloves
Models of human reproductive systems

## ⚠ SAFETY

- Wear disposable gloves when working on the cat dissection.
- Dispose of tissue remnants and gloves as instructed.
- Wash the dissecting tray and instruments as instructed.
- Wash your laboratory table.
- Wash your hands before leaving the laboratory.

## PURPOSE OF THE EXERCISE

To examine and identify the reproductive organs of the cat and to compare them with the corresponding organs of the human.

##  LEARNING OUTCOMES AP|R

After completing this exercise, you should be able to

①  Locate and identify the reproductive organs of a cat.

②  Compare the reproductive organs of the cat with those of the human.

③  Identify the corresponding organs in models of the human reproductive systems.

I n this laboratory exercise, you will dissect the reproductive system of the cat. If you have a female cat, begin with Procedure A. If you have a male cat, begin with Procedure B. After completing the dissection, exchange cats with someone who has dissected one of the opposite sex, and examine its reproductive organs.

As you observe the cat reproductive organs, compare them with the corresponding human organs by examining the models of the human reproductive systems.

##  PRACTICE

### PROCEDURE A—Female Reproductive System

1. Place the preserved female cat in a dissecting tray with its ventral surface up, and open its abdominal cavity.
2. Locate the small, oval *ovaries* just posterior to the kidneys (figs. 59.1 and 59.2).
3. Examine an ovary, suspended from the dorsal body wall by a fold of peritoneum called the *mesovarium*. Another attachment, the *ovarian ligament,* connects the ovary to the tubular *uterine horn* (see fig. 59.1).
4. Near the anterior end of the ovary, locate the funnel-shaped *infundibulum,* which is at the end of a very small *uterine tube* (*oviduct*). Note the tiny projections, or *fimbriae,* that create a fringe around the edge of the infundibulum. Trace the uterine tube around the ovary to its nearby connection with the uterine horn near the attachment of the ovarian ligament.
5. Examine the uterine horn. It is suspended from the body wall by a fold of peritoneum, the *broad ligament,* that is continuous with the mesovarium of the ovary. Also note the fibrous *round ligament,* which extends from the uterine horn laterally and posteriorly to the body wall (figs. 59.1 and 59.2).
6. To observe the remaining organs of the reproductive system more easily, remove the left hindlimb by severing it near the hip with bone cutters. Then use the bone cutters to remove the left anterior portion of the pelvic girdle. Also remove the necessary muscles and connective tissue to expose the structures shown in figure 59.3.
7. Trace the relatively long uterine horns posteriorly. They unite to form the *uterine body,* which is located between the urethra and rectum. This Y-shaped uterus allows ample space for several offspring to develop at the same time in the uterine horns (see fig. 59.2). The uterine body is continuous with the *vagina,* which leads to the outside.

8. Trace the *urethra* from the urinary bladder posteriorly. The urethra and the vagina open into a common chamber, called the *urogenital sinus.* The opening of this chamber, ventral to the anus, is called the *urogenital aperture.* Locate the *labia majora,* which are folds of skin on either side of the urogenital aperture.

9. Use scissors to open the vagina along its lateral wall, beginning at the urogenital aperture and continuing to the body of the uterus. Note the *urethral orifice* in the ventral wall of the urogenital sinus, and locate the small, rounded

*cervix* of the uterus, which projects into the vagina at its deep end. The *clitoris* is located in the ventral wall of the urogenital sinus near its opening to the outside.

10. Review the locations of the reproductive organs without the aid of the figures. **1**

11. Examine the models of the human reproductive systems along with figures 59.1–59.3. Identify the reproductive organs of the cat that correspond with those of the human. **3**

12. Complete Part A of Laboratory Report 59.

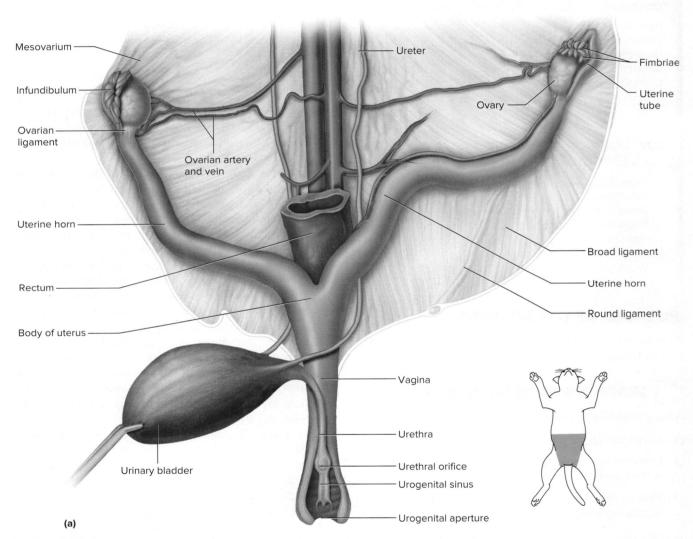

**(a)**

**FIGURE 59.1** Ventral view of the female cat's reproductive system: (*a*) illustration; (*b*) dissection. (©McGraw-Hill Education/APR)

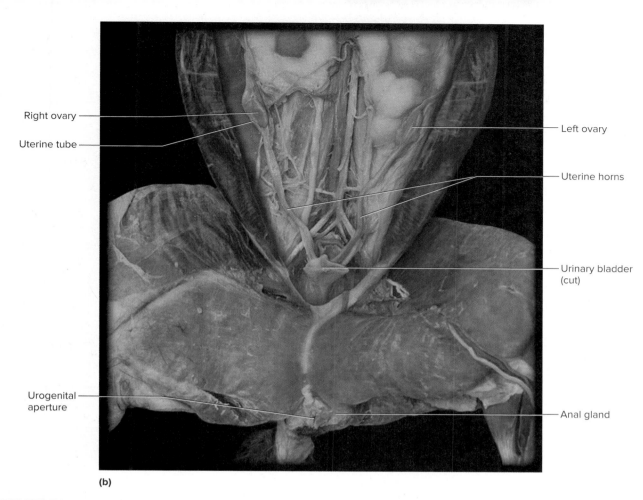

Right ovary

Uterine tube

Left ovary

Uterine horns

Urinary bladder (cut)

Urogenital aperture

Anal gland

(b)

**FIGURE 59.1** *Continued.*

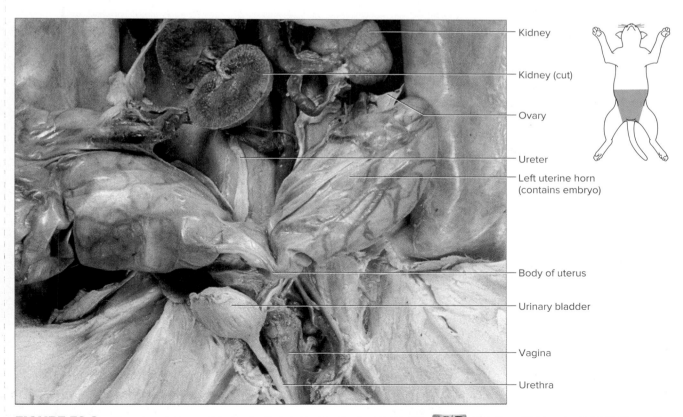

Kidney

Kidney (cut)

Ovary

Ureter

Left uterine horn (contains embryo)

Body of uterus

Urinary bladder

Vagina

Urethra

**FIGURE 59.2** Female urinary and reproductive systems of a pregnant cat. AP|R (©McGraw-Hill Education/Ralph W. Stevens III, Ph.d., photographer)

**475**

Colon

Urinary bladder

Ureter

Urethra

Vagina

Urogenital sinus

Urogenital aperture

Uterine horn

Uterine body

Rectum

Anal gland

Anus

**FIGURE 59.3** Lateral view of the female cat's reproductive system.

 **PRACTICE**

**PROCEDURE B—Male Reproductive System**

1. Place the preserved male cat in a dissecting tray with its ventral surface up.
2. Locate the *scrotum* between the hindlimbs. Make an incision along the midline of the scrotum, and expose the *testes* inside (fig. 59.4). The testes are separated by a septum, and each testis is enclosed in a sheath of connective tissue.
3. Remove the sheath surrounding the testes, and locate the convoluted *epididymis* on the dorsal surface of each testis (fig. 59.5).
4. Locate the *spermatic cord* on the right side, leading away from the testis. This spermatic cord contains the *ductus (vas) deferens,* which is continuous with the epididymis, as well as with the nerves and blood vessels that supply the testis on that side. Trace the spermatic cord to the body wall, where its contents pass through the *inguinal canal* and enter the pelvic cavity (fig. 59.5).
5. Locate the *penis,* and identify the *prepuce,* which forms a sheath around the entire penis. Make an incision through the skin of the prepuce, and expose the *glans penis.* The glans penis has minute spines on its surface that assist copulation.
6. To observe the remaining structures of the reproductive system more easily, remove the left hindlimb by severing it near the hip with bone cutters. Then use the bone cutters to remove the left anterior portion of the pelvic girdle. Also remove the necessary muscles and connective tissues to expose the organs shown in figure 59.6.
7. Trace the ductus deferens from the inguinal canal to the penis. Note that the ductus deferens loops over the ureter within the pelvic cavity and passes downward behind the urinary bladder to join the urethra. Locate the *prostate gland,* which appears as an enlargement at the junction of the ductus deferens and urethra. The prostate gland is relatively small compared to that in humans and positioned along the urethra some distance from the urinary bladder.

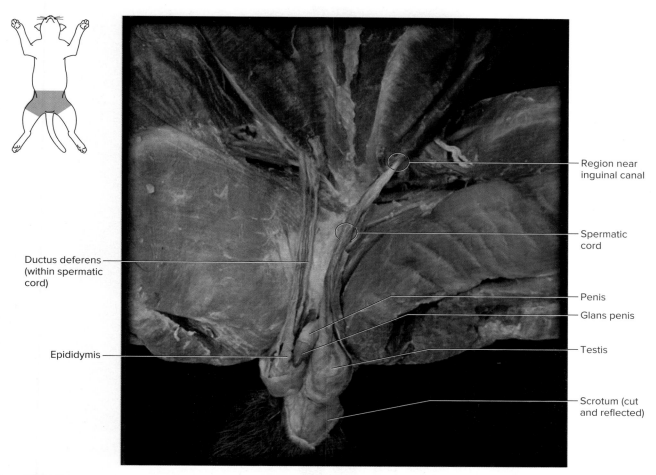

Ductus deferens
(within spermatic
cord)

Epididymis

Region near
inguinal canal

Spermatic
cord

Penis

Glans penis

Testis

Scrotum (cut
and reflected)

**FIGURE 59.4** Male reproductive system of the cat. **AP|R** (©McGraw-Hill Education/APR)

8. Trace the urethra to the penis. Locate the *bulbourethral glands,* which form small swellings on either side at the proximal end of the penis. Seminal vesicles are absent in cats.

9. Use a sharp scalpel to cut a transverse section of the penis. Identify the *corpora cavernosa,* each of which contains many blood spaces surrounded by a sheath of connective tissue.

10. Review the locations of the reproductive organs without the aid of the figures. **1**

11. Examine the models of the human reproductive systems along with figures 59.4–59.6. Identify the organs of the cat that correspond with those of the human. **3**

12. Complete Part B of the laboratory report.

**FIGURE 59.5** Ventral view of the male cat's reproductive system.

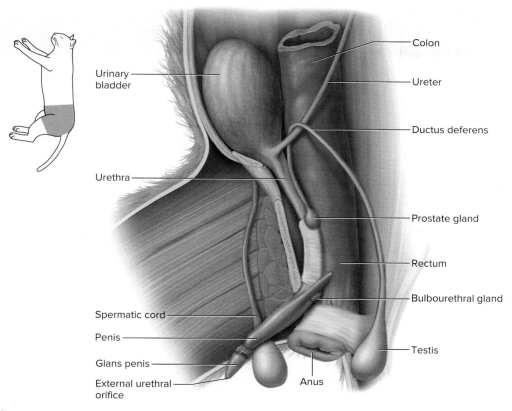

Urinary bladder

Urethra

Spermatic cord

Penis

Glans penis

External urethral orifice

Colon

Ureter

Ductus deferens

Prostate gland

Rectum

Bulbourethral gland

Testis

Anus

**FIGURE 59.6**    Lateral view of the male cat's reproductive system.

# Notes

Name _____

Date _____

Section _____

The ⚠ corresponds to the indicated Learning Outcome(s) found at the beginning of the laboratory exercise.

# Cat Dissection: Reproductive Systems

## ♻ PART A ASSESSMENTS

Complete the following:

1. Compare the relative lengths and paths of the uterine tubes (oviducts) of the cat and the human. ⚠2 _____

_____

_____

_____

_____

_____

2. How do the shape and structure of the uterus of the cat compare with that of the human? ⚠2 _____

_____

_____

_____

_____

_____

3. Assess the function of the uterine horns of the cat. ⚠1 _____

_____

_____

_____

_____

4. Compare the relationship of the urethra and the vagina in the cat and in the human. ⚠2 _____

_____

_____

_____

_____

Complete the following:

1. Compare the glans penis in the cat and in the human. 2 _____

_____

_____

_____

_____

2. How do the location and the relative size of the prostate gland of the cat compare with that of the human? 2

_____

_____

_____

_____

_____

3. What glands associated with the human ductus deferens are missing in the cat? 2 _____

_____

4. Compare the prepuce in the cat and in the human. 2 _____

_____

_____

Name_____

Date_____ Class_____

# Laboratory Safety Guidelines

*Carefully review the following safety guidelines before starting a lab exercise. Safety guideline reminders are also included in the appropriate lab exercise sections. Avoid tardiness to laboratory sessions because specific directions and precautions are often given at the beginning of the period. Your instructor should update you about safety procedures, as your school might have safety regulations or modifications in addition to those listed here.*

1. *Become familiar with all room exits and the location and operating procedures of all safety equipment (first aid kit, Material Safety Data Sheets (MSDS), eyewash station, safety shower, fire extinguisher, and fire blanket).*
2. *Read all laboratory exercises prior to starting the procedures. Do not work alone in the laboratory.*
3. *Do not smoke, chew, eat, drink, apply cosmetics or lip balm, handle contact lenses (regular eyeglasses should be worn), or work with open wounds in the laboratory. Consider all materials you use as potentially hazardous.*
4. *Use protective eyewear and laboratory coats during labs when handling dangerous chemicals or when heating dangerous materials. Volatile materials should never be heated using an open flame. When heating materials in a test tube, never point the test tube in the direction of a person or leave heated materials unattended.*
5. *Use a commercially prepared disinfectant or a 10% bleach solution to clean laboratory work surfaces before and after any procedures when using animals, body fluids, and dissection specimens.*
6. *Wear enclosed shoes and disposable gloves when working with dangerous chemicals, body fluids, and dissection specimens.*
7. *If body fluids are being studied, work only with your own. Special precautions to prevent contact with body fluids may include wearing disposable gloves, gowns, aprons, or masks, as directed by your instructor.*
8. *Special precautions during dissections may include wearing laboratory coats and protective eyewear. Always use cutting blades away from yourself and laboratory partners.*
9. *Restrain all loose clothing, long hair, and dangling jewelry during laboratory procedures. Keep all unnecessary materials away from the work area to reduce clutter and the possibility of an accident.*
10. *If you have special needs, are taking medications, experience allergic reactions, are pregnant, or are uncomfortable with the procedures, inform, your instructor.*
11. *Use only a mechanical pipetting device (never your mouth).*
12. *Immediately report any accidents, spills, or damaged equipment to your instructor.*
13. *Use only disposable lancets and needles, and never attempt to bend, cut, or recap them when finished. Place sharp items in a puncture-resistant container that is marked "Biohazard Container."*
14. *Dispose of chemicals, waste material, body fluids, and dissection specimens according to appropriate directions. Any reusable glassware or utensils that have been contaminated with body fluids should be placed in a disinfectant (10% bleach solution) and later autoclaved.*
15. *Thoroughly wash your hands with soap and warm water immediately after removing disposable gloves and before leaving the laboratory.*
16. *Practice any modified or additional safety guidelines required by your instructor.*

# Periodic Table of the Elements

| 1 1A | | | | | | | | | | | | | | | | | 18 8A |
|---|---|---|---|---|---|---|---|---|---|---|---|---|---|---|---|---|---|
| **1**<br>**H**<br>Hydrogen<br>1.008 | 2 2A | | | | | | | | | | | 13 3A | 14 4A | 15 5A | 16 6A | 17 7A | **2**<br>**He**<br>Helium<br>4.003 |
| **3**<br>**Li**<br>Lithium<br>6.941 | **4**<br>**Be**<br>Beryllium<br>9.012 | | | | | | | | | | | **5**<br>**B**<br>Boron<br>10.81 | **6**<br>**C**<br>Carbon<br>12.01 | **7**<br>**N**<br>Nitrogen<br>14.01 | **8**<br>**O**<br>Oxygen<br>16.00 | **9**<br>**F**<br>Fluorine<br>19.00 | **10**<br>**Ne**<br>Neon<br>20.18 |
| **11**<br>**Na**<br>Sodium<br>22.99 | **12**<br>**Mg**<br>Magnesium<br>24.31 | 3 3B | 4 4B | 5 5B | 6 6B | 7 7B | 8 | 9 ─8B─ | 10 | 11 1B | 12 2B | **13**<br>**Al**<br>Aluminum<br>26.98 | **14**<br>**Si**<br>Silicon<br>28.09 | **15**<br>**P**<br>Phosphorus<br>30.97 | **16**<br>**S**<br>Sulfur<br>32.07 | **17**<br>**Cl**<br>Chlorine<br>35.45 | **18**<br>**Ar**<br>Argon<br>39.95 |
| **19**<br>**K**<br>Potassium<br>39.10 | **20**<br>**Ca**<br>Calcium<br>40.08 | **21**<br>**Sc**<br>Scandium<br>44.96 | **22**<br>**Ti**<br>Titanium<br>47.88 | **23**<br>**V**<br>Vanadium<br>50.94 | **24**<br>**Cr**<br>Chromium<br>52.00 | **25**<br>**Mn**<br>Manganese<br>54.94 | **26**<br>**Fe**<br>Iron<br>55.85 | **27**<br>**Co**<br>Cobalt<br>58.93 | **28**<br>**Ni**<br>Nickel<br>58.69 | **29**<br>**Cu**<br>Copper<br>63.55 | **30**<br>**Zn**<br>Zinc<br>65.39 | **31**<br>**Ga**<br>Gallium<br>69.72 | **32**<br>**Ge**<br>Germanium<br>72.59 | **33**<br>**As**<br>Arsenic<br>74.92 | **34**<br>**Se**<br>Selenium<br>78.96 | **35**<br>**Br**<br>Bromine<br>79.90 | **36**<br>**Kr**<br>Krypton<br>83.80 |
| **37**<br>**Rb**<br>Rubidium<br>85.47 | **38**<br>**Sr**<br>Strontium<br>87.62 | **39**<br>**Y**<br>Yttrium<br>88.91 | **40**<br>**Zr**<br>Zirconium<br>91.22 | **41**<br>**Nb**<br>Niobium<br>92.91 | **42**<br>**Mo**<br>Molybdenum<br>95.94 | **43**<br>**Tc**<br>Technetium<br>(98) | **44**<br>**Ru**<br>Ruthenium<br>101.1 | **45**<br>**Rh**<br>Rhodium<br>102.9 | **46**<br>**Pd**<br>Palladium<br>106.4 | **47**<br>**Ag**<br>Silver<br>107.9 | **48**<br>**Cd**<br>Cadmium<br>112.4 | **49**<br>**In**<br>Indium<br>114.8 | **50**<br>**Sn**<br>Tin<br>118.7 | **51**<br>**Sb**<br>Antimony<br>121.8 | **52**<br>**Te**<br>Tellurium<br>127.6 | **53**<br>**I**<br>Iodine<br>126.9 | **54**<br>**Xe**<br>Xenon<br>131.3 |
| **55**<br>**Cs**<br>Cesium<br>132.9 | **56**<br>**Ba**<br>Barium<br>137.3 | **57**<br>**La**<br>Lanthanum<br>138.9 | **72**<br>**Hf**<br>Hafnium<br>178.5 | **73**<br>**Ta**<br>Tantalum<br>180.9 | **74**<br>**W**<br>Tungsten<br>183.9 | **75**<br>**Re**<br>Rhenium<br>186.2 | **76**<br>**Os**<br>Osmium<br>190.2 | **77**<br>**Ir**<br>Iridium<br>192.2 | **78**<br>**Pt**<br>Platinum<br>195.1 | **79**<br>**Au**<br>Gold<br>197.0 | **80**<br>**Hg**<br>Mercury<br>200.6 | **81**<br>**Tl**<br>Thallium<br>204.4 | **82**<br>**Pb**<br>Lead<br>207.2 | **83**<br>**Bi**<br>Bismuth<br>209.0 | **84**<br>**Po**<br>Polonium<br>(210) | **85**<br>**At**<br>Astatine<br>(210) | **86**<br>**Rn**<br>Radon<br>(222) |
| **87**<br>**Fr**<br>Francium<br>(223) | **88**<br>**Ra**<br>Radium<br>(226) | **89**<br>**Ac**<br>Actinium<br>(227) | **104**<br>**Rf**<br>Rutherfordium<br>(257) | **105**<br>**Db**<br>Dubnium<br>(260) | **106**<br>**Sg**<br>Seaborgium<br>(263) | **107**<br>**Bh**<br>Bohrium<br>(262) | **108**<br>**Hs**<br>Hassium<br>(265) | **109**<br>**Mt**<br>Meitnerium<br>(266) | **110**<br>**Ds**<br>Darmstadtium<br>(269) | **111**<br>**Rg**<br>Roentgenium<br>(272) | **112**<br>**Cn**<br>Copernicium<br>(285) | **113**<br>**Nh**<br>Nihonium<br>(286) | **114**<br>**Fl**<br>Flerovium<br>(289) | **115**<br>**Mc**<br>Moscovium<br>(289) | **116**<br>**Lv**<br>Livermorium<br>(293) | **117**<br>**Ts**<br>Tennessine<br>(294) | **118**<br>**Og**<br>Oganesson<br>(294) |

Atomic number — **9 F** Fluorine 19.00 — Atomic mass

Metals / Metalloids / Nonmetals

| **58**<br>**Ce**<br>Cerium<br>140.1 | **59**<br>**Pr**<br>Praseodymium<br>140.9 | **60**<br>**Nd**<br>Neodymium<br>144.2 | **61**<br>**Pm**<br>Promethium<br>(147) | **62**<br>**Sm**<br>Samarium<br>150.4 | **63**<br>**Eu**<br>Europium<br>152.0 | **64**<br>**Gd**<br>Gadolinium<br>157.3 | **65**<br>**Tb**<br>Terbium<br>158.9 | **66**<br>**Dy**<br>Dysprosium<br>162.5 | **67**<br>**Ho**<br>Holmium<br>164.9 | **68**<br>**Er**<br>Erbium<br>167.3 | **69**<br>**Tm**<br>Thulium<br>168.9 | **70**<br>**Yb**<br>Ytterbium<br>173.0 | **71**<br>**Lu**<br>Lutetium<br>175.0 |
|---|---|---|---|---|---|---|---|---|---|---|---|---|---|
| **90**<br>**Th**<br>Thorium<br>232.0 | **91**<br>**Pa**<br>Protactinium<br>(231) | **92**<br>**U**<br>Uranium<br>238.0 | **93**<br>**Np**<br>Neptunium<br>(237) | **94**<br>**Pu**<br>Plutonium<br>(242) | **95**<br>**Am**<br>Americium<br>(243) | **96**<br>**Cm**<br>Curium<br>(247) | **97**<br>**Bk**<br>Berkelium<br>(247) | **98**<br>**Cf**<br>Californium<br>(249) | **99**<br>**Es**<br>Einsteinium<br>(254) | **100**<br>**Fm**<br>Fermium<br>(253) | **101**<br>**Md**<br>Mendelevium<br>(256) | **102**<br>**No**<br>Nobelium<br>(254) | **103**<br>**Lr**<br>Lawrencium<br>(257) |

The 1–18 group designation has been recommended by the International Union of Pure and Applied Chemistry (IUPAC).

Elements up to atomic number 118 have been confirmed and officially named.

Different sources may report different numbers of elements occurring naturally. According to Los Alamos National Laboratory (New Mexico, US), of elements 92 (Uranium) and below, only Promethium appears to be completely absent on earth outside of the particle physics laboratory. A few elements above and below Uranium are reportedly found naturally only in trace amounts on earth or in spectral analysis of certain stars. Including or excluding such elements would affect the stated total.

# APPENDIX 3

## Preparation of Solutions

### AMYLASE SOLUTION, 0.5%

Place 0.5 g of bacterial amylase in a graduated cylinder or volumetric flask. Add distilled water to the 100 mL level. Stir until dissolved. The amylase should be free of sugar for best results; a low-maltose solution of amylase yields good results. (Store amylase powder in a freezer until mixing this solution.)

### BENEDICT'S SOLUTION

Prepared solution is available from various suppliers.

### CAFFEINE, 0.2%

Place 0.2 g of caffeine in a graduated cylinder or volumetric flask. Add distilled water to the 100 mL level. Stir until dissolved.

### CALCIUM CHLORIDE, 2.0%

Place 2.0 g of calcium chloride in a graduated cylinder or volumetric flask. Add distilled water to the 100 mL level. Stir until dissolved.

### CALCIUM HYDROXIDE SOLUTION (LIMEWATER)

Add an excess of calcium hydroxide to 1 L of distilled water. Stopper the bottle and shake thoroughly. Allow the solution to stand for 24 hours. Pour the supernatant fluid through a filter. Store the clear filtrate in a stoppered container.

### EPSOM SALT SOLUTION, 0.1%

Place 0.5 g of Epsom salt in a graduated cylinder or volumetric flask. Add distilled water to the 500 mL level. Stir until dissolved.

### GLUCOSE SOLUTIONS

1. *1.0% solution.* Place 1 g of glucose in a graduated cylinder or volumetric flask. Add distilled water to the 100 mL level. Stir until dissolved.
2. *10% solution.* Place 10 g of glucose in a graduated cylinder or volumetric flask. Add distilled water to the 100 mL level. Stir until dissolved.

### IODINE-POTASSIUM-IODIDE (IKI SOLUTION)

Add 20 g of potassium iodide to 1 L of distilled water, and stir until dissolved. Then add 4.0 g of iodine, and stir again until dissolved. Solution should be stored in a dark, stoppered bottle.

### METHYLENE BLUE

Dissolve 0.3 g of methylene blue powder in 30 mL of 95% ethyl alcohol. In a separate container, dissolve 0.01 g of potassium hydroxide in 100 mL of distilled water. Mix the two solutions. (Prepared solution is available from various suppliers.)

### MONOSODIUM GLUTAMATE (MSG) SOLUTION, 1%

Place 1 g of monosodium glutamate in a graduated cylinder or volumetric flask. Add distilled water to the 100 mL level. Stir until dissolved.

### PHYSIOLOGICAL SALINE SOLUTION

Place 0.9 g of sodium chloride in a graduated cylinder or volumetric flask. Add distilled water to the 100 mL level. Stir until dissolved.

### POTASSIUM CHLORIDE, 5%

Place 5.0 g of potassium chloride in a graduated cylinder or volumetric flask. Add distilled water to the 100 mL level. Stir until dissolved.

### QUININE SULFATE, 0.5%

Place 0.5 g of quinine sulfate in a graduated cylinder or volumetric flask. Add distilled water to the 100 mL level. Stir until dissolved.

### RINGER'S SOLUTION (FROG)

Dissolve the following salts in 1 L of distilled water:

    6.50 g sodium chloride
    0.20 g sodium bicarbonate
    0.14 g potassium chloride
    0.12 g calcium chloride

## SODIUM CHLORIDE SOLUTIONS

1. *0.9% solution.* Place 0.9 g of sodium chloride in a graduated cylinder or volumetric flask. Add distilled water to the 100 mL level. Stir until dissolved.
2. *1.0% solution.* Place 1.0 g of sodium chloride in a graduated cylinder or volumetric flask. Add distilled water to the 100 mL level. Stir until dissolved.
3. *3.0% solution.* Place 3.0 g of sodium chloride in a graduated cylinder or volumetric flask. Add distilled water to the 100 mL level. Stir until dissolved.
4. *5.0% solution.* Place 5.0 g of sodium chloride in a graduated cylinder or volumetric flask. Add distilled water to the 100 mL level. Stir until dissolved.

## STARCH SOLUTIONS

1. *0.5% solution.* Add 5 g of cornstarch to 1 L of distilled water. Heat until the mixture boils. Cool the liquid, and pour it through a filter. Store the filtrate in a refrigerator.
2. *1.0% solution.* Add 10 g of cornstarch to 1 L of distilled water. Heat until the mixture boils. Cool the liquid, and pour it through a filter. Store the filtrate in a refrigerator.
3. *10% solution.* Add 100 g of cornstarch to 1 L of distilled water. Heat until the mixture boils. Cool the liquid, and pour it through a filter. Store the filtrate in a refrigerator.

## SUCROSE, 5% SOLUTION

Place 5.0 g of sucrose in a graduated cylinder or volumetric flask. Add distilled water to the 100 mL level. Stir until dissolved.

## WRIGHT'S STAIN

Prepared solution is available from various suppliers.

# APPENDIX 4

# Assessments of Laboratory Reports

Many assessment models can be used for laboratory reports. A rubric, which can be used for performance assessments, contains a description of the elements (requirements or criteria) of success to various degrees. The term *rubric* originated from *rubrica terra,* which is Latin for the application of red earth to indicate anything of importance. A rubric used for assessment contains elements for judging student performance, with points awarded for varying degrees of success in meeting the learning outcomes. The content and the quality level necessary to attain certain points are indicated in the rubric. It is effective if the assessment tool is shared with the students before the laboratory exercise is performed.

Following are two sample rubrics that can easily be modified to meet the needs of a specific course. Some of the elements for these sample rubrics may not be necessary for every laboratory exercise. The generalized rubric needs to contain the possible assessment points that correspond to learning outcomes for a specific course. The point value for each element may vary. The specific rubric example contains performance levels for laboratory reports. The elements and the point values can easily be altered to meet the value placed on laboratory reports for a specific course.

**Assessment: Generalized Laboratory Report Rubric**

| Element | Assessment Points Possible | Assessment Points Earned |
|---|---|---|
| 1. Figures are completely and accurately labeled. | | |
| 2. Sketches are accurate, contain proper labels, and are of sufficient detail. | | |
| 3. Colored pencils were used extensively to differentiate structures on illustrations. | | |
| 4. Matching and fill-in-the-blank answers are completed and accurate. | | |
| 5. Short-answer/discussion questions contain complete, thorough, and accurate answers. Some elaboration is evident for some answers. | | |
| 6. Data collected are complete, are accurately displayed, and contain a valid explanation. | | |

TOTAL POINTS: POSSIBLE _____ EARNED _____

**Assessment: Specific Laboratory Report Rubric**

| Element | Excellent Performance (4 points) | Proficient Performance (3 points) | Marginal Performance (2 points) | Novice Performance (1 point) | Points Earned |
|---|---|---|---|---|---|
| Figure labels | Labels completed with ≥ 90% accuracy. | Labels completed with 80%–89% accuracy. | Labels completed with 70%–79% accuracy. | Labels <70% accurate. | |
| Sketches | Accurate use of scale, details illustrated, and all structures labeled accurately. | Minor errors in sketches. Missing or inaccurate labels on one or more structures. | Sketch not realistic. Missing or inaccurate labels on two or more structures. | Several missing or inaccurate labels. | |
| Matching and fill-in-the-blanks | All completed and accurate. | One or two errors or omissions. | Three or four errors or omissions. | Five or more errors or omissions. | |
| Short-answer and discussion questions | Answers complete, valid, and contain some elaboration. No misinterpretations noted. | Answers generally complete and valid. Only minor inaccuracies noted. Minimal elaboration. | Marginal answers to the questions and contains inaccurate information. | Many answers incorrect or fail to address the topic. May be misinterpretations. | |
| Data collection and analysis | Data complete and displayed with a valid interpretation. | Only minor data missing or a slight misinterpretation. | Some omissions. Not displayed or interpreted accurately. | Data incomplete or show serious misinterpretations. | |

TOTAL POINTS EARNED _____

# INDEX

esophagus
- in cat, 454, 455f, 457f, 461, 462f, 463f
- location in human, 341f, 352f
- as major organ, 12
- structure, 336

estrogens, 272
ethmoidal air cells, 100f
ethmoidal sinus, 97, 351
ethmoid bone, 97, 98f
exercise effects on respiration, 370
exocrine cells, 273, 275f, 276f, 338, 340f
experiments, 1
expiratory reserve volume, 361, 362f
extension, 138t, 139f
extensor carpi radialis brevis muscle, 158, 160f
extensor carpi radialis longus muscle, 158, 160f
extensor carpi radialis muscle (cat), 430f, 431, 432t
extensor carpi ulnaris muscle, 158, 430f, 431, 432t
extensor digitorum communis muscle, 430f, 431, 432t
extensor digitorum lateralis muscle, 430f, 431, 432t
extensor digitorum longus muscle, 171
extensor digitorum muscle, 158
extensor group, 431
extensor retinacula, 175f
external acoustic meatus, 97, 99f, 182f, 239
external anal sphincter, 168f, 340
external carotid artery, 310
external iliac arteries, 310, 445, 446f, 447f, 468f
external iliac veins, 316, 448f, 449
external intercostals, 158f, 368f
external jugular veins, 316, 317f, 442f, 443f
external oblique muscle
- in cat, 423f, 425, 426f, 429f
- location in human, 165
- role in breathing, 367
- surface anatomy, 184f
external occipital protuberance, 182f
external urethral orifice, 387

extracellular matrix
- in bone, 85, 87f, 88
- in various connective tissues, 67, 68f, 69f
extraembryonic cavity, 403f
extrafollicular cells, 272
extrinsic muscles of eye, 253, 254f, 256t, 257
eyelids, 253, 256t
eyepiece lens systems, 30f, 31, 34f
eyes, 204, 253–258, 259f. *See also* vision

**F**

facets, 94, 113, 114f
facial artery, 310
facial bones, 91
facial expressions, 151, 156
facial muscles, 151, 152f
facial nerves, 204, 209f, 210, 240f
falciform ligament, 338, 454
false ribs, 113
false vocal cords, 351
falx cerebelli, 202
falx cerebri, 216f
farsightedness, 263, 408
fascia, 147f
fascicles, 146f
fascicles (nerve), 196
fasciculus cuneatus, 215f
fasciculus gracilis, 215f
F cells, 274
feet
- arches of foot, 186f
- bones of, 132, 135f
- muscles of, 171–172, 175f–177f
- reflexes, 22f, 220
female pelvic floor, 168f
female reproductive system
- cat, 469f, 473–474, 475f
- human, 393–398
- major organs, 12
female *versus* male pelvic girdles, 129
female vital capacities, 363t
femoral artery, 310
femoral vein, 316, 449
femur, 4, 91, 130, 186f
fertilization and early development, 401–403, 404f
fertilization membranes, 401
fetal skeleton, 102
fibroblasts, 68f
fibrocartilage, 69f
fibrous capsule, 389, 470
fibrous coat in ureter, 378
fibrous joints, 137

fibrous pericardium, 291
fibula, 91, 132, 186f, 187f
fibular collateral ligament, 138f
fibularis brevis muscle, 172
fibularis group (cat), 434, 435f
fibularis longus muscle, 172, 177f
fibularis muscle (cat), 436f
fibularis tertius muscle, 171, 176f, 436f
field of view (microscope), 32
filaments (skeletal muscle), 147f, 148f
filter paper, 50
filtration, 49–50
fimbriae, 393, 395f, 473, 474f
fine adjustment knob of microscope, 30f, 32
fingers. *See* phalanges (hand)
fish embryo, 55, 56f
fish, insulin shock in, 274
fissures
- in bone, 94
- in brain, 201, 207, 208f
flat bones, 86
flexion, 138t, 139f
flexor carpi radialis muscle, 158, 429f, 431, 432t
flexor carpi ulnaris muscle, 158, 160f, 429f, 431, 432t
flexor digitorum longus muscle, 172, 177f
flexor digitorum profundus muscle, 158, 429f, 432t
flexor digitorum superficialis muscle, 158
flexor group, 431
floating ribs, 113
focusing knobs, 30f, 34f
follicles, thyroid, 272
fontanels, 102
foramen lacerum, 99f, 101, 108f
foramen magnum, 97
foramen ovale, 99f, 101, 108f
foramen rotundum, 99f, 101
foramen spinosum, 99f, 101, 108f
foramina
- defined, 94
- in skull, 97, 99f, 101–102, 108f
- of vertebral column, 113
forearm muscles, 157
forelimb muscles of cat, 429–431, 432t
foreskin, 387
fornices, 393
fornix, 206f

fibrous pericardium, 291

fossae
- defined, 94
- in heart, 294f
- of scapulae, 117
- of upper limbs, 118
fossa ovalis, 294f
fourth ventricle, 202, 206f
fovea capitis, 130
fovea centralis, 255, 258f
foveae, 94
freckles, 408, 409f
free earlobes, 408, 410f
frenulum, 454
frog blood vessels, 308–309, 310f
frontal bone, 97
frontalis, 151, 182f
frontal lobe, 201, 207, 208f
frontal plane, 13f
frontal sinus, 97, 351
FSH, 270
functional residual capacity, 362
fundus, 337, 393, 454, 457f
funiculi, 213

**G**

gallbladder
- in cat, 445f, 454, 456f, 457f, 458f
- location in human, 11f, 338, 341f
- as major organ, 12
G1 and G2 phases, 55
ganglion cells, 255, 256f
gap phases, 55
gastric arteries, 312f, 445, 446f
gastric folds, 337, 455
gastric glands, 338
gastric pits, 338
gastric vein, 316
gastrocnemius muscle
- surface anatomy, 186f, 187f
gastrula, 402
general senses, 225–227
genetics, 407–412
geniohyoid muscle, 422, 422t
genome, 407–408
genotypes, 407, 412
germinal epithelium, 396f, 397
gingivae, 333
glans penis
- in cat, 477f, 478f, 479f
- in human, 387
glenoid cavity, 117
globus pallidus, 201
glomerular capsule, 373, 376, 377f
glomeruli, 376, 377f
glossopharyngeal nerves, 204, 209f, 210
glottis, 352, 461

systole, 299
systolic pressure, 324

**T**

tactile corpuscles, 225, 226f
taeniae coli, 338
talus, 132
tapetum fibrosum, 258, 259f
tarsal bones, 91, 132
taste buds, 231, 233f, 234f
taste hairs, 234
taste pores, 234
taste sense, 231, 233–234
tattoos, 80
teeth
    of cat, 454
    digestive system role, 12
    structure, 333, 335f
telophase, 56f
temperature
    blood vessel response to
        changes, 309
    effects on digestion,
        345, 347
    metric units, 2t
    sensory receptors, 227
temporal bones, 97
temporalis muscle, 151
temporal lobe, 201, 207, 208f
temporal process, 101
temporomandibular
        joint, 182f
tendons
    cell structure, 68f
    of lower limbs, 172f,
        177f, 187f
    as major part of muscular
        system, 12
    origin and insertion, 146
tensor fasciae latae muscle,
        171, 427f, 432, 433f,
        436f, 437t
tensor tympani muscle, 239
tentorium cerebelli, 216f
tenuissimus muscle, 434, 435f,
        436f, 437t
teres major muscle
    in cat, 421f, 426, 427f, 428t,
        429f, 430f
    human, 157
teres minor muscle, 157
tertiary bronchi, 352
testes
    in cat, 476, 477f,
        478f, 479f
    location in human, 270
    as major organs, 12
    structure, 387, 388f, 389
testosterone, 389
thalamus, 202, 210
thenar eminence, 185f
theories, 1

thigh muscles
    in cat, 434, 435f, 436f
    human, 171, 173f–174f
third ventricle, 202
thoracic aorta, 310, 444f
thoracic cage, 91, 109,
        113, 114f
thoracic cavity, 9, 11f
thoracic curvature, 109, 110f
thoracic duct, 327, 328f
thoracic muscles in cat, 421f,
        423, 423f, 424f, 425t
thoracic vertebrae, 109, 112f
thymus
    of cat, 442f, 443f, 461, 463f
    location, 270
    as major organ, 12
    structure, 329, 330f
thyrocervical artery, 313f,
        442f, 443
thyrohyoid muscle, 422, 422t
thyroid cartilage, 182f, 351,
        461, 463f
thyroid gland
    of cat, 461, 463f
    examining, 272
    location in human, 11f, 270
    as major organ, 12
thyroid hormone, 272
tibia, 91, 131f, 132, 175f
tibial arteries, 310
tibial collateral ligament, 138f
tibialis anterior muscle
    in cat, 433f, 434, 435f, 437t
    human, 171, 186f
tibialis posterior muscle, 175
tibial tuberosity, 132, 186f
tibial veins, 318f
tidal volume, 360–361, 362f,
        362t
time, metric units, 2t
tissues
    blood, 281–285
    epithelial, 63, 64f
    joint classification by, 137
    major types, 63
    muscle, 75–76, 145–146
    nervous, 75–76, 193, 197
    respiratory, 351–355, 356f
tongue
    of cat, 454, 455f, 462f
    digestive system role, 12
    rollers and nonrollers, 408,
        409f
    structure, 333
    taste buds on, 231, 233f
    testing gustation on,
        233–234
tonsils, 333, 454
total lung capacity, 362
total magnification, 32
touch receptors, 80f, 225–227

trabeculae, 87f
trachea
    of cat, 443f, 461, 462f, 463f,
        464f
    as major organ, 12
    structure, 336f, 352, 353f
tracheal rings, 461
traits, 407, 408–411
transitional epithelium, 378
transmembrane proteins,
        41f, 41f
transparency of urine, 382
transverse arch, 186f
transverse colon, 338, 456
transverse fissure, 201
transverse foramina, 111f,
        112f, 113
transverse plane, 13f
transverse processes, 113, 114f
transverse tubules, 148
transversus abdominis muscle
    in cat, 423f, 425, 426f
    location in human, 165
transversus costarum muscle,
        423, 424f, 425t
trapezium, 119, 142f
trapezius muscles
    basic functions, 157
    location, 158f
    surface anatomy, 182f, 183f
trapezoid, 119
triceps brachii muscle
    in cat, 421f, 427f, 429, 429f,
        430f, 431f, 432t
    location in human, 157,
        160f
    reflexes, 220
    surface anatomy, 184f
triceps reflex, 220, 222f
Trichomonas, 384f
tricuspid valve, 292, 295–296,
        300
trigeminal nerves, 204,
        209f, 210
triquetrum, 119, 122f
trochanters, 94, 132
trochlea, 118, 120f, 253, 254f
trochlear nerves, 204, 209f,
        210, 256t
trochlear notch, 118
trophoblasts, 403
true ribs, 113
true vocal cords, 351
trypsin, 346
TSH, 270
tubercles
    defined, 94
    of humerus, 118, 120f, 184f
    of ribs, 113, 114f
    of sacrum, 113
tuberosities, 94, 118, 132
tunica albuginea, 387, 389, 390

tunica externa, 309, 309f
tunica interna, 308, 309f
tunica media, 308, 309f
two-point threshold, 226–227
tympanic cavity, 239, 240f
tympanic membrane, 239
type 1 diabetes mellitus, 270
type 2 diabetes mellitus, 270
tyrosine crystals, 384f

**U**

ulna, 91, 118, 121f, 185f
ulnar artery, 310
ulnar notch, 118
ulnar vein, 316, 317f
umami sensation, 231
umbilical arteries and veins,
        403, 404f
umbilical cord, 403
umbilicus, 167f, 184f
unknown compounds,
        identifying, 24
upper limbs
    arteries, 310
    bones, 91, 117–119,
        120f–122f, 124–126f
    muscles, 157–159, 160f
    veins, 316, 317f
ureters
    in cat, 467, 468f, 469f
    location in human, 373,
        375, 388f
    as major organs, 12
    structure, 378
urethra
    in cat, 468f, 469f, 474, 474f
    location in human, 373,
        375, 390, 394f
    as major organ, 12
urethral orifice, 168f, 474
uric acid crystals, 384f
urinalysis, 381–384
urinary bladder
    in cat, 445f, 457f, 458f, 467,
        468f, 469f, 478f,
        479f
    in female, 394f
    location in human, 373, 375
    as major organ, 12
    in male, 388f
    structure, 378
urinary system
    cat, 467–470
    kidney structure, 373–376,
        377f
    urinalysis, 381–384
urochrome, 382, 383
urogenital aperture, 474
urogenital diaphragm, 165,
        167, 168f
urogenital sinus, 467, 474
uterine horns, 469f, 474f, 475f

# Notes

# Notes

# Notes

# Notes

# Notes

# Notes

**Notes**

# Notes